Innovative Materials for Construction

Innovative Materials for Construction

Editors

Mariaenrica Frigione
José Barroso de Aguiar

MDPI • Basel • Beijing • Wuhan • Barcelona • Belgrade • Manchester • Tokyo • Cluj • Tianjin

Editors
Mariaenrica Frigione
University of Salento
Italy

José Barroso de Aguiar
University of Minho
Portugal

Editorial Office
MDPI
St. Alban-Anlage 66
4052 Basel, Switzerland

This is a reprint of articles from the Special Issue published online in the open access journal *Materials* (ISSN 1996-1944) (available at: https://www.mdpi.com/journal/materials/special_issues/Innovative_Materials_for_Construction).

For citation purposes, cite each article independently as indicated on the article page online and as indicated below:

LastName, A.A.; LastName, B.B.; LastName, C.C. Article Title. *Journal Name* **Year**, *Volume Number*, Page Range.

ISBN 978-3-0365-0166-6 (Hbk)
ISBN 978-3-0365-0167-3 (PDF)

Cover image courtesy of Mariaenrica Frigione and José Barroso de Aguiar.

© 2021 by the authors. Articles in this book are Open Access and distributed under the Creative Commons Attribution (CC BY) license, which allows users to download, copy and build upon published articles, as long as the author and publisher are properly credited, which ensures maximum dissemination and a wider impact of our publications.

The book as a whole is distributed by MDPI under the terms and conditions of the Creative Commons license CC BY-NC-ND.

Contents

About the Editors vii

Mariaenrica Frigione and José Luís Barroso de Aguiar
Innovative Materials for Construction
Reprinted from: *Materials* **2020**, *13*, 5448, doi:10.3390/ma13235448 1

**Juan María Terrones-Saeta, Francisco Javier Iglesias-Godino,
Francisco Antonio Corpas-Iglesias and Carmen Martínez-García**
Study of the Incorporation of Ladle Furnace Slag in the Manufacture of Cold In-Place Recycling
with Bitumen Emulsion
Reprinted from: *Materials* **2020**, *13*, 4765, doi:10.3390/ma13214765 5

Md Al Imran, Sivakumar Gowthaman, Kazunori Nakashima and Satoru Kawasaki
The Influence of the Addition of Plant-Based Natural Fibers (Jute) on Biocemented Sand Using
MICP Method
Reprinted from: *Materials* **2020**, *13*, 4198, doi:10.3390/ma13184198 25

Michele Angiolilli, Marco Vailati and Amedeo Gregori
Lime-Based Mortar Reinforced by Randomly Oriented Short Fibers for the Retrofitting of the
Historical Masonry Structure
Reprinted from: *Materials* **2020**, *13*, 3462, doi:10.3390/ma13163462 45

**Mari Masanaga, Tsuyoshi Hirata, Hirokatsu Kawakami, Yuka Morinaga, Toyoharu Nawa
and Yogarajah Elakneswaran**
Effects of a New Type of Shrinkage-Reducing Agent on Concrete Properties
Reprinted from: *Materials* **2020**, *13*, 3018, doi:10.3390/ma13133018 65

Dominika Madej
Strontium Retention of Calcium Zirconium Aluminate Cement Paste Studied by NMR, XRD
and SEM-EDS
Reprinted from: *Materials* **2020**, *13*, 2366, doi:10.3390/ma13102366 77

**Antonella Sarcinella, José Luìs Barroso De Aguiar, Mariateresa Lettieri, Sandra Cunha and
Mariaenrica Frigione**
Thermal Performance of Mortars Based on Different Binders and Containing a Novel
Sustainable Phase Change Material (PCM)
Reprinted from: *Materials* **2020**, *13*, 2055, doi:10.3390/ma13092055 93

Dongmin Lee, Dongyoun Lee, Myungdo Lee, Minju Kim and Taehoon Kim
Analytic Hierarchy Process-Based Construction Material Selection for Performance
Improvement of Building Construction: The Case of a Concrete System Form
Reprinted from: *Materials* **2020**, *13*, 1738, doi:10.3390/ma13071738 109

Won Jong Chin, Young Hwan Park, Jeong-Rae Cho, Jin-Young Lee and Young-Soo Yoon
Flexural Behavior of a Precast Concrete Deck Connected with Headed GFRP Rebars and UHPC
Reprinted from: *Materials* **2020**, *13*, 604, doi:10.3390/ma13030604 131

**Mariaenrica Frigione, Mariateresa Lettieri, Antonella Sarcinella
and José Luìs Barroso de Aguiar**
Applications of Sustainable Polymer-Based Phase Change Materials in Mortars Composed by
Different Binders
Reprinted from: *Materials* **2019**, *12*, 3502, doi:10.3390/ma12213502 147

Tomasz Kopecki, Przemysław Mazurek and Tomasz Lis
Experimental and Numerical Analysis of a Composite Thin-Walled Cylindrical Structures with Different Variants of Stiffeners, Subjected to Torsion
Reprinted from: *Materials* **2019**, *12*, 3230, doi:10.3390/ma12193230 . . . . **161**

Sing Chuong Chuo, Sarajul Fikri Mohamed, Siti Hamidah Mohd Setapar, Akil Ahmad, Mohammad Jawaid, Waseem A. Wani, Asim Ali Yaqoob and Mohamad Nasir Mohamad Ibrahim
Insights into the Current Trends in the Utilization of Bacteria for Microbially Induced Calcium Carbonate Precipitation
Reprinted from: *Materials* **2020**, *13*, 4993, doi:10.3390/ma13214993 . . . . **175**

About the Editors

Mariaenrica Frigione, Ph.D. Leader of "Materials and Technologies for Constructions and Cultural Heritage (MaTech-CCult)" research group at the University of Salento, maintaining scientific collaborations with several Italian and International Universities, Research Centers, and Companies. She is/has been responsible for several competitive research projects. Since 2018 she has been the Secretary of "ICPIC—International Congress of Polymers in Concrete" (Member of Board of Directors in the sub-committee "International Exchange" since 2013). She is a member of the Board of Experts of the TICHE Foundation (Technological Innovation in Cultural Heritage—Italy). She is an Associate Editor for Polymers MDPI, "Polymer Applications" Section. She is a member of the Editorial Board of the ASCE *Journal of Composite for Constructions, Polymers* MDPI, and *Coatings* MDPI. She has been the supervisor of more than 70 Ph.D., bachelor, and master students. She is the coordinator for the University of Salento of the Ph.D. Program with Al Akhawayn University (Ifrane, Morocco). In 2015, she was awarded the Industria Felix Prize. From 2013 to 2019, she has been Vice-Rector of the University of Salento for the Technical Scientific Area and a Delegate of the Rector for Internationalization.

José Barroso de Aguiar, Ph.D. Leader of the Research Center for Territory, Environment and Construction at the University of Minho, maintaining scientific collaborations with several Portuguese and International Universities, Research Centers, and Companies. He is the Director of the Doctoral Program in Management and Treatment of Residues at the University of Minho. He has been responsible for several competitive research projects. He is a member of the Board of Directors of "ICPIC—International Congress of Polymers in Concrete". He is a member of the Editorial Boards of *Advances in Civil Engineering, Review of Industrial Engineering Letters, Exploratory Materials Science Research, and Advances in Materials*. He has been the supervisor of more than 60 Ph.D. and master students. In 2018, he was awarded the Owen Nutt Award for outstanding research in the field of polymers in concrete.

Editorial

Innovative Materials for Construction

Mariaenrica Frigione [1,*] and José Luís Barroso de Aguiar [2]

1. Innovation Engineering Department, University of Salento, Prov.le Lecce-Monteroni, 73100 Lecce, Italy
2. Civil Engineering Department, University of Minho, Campus de Azurém, 4800-058 Guimarães, Portugal; aguiar@civil.uminho.pt
* Correspondence: mariaenrica.frigione@unisalento.it

Received: 14 November 2020; Accepted: 30 November 2020; Published: 2 December 2020

Academic and industrial efforts around the world are continuously engaged to develop new smart materials that can provide efficient alternatives to conventional construction materials and improve the energy-efficiency in buildings or are able to upgrade, repair, and protect existing infrastructures. This new generation of materials, before an actual market entry, needs to be analyzed, validated, tested on in the field, and, possibly, modeled to enable predictions as to their long-term behavior and performance. To this regard, the valuable contributions in the Special Issue "Innovative Materials for Construction" provide a collection of original research and new trends in the field of innovative materials and technologies proposed for the construction sector, with a special focus on sustainable materials for the building industry of the future, keeping attention on innovative solutions suitable for ancient constructions and cultural heritage.

The first stage of any construction design relies on the selection of the best-performing materials able to satisfy the project goal. Although this is a very important task in the design of a construction project, since mechanical, functional, and physical properties of construction materials greatly affect the overall project performance, there are currently no systematic methods to guide designers in this selection. The work by Lee and co-workers [1] would propose an original approach to fill this gap based on an analytic hierarchy process (AHP) method. The authors validated the AHP method in the design of a composite system form (CSF) panel, proving the efficacy of the proposed model able to identify the best combination of materials and ensuring the greatest performance.

In concrete technology, the shrinkage occurring during the cure of concrete represents a major problem, hampering appropriate performance and limiting durability of concrete structures. To address this issue, Masanaga et al. [2] investigated the introduction of a new shrinkage-reducing agent (SRA) to a concrete mix based on Portland cement, comparing its properties and performance with those of a conventional SRA. The superiority of the new shrinkage-reducing agent was confirmed in terms of improved durability of concrete towards freeze-thaw cycles. Additionally, the authors of the study analyzed the mechanism, through which the proposed SRA was able to reduce shrinkage in concrete.

The incorporation of industrial by-products in construction materials is a sustainable and convenient way to exploit waste to produce new materials, eliminating the problem of waste treatment and, at the same time, avoiding the use and depletion of new natural resources. To this regard, the research presented by Terrones-Saeta et al. [3] proposed the incorporation of ladle furnace slag in reclaimed asphalt pavements (RAPs) in the cold in-place recycling, with the bitumen emulsion manufacturing technique. The authors demonstrated that, with a proper selection of the asphalt mix components (RAP, ladle furnace slag, water, and emulsion), it is possible to manufacture road pavements characterized by good mechanical properties and durability.

The development of interior mortars, based on different binders (aerial and hydraulic lime, gypsum and cement), containing an original sustainable phase change material (PCM) able to improve the energy efficiency of buildings, was the object of two publications resulting from an Italian–Portuguese scientific collaboration [4,5]. Since PCMs have the ability to change their physical state according to the environmental temperature, the incorporation of a suitable PCM in mortars can reduce indoor temperature fluctuations, leading to improvement in human comfort and reductions in energetic consumption. The selection of the components to produce a "green" composite PCM system was based on the use of non-toxic, low environmental impact materials, and waste materials/by-products from other industries. The authors of this study demonstrated that it was possible to obtain mortars with suitable mechanical properties with an optimization of their composition, depending on the kind of binder [4]. Subsequently, the measurements of thermal performance of the best-performing mortars, i.e., those based on cement and gypsum, confirmed that the addition of the experimented PCM in mortars leads to a decrease of the maximum achievable temperatures in the hot season and an increase of the minimum temperatures in the cold season, with a reduction of heating and cooling indoor needs, thus confirming the capability of this new PCM material to achieve energy savings.

The microbial-induced calcium carbonate precipitation (MICP) has received great attention for its potential in construction applications. This technique has been used in biocementation of sand, consolidation of soil, production of self-healing concrete or mortar, and removal of heavy metal ions from water. The self-healing ability is very important for repair of concrete structures. The presence of bacteria is essential for MICP to occur. To this regard, the paper presented by Chuo et al. [6] reviews the bacteria used for MICP in some of the most recent studies. Several factors that affect MICP performance are bacterial strain, bacterial concentration, nutrient concentration, calcium source concentration, addition of other substances, and methods to distribute bacteria. The paper presented by Imran et al. [7] proposed the addition of plant-based natural jute fibers to MICP-treated sand. The results of this study showed that the added jute fibers improved the engineering properties (ductility, toughness, and brittleness behavior) of the biocemented sand using the MICP method. The fibers facilitated the MICP process by bridging the pores in the calcareous sand, reduced the brittleness of the treated samples, and increased the mechanical properties of the biocemented sand. The results of this study could significantly contribute to further improvement of fiber-reinforced biocemented sand in geotechnical engineering field applications.

The study of materials that can influence the cement hydration is very significant. These admixtures modify the hydration behavior, contributing to the adaptation of the cementitious materials to the site conditions. Madej [8] showed that the substitution of Sr^{2+} for Ca^{2+} in the $Ca_7ZrAl_6O_{18}$ lattice decreased the reactivity of Sr-substituted $Ca_7ZrAl_6O_{18}$ in the presence of water. Therefore, strontium can be considered an inhibition agent for cement hydration. The techniques NMR, XRD, and SEM-EDS were employed for the justification of this behavior. The authors studied calcium zirconium aluminate cement pastes and detected important changes during the first 24 h or 7 days of hydration.

The use of fiber reinforced polymer (FRP) technologies is gaining great success worldwide as an efficient alternative to steel reinforcements to improve load-bearing and performance of modern constructions, as well as ancient buildings. The complete study of these materials should include experimental and numerical analysis.

Chin and co-workers [9] developed rebars with headed ends with the aim to improve their anchorage to concrete. The authors of this study systematically evaluated the mechanical properties of FRP rebars and the bond strength developed with concrete, and these parameters were fundamental in the design of FRP structural members. Pull-out tests were performed by changing the GFRP rebar diameter, the concrete strength, and the head type; precast concrete decks connected with different headed GFRP rebars were also tested in flexural mode to estimate the flexural behavior of the connected decks. The results confirmed the effectiveness of the headed GFRP rebars, evidencing the effects of the different parameters analyzed.

Kopecki et al. [10] used a non-linear numerical analysis of the examined structures by comparing them with the results of the model experiment. The study contains the results of the experimental research using models made of glass epoxy composites. The results of the research allowed the creation of the concept of an adequate numerical model in terms of the finite element method, allowing to determine the distribution of stress and strain in the components of the studied structures.

The presented research allows to determine the nature of the deformation of composite thin-walled structures, in which local loss of stability of the covering is acceptable in the area of post-critical loads.

The mechanical behavior (in flexural, tensile, and compressive modes and the energy fracture) of lime-based mortars reinforced by randomly oriented short glass fibers, with different contents and aspect ratios, was investigated by Angiolilli et al. [11]. The aim of their research was the assessment of the effect of the introduction of various types/amounts of fibers in binders, reproducing the compositions of historical lime-based mortars. The obtained results highlighted that the fiber reinforced mortar composites ensures increased strength and excellent ductility capacity, thus making them a promising alternative to traditional fiber reinforcement systems, even for ancient constructions.

Funding: This research received no external funding.

Acknowledgments: The Guest Editors of this Special Issue would like to thank all the Authors from all over the world (India, Italy, Japan, Korea, Malaysia, Poland, Portugal, Spain), who contributed with their valuable works to the accomplishment of the Special Issue. Special thanks are due to the Reviewers for their constructive comments and thoughtful suggestions. Finally, the Authors are grateful to the Materials Editorial Office, particularly to Jason Huang, for their kind assistance.

Conflicts of Interest: The authors declare no conflict of interest.

References

1. Lee, D.; Lee, D.; Lee, M.; Kim, M.; Kim, T. Analytic Hierarchy Process-Based Construction Material Selection for Performance Improvement of Building Construction: The Case of a Concrete System Form. *Materials* **2020**, *13*, 1738. [CrossRef]
2. Masanaga, M.; Hirata, T.; Kawakami, H.; Morinaga, Y.; Nawa, T.; Elakneswaran, Y. Effects of a New Type of Shrinkage-Reducing Agent on Concrete Properties. *Materials* **2020**, *13*, 3018. [CrossRef]
3. Terrones-Saeta, J.M.; Francisco Javier Iglesias-Godino, F.J.; Corpas-Iglesias, F.A.; Martínez-García, C. Study of the Incorporation of Ladle Furnace Slag in the Manufacture of Cold In-Place Recycling with Bitumen Emulsion. *Materials* **2020**, *13*, 4765. [CrossRef]
4. Frigione, M.; Lettieri, M.; Sarcinella, A.; Barroso de Aguiar, J.L. Applications of Sustainable Polymer-Based Phase Change Materials in Mortars Composed by Different Binders. *Materials* **2019**, *12*, 3502. [CrossRef]
5. Sarcinella, A.; Barroso de Aguiar, J.L.; Lettieri, M.; Cunha, S.; Frigione, M. Thermal Performance of Mortars Based on Different Binders and Containing a Novel Sustainable Phase Change Material (PCM). *Materials* **2020**, *13*, 2055. [CrossRef]
6. Chuo, S.C.; Mohamed, S.F.; Mohd Setapar, S.H.; Ahmad, A.; Jawaid, M.; Wani, W.A.; Yaqoob, A.A.; Mohamad Ibrahim, M.N. Insights into the Current Trends in the Utilization of Bacteria for Microbially Induced Calcium Carbonate Precipitation. *Materials* **2020**, *13*, 4993. [CrossRef]
7. Al Imran, M.; Gowthaman, S.; Nakashima, K.; Kawasaki, S. The Influence of the Addition of Plant-Based Natural Fibers (Jute) on Biocemented Sand Using MICP Method. *Materials* **2020**, *13*, 4198. [CrossRef] [PubMed]
8. Madej, D. Strontium Retention of Calcium Zirconium Aluminate Cement Paste Studied by NMR, XRD and SEM-EDS. *Materials* **2020**, *13*, 2366. [CrossRef]
9. Chin, W.J.; Park, Y.H.; Cho, J.-R.; Lee, J.-Y.; Yoon, Y.-S. Flexural Behavior of a Precast Concrete Deck Connected with Headed GFRP Rebars and UHPC. *Materials* **2020**, *13*, 604. [CrossRef]
10. Kopecki, T.; Mazurek, P.; Lis, T. Experimental and Numerical Analysis of a Composite Thin-Walled Cylindrical Structures with Different Variants of Stiffeners, Subjected to Torsion. *Materials* **2019**, *12*, 3230. [CrossRef] [PubMed]

11. Angiolilli, M.; Gregori, A.; Vailati, M. Lime-Based Mortar Reinforced by Randomly Oriented Short Fibers for the Retrofitting of the Historical Masonry Structure. *Materials* **2020**, *13*, 3462. [CrossRef]

Publisher's Note: MDPI stays neutral with regard to jurisdictional claims in published maps and institutional affiliations.

© 2020 by the authors. Licensee MDPI, Basel, Switzerland. This article is an open access article distributed under the terms and conditions of the Creative Commons Attribution (CC BY) license (http://creativecommons.org/licenses/by/4.0/).

Article

Study of the Incorporation of Ladle Furnace Slag in the Manufacture of Cold In-Place Recycling with Bitumen Emulsion

Juan María Terrones-Saeta *, Francisco Javier Iglesias-Godino, Francisco Antonio Corpas-Iglesias and Carmen Martínez-García

Department of Chemical, Environmental, and Materials Engineering, Higher Polytechnic School of Linares, University of Jaen, Scientific and Technological Campus of Linares, 23700 Linares, Spain; figodino@ujaen.es (F.J.I.-G.); facorpas@ujaen.es (F.A.C.-I.); cmartin@ujaen.es (C.M.-G.)
* Correspondence: terrones@ujaen.es; Tel.: +034-675-201-939

Received: 30 August 2020; Accepted: 23 October 2020; Published: 26 October 2020

Abstract: Cold in-place recycling with bitumen emulsion is a good environmental option for road conservation. The technique produces lower CO_2 emissions because the product is manufactured and spread in the same location as the previous infrastructure, and its mixing with bitumen emulsion occurs at room temperature. Adding materials with cementitious characteristics gives the final mixture greater resistance and durability, and incorporating an industrial by-product such as ladle furnace slag (of which cementitious characteristics have been corroborated by various authors) enables the creation of sustainable, resistant pavement. This paper describes the incorporation of ladle furnace slag in reclaimed asphalt pavements (RAP) to execute in-place asphalt pavement recycling with bitumen emulsion. Various test groups of samples with increasing percentages of emulsion were created to study both the density of the mixtures obtained, and their dry and post-immersion compressive strength. To determine these characteristics, the physical and chemical properties of the ladle furnace slag and the reclaimed asphalt pavements were analyzed, as well as compatibility with the bitumen emulsion. The aforementioned tests define an optimal combination of RAP (90%), ladle furnace slag (10%), water (2.6%), and emulsion (3.3%), which demonstrated maximum values for compressive strength of the dry and post-immersion bituminous mixture. These tests therefore demonstrate the suitability of ladle furnace slag for cold in-place recycling with bitumen emulsion.

Keywords: ladle furnace slag; reclaimed asphalt pavements; cold in-place recycling; simple compressive strength; bitumen emulsion; waste; circular economy

1. Introduction

The construction industry is one of the sectors with the greatest environmental impact [1], as well as being crucial for the development of a region's social welfare and economy. Civil constructions contribute to the progress of a population and the economic development of a nation. Building such constructions is therefore essential, even if large amounts of materials are consumed and the environment is affected [2,3].

A high proportion of the materials consumed by the construction of civil infrastructure projects are natural aggregates from nearby quarries. Taken together, the stages of extracting these natural aggregates and the binder used, transporting them to the factory, and designing the asphalt mix, generate a significant emission of greenhouse gases and a high environmental cost. Given the critical importance of roads and the current need to build them, the main countries involved in such infrastructure development stipulate regulations to mitigate environmental impacts throughout a road's life cycle [4,5].

The ideal solution—building high-quality structures while reducing environmental impact—requires techniques that are less damaging to the environment, including designing less-polluting processes for mixing the asphalt, and using waste from other industries in the development of road infrastructure [6]. All of these ideas are inherent in the so-called "circular economy," a strategy to reduce consumption of virgin materials, optimize industrial processes, and ultimately reduce waste. The circular economy enables industries to close economic and ecological flows of resources.

Based on the above, and following the principles of the circular economy, a trend toward less-polluting techniques with a similar quality of results is essential for preserving road infrastructure. One of the best solutions is the recycling of asphalt pavement with bitumen emulsion.

Road pavement made with an asphalt mix has good strength, adhesion, and comfort for drivers, but these characteristics can decrease significantly at the end of its useful life. However, the aggregates' standardized and binding characteristics give the aged material quality for use in new asphalt mixes. Pavement aging is characterized by the formation of cracks of irregular shape and the loss of macroscopic adhesion for vehicles. Merely disposing of the aged pavement into landfill without any reuse wastes significant amounts of material, which is a strong negative environmental impact [7,8]. Ideally, mixtures would be developed that use this aged material to achieve the qualities required in the construction of the new infrastructure, reducing the use of new virgin materials in the construction of infrastructure, and of fossil fuels to produce and transport the new mixture [9,10]—in short, a lower emission of greenhouse gases and lower environmental impact. Some authors call this result sustainable pavement, thanks to the increase in the useful life of the infrastructure, and savings in terms of economic and environmental costs [11–13].

Lack of sustainability is a problem in the construction sector, albeit one which may slightly be mitigated through the reclamation of aged pavement to create a new asphalt mix. Disposing of the aged layer in landfill and producing a new asphalt mix should be rejected as an idea, due to its high impact on the environment and the contemporary imperative to optimize resources.

Various techniques have been used in this process. Every technique has its advantages and disadvantages, but each use all or part of the milled and aged road material.

These techniques are classified into two main types: hot and cold, depending on whether or not the asphalt mix is heated. In hot in-plant recycling, the road's aged asphalt mix is milled, transported to the manufacturing plant, mixed with virgin material, and manufactured at temperatures of 180 °C. It is then transported to the site of the infrastructure construction project, spread, and compacted. Unlike hot recycling, cold recycling is performed at room temperature, reducing the consumption of fossil fuels and the emission of CO_2 required in the heating of the mixture. Cold central plant pavement recycling-based processes follow implementation stages similar to those of hot recycling. One difference is the much lower temperature, cold in-place recycling does not require transportation of the milled mixture to a manufacturing plant, and the new mix is manufactured with 100% of the aged pavement reused in the new infrastructure [14].

Cold in-place recycling, on which this study is based, has a number of obvious advantages over other techniques. Environmental advantages include the reduced transport by heavy vehicles, lower CO_2 emissions, and a lower consumption of fossil fuels. Operational advantages include [15,16] low traffic influence, the maintenance of road geometry, a high pace of construction, and safety. These advantages stem mainly from the efficient practices involved in the execution of all pavement manufacturing operations, from milling and mixing with emulsion, water, and other additives to spreading and compaction.

Based on the above, cold in-place recycling with bitumen emulsion decreases the environmental impact, mainly due to a reduced extraction of virgin materials, and less use of machinery in the transport and manufacture of the asphalt mix. These improvements significantly reduce the emission of greenhouse gases such as CO_2 and NO_X [17].

Recycling pavement also has some disadvantages that must be taken into account during design. Firstly, not all pavements can be recycled; some pavements have suffered plastic deformation and are not recommended for regeneration, as the problems are likely to re-impact the new pavement created. Furthermore, since the new recycled pavement with bitumen emulsion is manufactured, curing time is required to achieve optimal mechanical properties. Therefore, even if it is permissible to open the road to traffic after the reclamation of the pavement, sufficient time is required before the final mechanical characteristics are obtained [18]. It is therefore essential to choose a bitumen emulsion compatible with both the material to be treated, and the expected breaking and curing times.

Secondly, the difficulty of reproducing the final characteristics of the projected pavement in the laboratory [19,20], variability of materials, and the dependence on proper execution make it difficult to determine the final behavior of the cold in-place recycling mix in advance, thus limiting its use to low traffic-volume roads [21].

In several cases, however, proper laboratory study and accurate execution of the technique has led to the creation of pavement with characteristics notably better than expected, producing a sustainable mix with good mechanical characteristics, even for high traffic [22–24]. Additives such as cement [25], and even industrial by-products such as fly ash from coal-fired thermal plants [26], have been used with good results.

Based on the preceding research, obtaining optimal mechanical characteristics in cold in-place recycling with bitumen emulsion requires the incorporation of an additive with cementitious characteristics to achieve three goals: proper pavement strength during the curing time; granulometric adjustment of the reclaimed asphalt pavement; and material that maximizes final resistance of the mix. This project therefore uses ladle furnace slag as a grading corrector and additive to improve mechanical qualities in the short and long term.

The ladle furnace slag comes from the steel industry. It is a by-product of the process of obtaining quality steel from scrap steel—more specifically, from the refining stage. In the first stage of meltdown performed in an electric arc furnace, oxidation eliminates manganese and silicon and achieves dephosphorization, creating a foamed slag in which all the dross accumulates. These slags are called electric arc furnace slags. The subsequent refining stage is for the removal of metal oxides, desulfurization, and decarbonization of the steel [27]. The main purpose of the refining stage is to obtain steel with a low oxygen and sulfur content, so the refining furnace or ladle furnace is fed with melted liquid from the previous stage and then covered with a reduction slag that is formed of lime, fluorspar, coke or graphite in appropriate amounts. Deoxidation is achieved by simple contact of the molten liquid with the slag. Full deoxidation occurs, however, through the addition of silicon and manganese ferroalloys, leading to liquid particles retained in the slag forming in the metal bath. Desulfurization minimizes the amount of sulfur in steel thanks to the presence of calcium oxide and carbon. This process produced the ladle furnace slag used in this project in a portion of 20–30 kg per final ton of steel. Although the proportion is lower than that obtained in the production of electric arc furnace slag, the landfill of ladle furnace slag produces serious environmental problems due to its chemical composition and small particle size.

Unlike electric arc furnace slags, with resistant characteristics, angularity, and hardness making them ideal substitutes for virgin aggregate in road infrastructure diversity (mainly in asphalt mix), ladle furnace slag has not been reused abundantly in the creation of new materials [28–31]. Ladle furnace slags have a fine grading which is suitable for various purposes [32,33]—among them, the addition to reclaimed asphalt pavements for the execution of cold in-place recycling with bitumen emulsion, the focus of this project. The presence of some metal oxides can cause volumetric expansion of the material when hydrated, however, and laboratory monitoring was required during the study [34,35].

Among the few applications of ladle furnace slag is its addition to cement [36,37], as a substitute for sand in cement mortars [38–41], as a replacement in concrete [42], and as a stabilizer of clayey materials due either to its high lime content [43] or its treatment of water [44,45]. Its appropriateness as a material with cementitious characteristics has been confirmed by the success stories involving its addition to cements, mortars, and concretes.

On the other hand, the use of ladle furnace slag in pavements has been limited, and only practiced at the research level. In the incorporation of the slags into low quality soils for its stabilization [46], acceptable results were obtained, with no problems of expansiveness in the treated material, and an improved quality of the soil. In turn, ladle furnace slags were used for the manufacture of concrete pavements [47], reflecting adequate strengths after testing, and even used as road bases, or subbases [48]. In the field of hot mix asphalt, research has been carried out in which the filler was replaced by ladle furnace slag [49], obtaining good properties of rigidity, tensile splitting strength and resistance to repeated loads. At the same time, the substitution of a high proportion of calcareous aggregates by slags in warm mix asphalt [50] showed a considerable improvement in mechanical properties compared with a traditional bituminous mix. Finally, the development of porous asphalt with ladle furnace slag and bitumen [51] showed excellent adhesion of the bitumen with the slag, resulting in good mechanical properties. However, there is no research referring to the use of ladle furnace slag in bituminous mixtures with bitumen emulsion and reclaimed asphalt pavements (RAP), the process developed in this research.

Based on the above, this study analyzed cold in-place recycling with bitumen emulsion and the incorporation of ladle furnace slag as an additive to improve grading and to provide special characteristics of resistance. The study therefore explored the use of industrial by-products as a substitute for virgin material, and also to lengthen the life cycle of a pavement. In addition to reducing the use of virgin materials in the formation of the asphalt mix, use of cold in-place techniques significantly decrease greenhouse gas emissions and fossil fuel consumption. This project sought to optimize the use of resources and manufacturing techniques, while reducing industrial waste. The study was therefore performed within the framework of a circular economy.

Regulations for studying the suitability of cold in-place recycling not only vary by country, but are somewhat inaccurate and depend on the empirical conditions of sampling, compaction, and testing [52]. Given the proliferation of such techniques in Spain for over 30 years, and the success achieved in their implementation, this study follows the Spanish regulations outlined in Circular 8/2001 [53], which provides a series of warnings concerning the treatment of materials and a series of standardized tests to be performed to achieve results that conform to the various minimum values established.

This regulation was applied for the in-place implementation of a bituminous mixture with bitumen emulsion, ladle furnace slag and milled pavement as the most superficial layer. Subsequently, a bituminous mixture reduced in thickness was applied on this pavement, which would improve the friction of a tyre with the road and provide a comfortable and safe wearing layer.

2. Materials and Methods

This section details the starting materials and the methodology used to study the suitability of in-place recycling of asphalt pavement with bitumen emulsion, using ladle furnace slag as an additive.

2.1. Materials

The following subsections detail the materials used in this study, highlighting their nature and origin, as well as specific noteworthy characteristics.

2.1.1. Reclaimed Asphalt Pavements

The reclaimed asphalt pavement came from the surface layer of the road joining the towns of Linares and Jabalquinto, located in Spanish territory. The road on which the pavement was located had a medium volume of heavy-vehicle traffic, and the surface layer was cracked with an irregular shape. This cracking reflected structural depletion of the layer due to aging of the bitumen. The layer did not, however, show significant deformation due to poor design of the initial mix or poor execution of the subgrade. The pavement composed of hot mix asphalt was milled with machinery similar to that which would be used on-site, to obtain an appropriate sample for study. The tests performed to confirm the pavement's suitability for recycling are detailed in the methodology.

2.1.2. Bitumen Emulsion

Bitumen emulsion plays a key role in achieving the mechanical and durability characteristics of the final asphalt mix. As an element to bond the various particles of the reclaimed asphalt pavements and the ladle furnace slag, bitumen emulsion can provide the asphalt mix with both tensile strength and durability.

Choosing the right bitumen emulsion was essential for a variety of reasons, including achieving adequate adhesivity between the aggregates and the bitumen emulsion, and breaking and coating times suitable for execution on-site. To achieve these properties, a slow-breaking bitumen emulsion must be used, since longer breaking time favors coating of the reclaimed asphalt pavements particles. To achieve chemical compatibility between aggregates, reclaimed asphalt pavements, and ladle furnace slag, a cationic emulsion with characteristics appropriate to the nature of ladle furnace slag was chosen; this emulsion did not cause problems in the coating of the reclaimed asphalt pavements. European regulations label this emulsion C60B5 REC. Table 1 lists its technical characteristics.

Table 1. Technical details of the bitumen emulsion C60B5 REC.

Characteristics	Unit	Standard	Min.	Max.
Original Emulsion				
Particle polarity	-	UNE EN 1430 [54]	Positive	
Breaking value	g	UNE EN 13075-1 [55]	170	
Binder content (per water content)	%	UNE EN 12846-1 [56]	58	62
Efflux time (2 mm, 40 °C)	s	UNE EN 12,846 [57]	15	70
Residue on sieving (0.5 mm)	%	UNE EN 1429 [58]	-	0.10
Setting tendency (7 days storage)	%	UNE EN 12,847 [59]	-	10
Water effect of binder adhesion	%	UNE EN 13,614 [60]	90	-
Binder after Distillation (UNE EN 1431 [61])				
Penetration (25 °C; 100 g; 5 s)	0.1mm	UNE EN 1426 [62]	-	270
Softening point	°C	UNE EN 1427 [63]	35	-
Evaporation Residue (UNE EN 13074-1 [64])				
Penetration (25 °C; 100 g; 5 s)	0.1mm	UNE EN 1426 [62]	-	330
Softening point	°C	UNE EN 1427 [63]	35	-
Stabilizing Residue (UNE EN 13074-2 [65])				
Penetration (25 °C; 100 g; 5 s)	0.1mm	UNE EN 1426 [62]	-	270
Softening point	°C	UNE EN 1427 [63]	35	-

2.1.3. Ladle Furnace Slag

The ladle furnace slag used in this study came from the steel manufacturing industry. As mentioned above, ladle furnace slag is found in steel refining or ladle furnaces. The sample was taken representatively, such that it contained all particle sizes typical of the unaltered by-product. The following sections detail the tests performed to characterize and study the by-product.

2.2. Methodology

A clear, objective methodology was used to confirm the suitability of the mechanical and resistance-related characteristics produced by cold in-place recycling with bitumen emulsion and ladle furnace slag.

Firstly, both of the starting materials, reclaimed asphalt pavement and ladle furnace slag, were analyzed to determine their physical, chemical, and mechanical characteristics. Consideration was given to differences in the materials, as well as to the role each played in the final asphalt mix.

Individualized study best determined the critical characteristics that could pose problems in executing the job.

After analyzing the properties of the various materials and their suitability for use in asphalt pavement recycling, the different specimen families were manufactured. Firstly, however, the percentages of each material to be added—reclaimed asphalt pavement, bitumen emulsion, ladle furnace slag, and water—was determined.

It should be noted that the mixing sequence was as follows: first, the pavement was milled and added to the ladle furnace slag; secondly, the appropriate percentage of precoating water was incorporated to facilitate the mixing process; and finally, the bitumen emulsion was added by re-mixing to achieve a homogeneous mix of materials.

According to Spanish Circular 8/2001 [53], these percentages were determined through a series of steps outlined in the NLT-389/00 [66] standard. This regulation primarily describes the grading envelope for in-place recycling with bitumen emulsion. The grading curve obtained by combining different percentages of the recycled asphalt pavement and ladle furnace slag must be contained in the grading envelope. After defining the percentages of each material, the percentages of fluids (bitumen emulsion and pre-coating water) required for the final mix were identified. NLT-389/00 [66] establishes the margins for both materials, and the addition of fluids varied based on these margins. These margins are determined by the UNE 103,501 [67] Modified Proctor Compaction Test and the NLT-196/84 [68] test for bitumen coating. Only in this way could a final mix with the greatest resistance be created, due to the higher density determined by the Modified Proctor Compaction Test. Good mechanical characteristics depend on the coating and adhesion of the aggregates with the emulsion, which was quantified by the Coating Test.

After determining the percentage of each material to be added to obtain the different families, the same tests were performed: simple compression strength NLT-161/98 [69] and immersion-compression NLT-162/00 [70].

With the data obtained from dry and post-immersion resistance, as well as preserved resistance, the optimal formula for the job was calculated mathematically, providing a glimpse of the mechanical characteristics of this job mix formula to be verified later.

The following defines the steps detailed in each section.

2.2.1. Analysis of Starting Materials

Industrial by-products have an environmental advantage over virgin materials, as using the former reduces the environmental impact by decreasing the extraction rate of other materials. These by-products must be studied in detail, however, as most have special characteristics that could cause revalorization to fail.

Among the by-products used in this study, reclaimed asphalt pavement was studied to determine its physical, chemical, and mechanical characteristics. Firstly, particle size was analyzed using the test UNE-EN 933-1 [71], to assess suitability for the grading envelope stipulated by this standard, as well as the percentage of ladle furnace slag to be added. After analyzing particle size, the binder and coarse and fine aggregates were separated according to UNE-EN 12697-1 [72]. The aged binder was then studied through the UNE-EN 1426 [62] penetration and UNE-EN 1427 [63] softening point tests, to evaluate the aging point of the binder. Next, since the coarse aggregate of the reclaimed asphalt pavement is responsible for providing mineral skeleton to the mix, the following tests are compulsory: determination of resistance to fragmentation, UNE-EN 1097-2 [73]; determination of percentage of crushed and broken surfaces in coarse aggregate particles, UNE-EN 933-5 [74]; and flakiness index, UNE-EN 933-3 [75]. The fine aggregate was tested using the sand equivalent test UNE-EN 933-8 [76] and the plasticity index (UNE 103,103 [77] and UNE 103,104 [78]) to identify the presence of clay elements that could impair the mix.

Next, the ladle furnace slag was studied chemically and physically. An X-ray fluorescence test provided the elemental composition of the sample, enabling evaluation of its chemical aptitude for the final mix and expected cement characteristics, as well as chemical elements that could pose problems in the mixture once the road is finished.

The physical properties were then evaluated through particle size analysis, UNE-EN 933-1 [71], a determination of particle density, UNE-EN 1097-7 [79], and the bulk density of filler in kerosene, UNE-EN 1097-3 [80]. These typical civil engineering tests are essential in the study of waste, because variation from the density of the usual virgin material necessitates appropriate volumetric corrections. As with the fine aggregate, the material's plasticity was analyzed (UNE 103,103 [77] and 103,104 [78]) to prevent it from impairing the final mix due to expansivity problems.

2.2.2. Manufacture of the Different Sample Families and Tests

After analyzing the starting materials and evaluating their suitability for cold in-place recycling with bitumen emulsion, the corresponding grading curve was adjusted by adding the ladle furnace slag to the reclaimed asphalt pavement. The grading curve obtained by combining the two materials must conform to the grading envelope defined by Circular 8/2001 [53], more specifically RE2. The type of mix (RE2) was chosen for its reduced layer thickness, and thus its more suitable mechanical characteristics.

Once the combination percentage of reclaimed asphalt pavement and ladle furnace slag was determined, the percentages for the addition of fluids (bitumen emulsion and precoating water) were studied. Following the NLT-389/00 [66] standard, we first calculated the Theoretical Content of Fluids (TCF), that is, the percentage of optimal humidity obtained from the Modified Proctor Compaction test UNE 103,501 [67] for combining reclaimed asphalt pavement and ladle furnace slag. The value is the reference for the percentage of precoating water, plus the percentage of bitumen emulsion that must be added to the aggregates to achieve maximum density and approximate highest resistance.

The definitive percentage, termed Optimal Fluid Content (OFC), was determined by performing several coating tests with different percentages of precoating water, while keeping the emulsion percentage constant. This procedure should produce a total fluid content (water plus emulsion) ranging from TCF−2% to TCF. The coating test was performed in accordance with NLT-196/84 [68], adding no calcium carbonate and only the combined reclaimed asphalt pavement and ladle furnace slag, as well as the emulsion and precoating water stipulated. A coating test then evaluated the suitability of the emulsion in combination with the aggregates, enabling selection of the percentage corresponding to the Optimal Fluid Content (OFC). This value had to be between TCF−1% and TCF%.

Following the determination of the Optimal Fluid Content (OFC), the percentages of precoating water and bitumen emulsion for the different families were determined, based on the knowledge that the emulsion percentage should be within a range of 2.5–4%. Three families were created with different percentages of bitumen emulsion-to-aggregate (recovered asphalt pavement plus ladle kiln slag) of 3%, 3.5% and 4%. Each percentage of emulsion corresponded to a percentage of precoating water equal to the difference between the Optimum Fluid Content (OFC) and the percentages of emulsion.

After determining the composition of the different families, 12 samples were manufactured for each family, designed according to the norm, NLT-161/98 [69]. The compaction process according to the detailed standard consisted of: first, pouring the bituminous mixture into standardized molds; next, an initial load of 1 MPa was then applied; finally, a load of 21 MPa was applied over a time of 2 to 3 min at constant speed. After manufacturing, these samples had to be cured in a forced air oven at 50 ± 2 °C until they reached constant mass, for no less than 3 days and not more than 7 days.

Upon completion of this process, maximum density, UNE-EN 12697-5 [81], and bulk density of the specimens, UNE-EN 12697-6 [82], as well as the void content, UNE-EN 12697-8 [83], of the different families was determined. The goal was to divide each family into two groups to study the effect of water on cohesion of the compacted asphalt mix (immersion-compression test NLT-162/00 [70]). A group from each family was subjected to the action of water to study cohesion. The samples were submerged in a water bath regulated at 49 ± 1 °C for 4 days and then tested following NLT-161/98 [69] to evaluate the difference in resistance between the sample subjected to the action of the water and the sample kept dry.

2.2.3. Determination of the Optimal Job Mix Formula

Once the results for densities, dry resistance, post-immersion resistance, and preserved resistance were obtained for the different families, they were evaluated mathematically and graphically to obtain maximum dry and post-immersion resistance results. These values had to be higher than the minimums stipulated in Circular 8/2001 [53]—a dry resistance value of 3 MPa and post-immersion resistance of 2.5 MPa. Once the maximum and expected resistance values were calculated by mathematical correlation, the specimen family was created with the optimal job mix formula in order to corroborate the approximate values. Ultimately, once resistance was confirmed, this step yielded the ideal combination of reclaimed asphalt pavement, ladle furnace slag, bitumen emulsion, and water to manufacture cold in-place recycled pavement.

It should be noted that the optimum combination of materials was obtained through the dry resistance test because it is the test that is most limited by the regulations, and which also best characterizes the bituminous mixture, as it provides the resistance of this structural course.

3. Results and Discussions

The following subsections present the results of the methodology described above.

3.1. Results of the Analysis of the Starting Materials

Particle size of the reclaimed asphalt pavement was analyzed to study its grading curve. This analysis enabled the adjustment of the percentage of ladle furnace slag added, to comply with the grading envelope established in the regulations. Figure 1 displays the results of the particle size analysis and the correspondence of the particle size distribution obtained for the reclaimed asphalt pavement to the RE2 grading envelope stipulated by Circular 8/2001 [53].

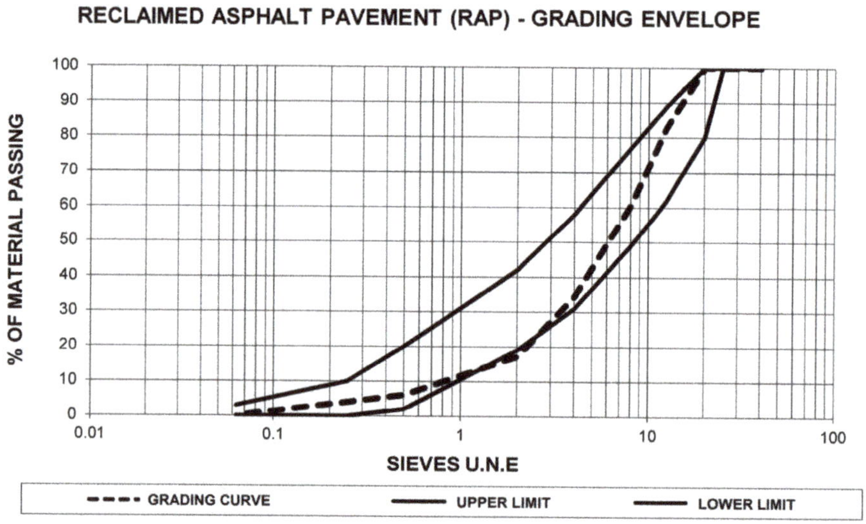

Figure 1. Graph of the grading curve of reclaimed asphalt pavement referenced to the RE2 grading envelope of Circular 8/2001 [53].

As Figure 1 shows, the grading curve of the reclaimed asphalt pavement reflects a majority composition of coarse aggregate, with a maximum size of less than 20 mm. Furthermore, the low proportion of fine aggregate led to the addition of ladle furnace slag adjusting the grading curve to the established grading envelope, as well as to reveal its cementitious characteristics.

Once the binder was extracted from the reclaimed asphalt pavement, separation of the aggregate enabled determination of the existing amount of the binder in the aggregate as 4.3%. This percentage is common in semi-dense asphalt mix and continuous grading used for medium- or low-traffic roads. The extracted bitumen binder was then analyzed (see Table 2 for results).

Table 2. Tests of binder extracted from reclaimed asphalt pavement.

Test	Standard	Value/Unit
Penetration (25 °C; 100 g; 5 s)	UNE-EN 1426 [62]	7 ± 0 (1/10) mm
Softening point	UNE-EN 1427 [63]	88 ± 2 °C

The results for penetration and softening point confirmed expectations. The pavement had aged by exhaustion, causing the binder to mostly lose its elasticity and become excessively hard. It should be noted that the hardness values were also influenced by the type of bitumen used in the area from which the pavement was reclaimed, where very hard bitumen (type B 40/50) is used due to the warm climate.

Moreover, the aggregate extracted from the pavement, unmilled to avoid alteration, was analyzed to confirm its suitability. The results of the tests are detailed in Table 3.

Table 3. Tests of the coarse aggregate and fine aggregate of the reclaimed asphalt pavement.

Coarse Aggregate		
Test	Standard	Value/Unit
Determination of percentage of crushed and broken surfaces	UNE-EN 933-5 [74]	93 ± 2%
Flakiness index	UNE-EN 933-3 [75]	88 ± 2 °C
Los Angeles Test method	UNE-EN 1097-2 [73]	19 ± 1%
Fine Aggregate		
Test	Standard	Value/Unit
Plasticity index	UNE 103103/UNE 103,104 [77,78]	2.9 ± 0.1%
Sand equivalent	UNE-EN 933-8 [76]	79 ± 2%

The results shown in Table 3 reflect a coarse aggregate of acceptable mechanical resistance and particle shape, more than adequate for use as aggregate in an asphalt mix for roads with intermediate traffic volume. The low plasticity index and sand equivalent value greater than 75 rule out the possible presence of clay particles that could cause problems due to expansiveness.

In line with the previous assumptions, reclaimed asphalt pavement can be classified as suitable for use in cold in-place recycling, but not without first correcting the grading through the addition of ladle furnace slag, and studying its compatibility with bitumen emulsion.

The ladle furnace slag was then analyzed elementally to detect its composition and to identify elements that could cause problems. Table 4 presents the results of the X-ray fluorescence.

The results show that the ladle furnace slag's composition derives directly from its nature and the production process, highlighting the percentages of calcium oxide so necessary for achieving primary resistance of the shaped mixture, and thus its suitability for supporting traffic. Percentages of silicon oxides are also necessary to achieve good cementitious characteristics over time.

The grading of the ladle furnace slag obtained in the physical tests is presented in Figure 2.

The results show a significant percentage of fine aggregates and a lower proportion of coarse aggregates, with a maximum aggregate size of 12.5 mm.

The remaining physical tests of density and plasticity for the fine portion of the ladle furnace slag are detailed in Table 5.

Table 4. Results of the X-ray fluorescence of ladle furnace slag.

Compound	wt,%	Est. Error
CaO	40.19	0.25
MgO	19.38	0.20
SiO_2	12.49	0.17
Al_2O_3	7.29	0.13
Fe_2O_3	2.38	0.08
MnO	0.936	0.047
S	0.548	0.027
TiO_2	0.486	0.024
BaO	0.240	0.012
Na_2O	0.118	0.042
Cr_2O_3	0.1100	0.0055
Cl	0.0833	0.0042
SrO	0.0733	0.0037
ZnO	0.0681	0.0034
K_2O	0.0506	0.0025
ZrO_2	0.0425	0.0021
V_2O_5	0.0179	0.0017
P	0.0138	0.0012
CuO	0.0117	0.0010
NiO	0.0082	0.0011
PbO	0.0048	0.0010
Nb_2O_5	0.0046	0.0006
MoO_3	0.0028	0.0009
Co_3O_4	0.0021	0.0009
SeO_2	0.0012	0.0005

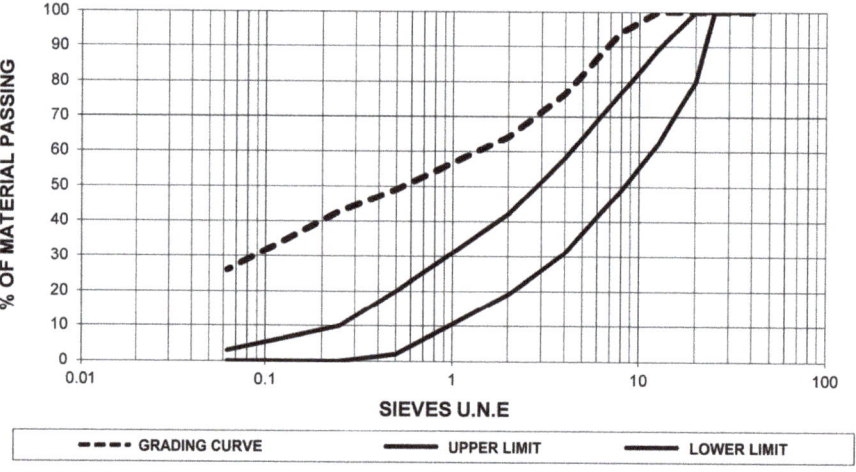

Figure 2. Graph of the grading curve of ladle furnace slag referenced to the RE2 grading envelope of Circular 8/2001 [53].

Table 5. Density and plasticity tests for the fine portion of ladle furnace slag.

Test	Standard	Value/Unit
Particle density	UNE-EN 1097-7 [79]	2.71 ± 0.07 t/m^3
Bulk density	UNE-EN 1097-3 [80]	0.75 ± 0.01 t/m^3
Plasticity index	UNE 103,103/UNE 103,104 [77,78]	No plasticity

Density values did not vary from those of typical virgin material filler; they showed a bulk density value of less than 0.8 t/m^3, indicating non-pulverulent behavior that would not impair operability. Plasticity is clearly zero, as this material had a significant percentage of calcium oxide and silicon oxide.

Therefore, after the study of both materials, reclaimed asphalt pavement and ladle furnace slag, it can be concluded that both are suitable for use in new bituminous mixtures. The reclaimed asphalt pavement had aged bitumen, albeit with an acceptable quality of aggregate for its reuse, while the ladle furnace slag had a chemical composition suitable for the development of the expected cementitious characteristics, and a fine grain size ideal for combination with the reclaimed asphalt pavement.

3.2. Test Results of the Different Sample Families

Based on the particle size analysis of the reclaimed asphalt pavement and ladle furnace slag, the percentage of each element to be added was calculated to confirm achievement of the grading envelope stipulated in Circular 8/2001 [53], and to incorporate an adequate percentage of ladle furnace slag to produce the cementitious characteristics that make the pavement resistant. The proportion of elements in the combination was 90% reclaimed asphalt pavement and 10% ladle furnace slag, a percentage chosen based on detailed adjustments. Figure 3 displays the corresponding grading curve for the mixture of the two materials. In the following sections, aggregate will be used to indicate the combination of these materials in the percentages specified, with the bitumen emulsion and water percentages referring to the mass of the two together.

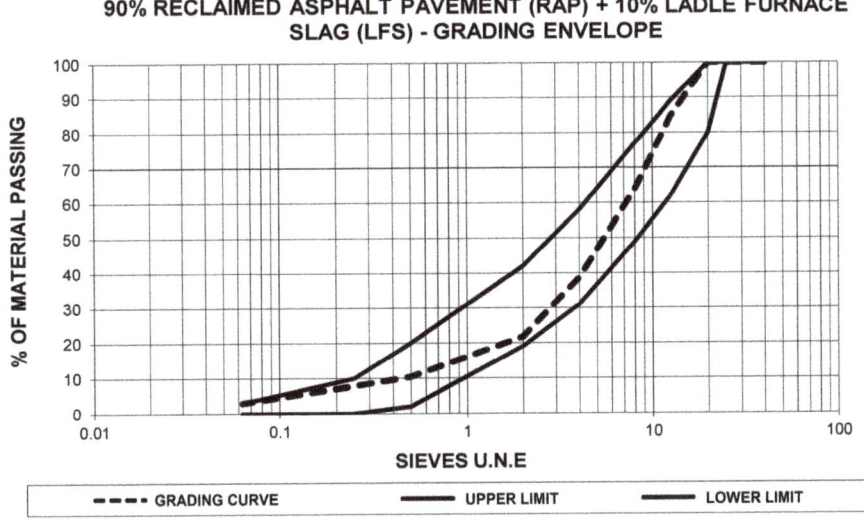

Figure 3. Graph of the grading curve for the combination of 90% reclaimed asphalt pavement and 10% ladle furnace slag, referenced to the RE2 grading envelope of Circular 8/2001 [53].

Once the percentage of the combined materials was determined, the Modified Proctor Compaction Test UNE 103,501 [67] was performed to determine the optimal humidity to obtain maximum compaction density. Optimal humidity, also termed the Theoretical Content of Fluids (TCF), corresponds to precoating water plus emulsion.

The Modified Proctor Compaction Test was performed for water percentages of 0%, 2.5%, 5%, 7.5%, and 10% in the mixture, establishing a clear maximum density value (1.72 t/m^3) with a humidity of 5.9%. The results of the Proctor Test can be seen in Figure 4.

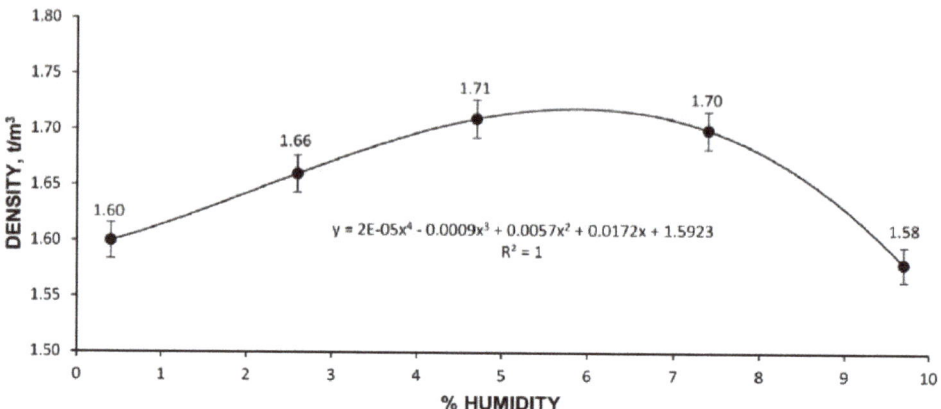

Figure 4. Graph of the Modified Proctor Compaction Test UNE 103,501 [67] for the combination of 90% reclaimed asphalt pavement and 10% ladle furnace slag.

After determining the Theoretical Content of Fluids (TCF) (5.9% to aggregate), Optimal Fluid Content (OFC) was calculated, starting with an essential test, the coating test NLT-196/84 [68]. An emulsion percentage of 3% to aggregate (reclaimed asphalt pavement plus ladle furnace slag) was established, as well as increasing percentages of precoating water from TCF–2% to TCF—that is, 0.9%, 1.9%, and 2.9% water-to-aggregate. Figure 5 presents images of adhesion of the emulsion with different percentages of precoating water.

Figure 5. Coating Test NLT-196/84 [68] for reclaimed asphalt pavement, ladle furnace slag, and 3% emulsion with different percentages of precoating water. (**a**) 0.9% water-to-aggregate. (**b**) 1.9% water-to-aggregate. (**c**) 2.9% water-to-aggregate.

Examination of the Coating Tests identified the best adhesion and coating of the emulsion, which occurred in the mix with 2.9% precoating water and 3% emulsion-to-aggregate, and was classified as good. This material's good dry behavior was easily observable, as it coincided in this case with the Theoretical Content of Fluids (TCF). It should be noted that the wrapping times were less than 60 s, and the breaking times of the emulsion were less than 300 s, reliable proof of compatibility of the emulsion with the aggregate.

Once Optimal Fluid Content (OFC) was determined, the different families of samples were manufactured with increasing percentages of emulsion. Circular 8/2001 [53] establishes a percentage of emulsion for cold in-place recycling of 2.5–4% to aggregate. To cover the entire possible range of combinations, therefore, three families were made, with increasing percentages of emulsion in increments of 0.5%, from 3% to 4% emulsion-to-aggregate. The 2.5% emulsion family was eliminated based on the assumption that this percentage of emulsion would be insufficient due to the large quantity of fine particles present. The precoating water for each family was calculated as the difference between Optimal Fluid Content (OFC) and the corresponding emulsion percentage. The percentage additions of precoating water and emulsion for each family are displayed in Table 6.

Table 6. Group of specimens manufactured with different percentages of emulsion and precoating water-to-aggregate.

Group	1	2	3
Precoating water-to-aggregate, %	2.9	2.4	1.9
Emulsion-to-aggregate, %	3	3.5	4

In total, 12 specimens were manufactured for each group following the standard NLT-161/98 [69]. After curing to a constant mass at a temperature of 50 ± 2 °C in a forced air stove, the specimens from each family were subdivided into two groups, which were subjected to different conditions. One group was immersed in water and the other kept dry, to assess the effect of water on the cohesion of the bituminous mixture. Table 7 shows the average values of particle density, bulk density, void content, dry compressive strength, post-immersion compressive strength, and preserved resistance.

Table 7. Test of maximum density, bulk density, void content, dry compressive strength, post-immersion compressive strength, and preserved resistance for the different groups of specimens.

Test	Standard	1	2	3
Maximum density, t/m^3	UNE-EN 12697-5 [81]	2.34 ± 0.04	2.31 ± 0.03	2.31 ± 0.08
Apparent density, t/m^3	UNE-EN 12697-6 [82]	2.14 ± 0.08	2.15 ± 0.07	2.14 ± 0.04
Void content, %	UNE-EN 12697-8 [83]	8.37 ± 0.31	6.81 ± 0.15	7.04 ± 0.25
Dry compressive strength, MPa	NLT-162/00 [70]	3.22 ± 0.12	3.51 ± 0.12	2.81 ± 0.09
Immersion compressive strength, MPa	NLT-162/00 [70]	2.90 ± 0.04	2.63 ± 0.03	1.99 ± 0.03
Preserved Resistance Index, %	NLT-162/00 [70]	90 ± 2	75 ± 2	71 ± 1

Based on the minimum values set forth in the Spanish regulations (3 MPa for dry resistance and 2.5 MPa for post-immersion compressive resistance), the values of 3% emulsion and 2.9% precoating water were acceptable in principle, as were 3.5% emulsion and 2.4% of precoating water-to-aggregate. The values obtained showed a clear decrease in the index of voids in the mixture with a higher percentage of emulsion, due to the compactability conditions provided by the emulsion. Based on dry compressive strength, the results identified an optimum point of 3–3.5% of emulsion-to-aggregate, as well as a decrease in post-immersion resistance with higher percentages of emulsion. The mathematically calculated optimal job mix formula provided maximum strength based on the results analyzed, showing the optimum combination of reclaimed asphalt pavement, ladle furnace slag, bitumen emulsion and water.

3.3. Optimal Job Mix Formula

Once the different specimen families were evaluated, the percentage of emulsion to obtain maximum resistance was studied mathematically. This percentage depended on the dry resistance and the points obtained from the different families. The maximum of this function, illustrated in Figure 6, coincided with 3.3% emulsion-to-aggregate and 2.6% precoating water.

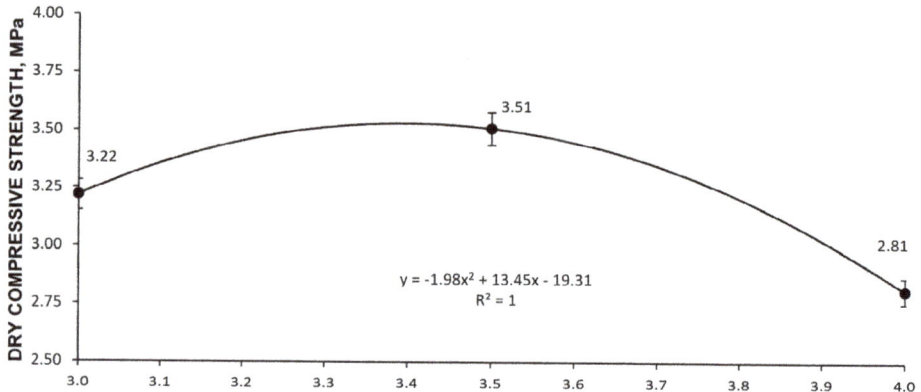

Figure 6. Test of dry compressive strength, NLT-162/00 [70], for the different emulsion percentages of each family of specimens.

Using the maximum obtained mathematically, and the percentages of the different materials to be added, 12 specimens were manufactured to confirm the optimal properties obtained with this job mix formula.

The process of manufacturing, curing, and studying the effect of water was similar to that performed on the other families. Table 8 presents the results of the trial for this optimal job mix formula.

Table 8. Tests for family of specimens made with the optimal job mix formula: 3.3% emulsion and 2.6% precoating water-to-aggregate.

	Optimal job mix formula	
Test	Standard	Value/Unit
Precoating water, % of aggregate	-	2.6
Emulsion, % of aggregate	-	3.3
Maximum density, t/m^3	UNE-EN 12697-5 [81]	2.32 ± 0.07
Bulk density, t/m^3	UNE-EN 12697-6 [82]	2.15 ± 0.04
Void content, %	UNE-EN 12697-8 [83]	7.30 ± 0.27
Dry compressive strength, MPa	NLT-162/00 [70]	3.65 ± 0.06
Immersion compressive strength, MPa	NLT-162/00 [70]	2.91 ± 0.08
Preserved Resistance Index, %	NLT-162/00 [70]	80 ± 2

The combination of 90% reclaimed asphalt pavement to 10% ladle furnace slag, plus 2.6% precoating water and 3.3% emulsion, produced results superior to those of the other families. The values for dry compressive strength, post-immersion compressive strength, and preserved resistance were higher than those required by the relevant regulations. The minimum values established by Circular 8/2001 [53] are 3 MPa for dry compressive strength and 2.5 MPa for post-immersion compressive strength, with a Preserved Resistance Index maintained as greater than 75%.

The following should be highlighted: a high percentage of RAP was utilized. This percentage of 90% RAP would have been unacceptable in other techniques that do not use bitumen emulsion and just use bitumen (e.g., hot mixes asphalt). This percentage of RAP was adequate, as reflected in the compressive strength tests. These results were obtained thanks to its cementitious properties and by the addition of ladle furnace slag. Finally, it should be pointed out that the same milling equipment was used for the laboratory tests as would later be used in the execution of the bituminous mix, since otherwise the particle size distribution could vary.

4. Conclusions

The following summarizes the partial conclusions that can be drawn from the results of the tests described in the methodology:

- Elementary composition of the ladle furnace slag studied showed a majority percentage of calcium oxide and a lower percentage of silicon oxide. Both compounds are essential for developing the desired cementitious characteristics of the ladle furnace slag and providing the strength of the asphalt mix manufactured with them.
- The ladle furnace slag had a maximum aggregate size of less than 12.5 mm, with primarily fine grading. Its particle and bulk density were comparable to those of a conventional aggregate and did not show plasticity.
- The slow-breaking cationic bitumen emulsion C60B5 REC showed good compatibility with the combination of ladle furnace slag and reclaimed asphalt pavement, as shown by the coating test. Both the adhesion of the emulsion to the aggregate and its cohesion and breaking times are suitable for use with ladle furnace slag and reclaimed asphalt pavement, with maximum precoating amounts of water-to-aggregate of 2.9% and emulsion-to-aggregate of 3%.
- The tests of both simple dry and post-immersion compressive strength showed good results (superior to Spanish regulations) for 3–3.5% emulsion-to-aggregate, corresponding to 2.9–2.4% precoating water-to-aggregate, respectively.
- The optimal combination of the different materials—considering an aggregate mixture of 90% reclaimed asphalt pavement and 10% ladle furnace slag, emulsion percentages of 3.3% to aggregate, and 2.6% precoating water—showed values of simple dry compressive strength and post-immersion compressive strength higher than those established by Spanish regulations and those of the other families tested.

Based on these detailed partial conclusions, ladle furnace slag has good characteristics for manufacturing pavement through cold in-place recycling with bitumen emulsion. The addition of ladle furnace slag achieved, on the one hand, the appropriate adjustment of the particle size of the reclaimed asphalt pavement, and on the other hand, provided resistance characteristics observable in the results obtained. It was therefore an ideal solution that created a sustainable asphalt mix: by considerably reducing CO_2 emissions; that used by-products in its composition therefore avoiding their disposal in landfills; and that was in-keeping with new environmental and circular economy trends.

Author Contributions: Conceptualization, F.A.C.-I., F.J.I.-G., J.M.T.-S. and C.M.-G.; methodology, F.A.C.-I., F.J.I.-G., J.M.T.-S. and C.M.-G.; software, J.M.T.-S. and C.M.-G.; validation, F.A.C.-I. and F.J.I.-G.; formal analysis, F.A.C.-I. and F.J.I.-G.; investigation, J.M.T.-S. and C.M.-G.; resources, F.A.C.-I.; data curation, F.J.I.-G.; writing—original draft preparation, C.M.-G.; writing—review and editing, J.M.T.-S.; visualization, J.M.T.-S.; supervision, F.A.C.-I.; project administration, C.M.-G.; funding acquisition, F.A.C.-I. All authors have read and agreed to the published version of the manuscript.

Funding: This research received no external funding.

Acknowledgments: Technical and human support provided by CICT of Universidad de Jaén (UJA, MINECO, Junta de Andalucía, FEDER) is gratefully acknowledged.

Conflicts of Interest: The authors declare no conflict of interest.

References

1. Plati, C. Sustainability factors in pavement materials, design, and preservation strategies: A literature review. *Constr. Build. Mater.* **2019**, *211*, 539–555. [CrossRef]
2. Shi, X.; Mukhopadhyay, A.; Zollinger, D.; Grasley, Z. Economic input-output life cycle assessment of concrete pavement containing recycled concrete aggregate. *J. Clean. Prod.* **2019**, *225*, 414–425. [CrossRef]
3. Arabani, M.; Azarhoosh, A.R. The effect of recycled concrete aggregate and steel slag on the dynamic properties of asphalt mixtures. *Constr. Build. Mater.* **2012**, *35*, 1–7. [CrossRef]
4. Jin, R.; Li, B.; Zhou, T.; Wanatowski, D.; Piroozfar, P. An empirical study of perceptions towards construction and demolition waste recycling and reuse in China. *Resour. Conserv. Recycl.* **2017**, *126*, 86–98. [CrossRef]
5. Menaria, Y.; Sankhla, R. Use of Waste Plastic in Flexible Pavements-Green Roads. *Open J. Civ. Eng.* **2015**, *05*, 299–311. [CrossRef]
6. Simone, A.; Vignali, V.; Lantieri, C. A new "frugal" approach to road maintenance: 100% Recycling of a deteriorated flexible pavement. In Proceedings of the 7th International Conference on Maintenance and Rehabilitation of Pavements and Technological Control, MAIREPAV 2012, Auckland, New Zealand, 28–30 August 2012.
7. Turk, J.; Mauko Pranjić, A.; Mladenovič, A.; Cotič, Z.; Jurjavčič, P. Environmental comparison of two alternative road pavement rehabilitation techniques: Cold-in-place-recycling versus traditional reconstruction. *J. Clean. Prod.* **2016**, *121*, 45–55. [CrossRef]
8. Zhang, W.; Yang, J.; Fan, X.; Yang, R.; Yu, B. Life-Cycle Cost Analysis of Base Course Using Cold In-Place Recycling: Case Study. In Proceedings of the Geo-Frontiers 2011, Dallas, TX, USA, 13–16 March 2011.
9. Liu, M.H. Research and Application Prospect on Cold Recycling Technology of Asphalt Pavement. *Appl. Mech. Mater.* **2012**, *204–208*, 1909–1913. [CrossRef]
10. Alkins, A.E.; Lane, B.; Kazmierowski, T. Sustainable Pavements: Environmental, Economic, and Social Benefits of In Situ Pavement Recycling. *Transp. Res. Rec.* **2008**, *2084*, 100–103. [CrossRef]
11. Martnez-Echevarría, M.J.; Rubio, M.C.; Menendez, A. The reuse of waste from road resurfacing: Cold in-place recycling of bituminous pavement, an environmentally friendly alternative to conventional pavement rehabilitation methods. *WIT Trans. Ecol. Environ.* **2008**, *109*, 459–469.
12. Thenoux, G.; González, Á.; Dowling, R. Energy consumption comparison for different asphalt pavements rehabilitation techniques used in Chile. *Resour. Conserv. Recycl.* **2007**, *49*, 325–339. [CrossRef]
13. Modarres, A.; Rahimzadeh, M.; Zarrabi, M. Field investigation of pavement rehabilitation utilizing cold in-place recycling. *Resour. Conserv. Recycl.* **2014**, *83*, 112–120. [CrossRef]
14. Cross, S.A.; Chesner, W.H.; Justus, H.G.; Kearney, E.R. Life-Cycle Environmental Analysis for Evaluation of Pavement Rehabilitation Options. *Transp. Res. Rec. J. Transp. Res. Board* **2011**, *2227*, 43–52. [CrossRef]
15. Euch Khay, S.E.; Euch Ben Said, S.E.; Loulizi, A.; Neji, J. Laboratory Investigation of Cement-Treated Reclaimed Asphalt Pavement Material. *J. Mater. Civ. Eng.* **2015**, *27*, 04014192. [CrossRef]
16. Lee, K.W.; Brayton, T.E.; Mueller, M.; Singh, A. Rational Mix-Design Procedure for Cold In-Place Recycling Asphalt Mixtures and Performance Prediction. *J. Mater. Civ. Eng.* **2016**, *28*, 04016008. [CrossRef]
17. Li, X.; Wen, H.; Edil, T.B.; Sun, R.; VanReken, T.M. Cost, energy, and greenhouse gas analysis of fly ash stabilised cold in-place recycled asphalt pavement. *Road Mater. Pavement Des.* **2013**, *14*, 537–550. [CrossRef]
18. Kim, Y.; Lee, H. "David" Development of Mix Design Procedure for Cold In-Place Recycling with Foamed Asphalt. *J. Mater. Civ. Eng.* **2006**, *18*, 116–124. [CrossRef]
19. Stimilli, A.; Ferrotti, G.; Graziani, A.; Canestrari, F. Performance evaluation of a cold-recycled mixture containing high percentage of reclaimed asphalt. *Road Mater. Pavement Des.* **2013**, *14*, 149–161. [CrossRef]
20. Diefenderfer, B.K.; Bowers, B.F.; Apeagyei, A.K. Initial Performance of Virginia's Interstate 81 In-Place Pavement Recycling Project. *Transp. Res. Rec. J. Transp. Res. Board* **2015**, *2524*, 152–159. [CrossRef]
21. Maurer, G.; Bemanian, S.; Polish, P. Alternative Strategies for Rehabilitation of Low-Volume Roads in Nevada. *Transp. Res. Rec. J. Transp. Res. Board* **2007**, *1989–2*, 309–320. [CrossRef]
22. Cox, B.C.; Howard, I.L.; Battey, R. In-Place Recycling Moisture-Density Relationships for High-Traffic Applications. In Proceedings of the IFCEE 2015, San Antonio, TX, USA, 17–21 March 2015.
23. Diefenderfer, B.K.; Apeagyei, A.K.; Gallo, A.A.; Dougald, L.E.; Weaver, C.B. In-Place Pavement Recycling on I-81 in Virginia. *Transp. Res. Rec. J. Transp. Res. Board* **2012**, *2306*, 21–27. [CrossRef]

24. Cox, B.C.; Howard, I.L.; Campbell, C.S. Cold In-Place Recycling Moisture-Related Design and Construction Considerations for Single or Multiple Component Binder Systems. *Transp. Res. Rec. J. Transp. Res. Board* **2016**, *2575*, 27–38. [CrossRef]
25. Recasens, R.M.; Pérez Jiménez, F.E.; Aguilar, S.C. Mixed recycling with emulsion and cement of asphalt pavements. Design procedure and improvements achieved. *Mater. Struct.* **2000**, *33*, 324–330. [CrossRef]
26. Wen, H.; Tharaniyil, M.P.; Ramme, B. Investigation of Performance of Asphalt Pavement with Fly-Ash Stabilized Cold In-Place Recycled Base Course. *Transp. Res. Rec. J. Transp. Res. Board* **2003**, *1819*, 27–31. [CrossRef]
27. Escorias de Acería de Horno de Arco Eléctrico | CEDEX. Available online: http://www.cedexmateriales.es/catalogo-de-residuos/25/escorias-de-aceria-de-horno-de-arco-electrico/ (accessed on 29 April 2020).
28. Motz, H.; Geiseler, J. Products of steel slags an opportunity to save natural resources. *Waste Manag. Ser.* **2000**, *1*, 207–220.
29. Pioro, L.; Pioro, I. Reprocessing of metallurgical slag into materials for the building industry. *Waste Manag.* **2004**, *24*, 371–379. [CrossRef]
30. Dippenaar, R. Industrial uses of slag (the use and re-use of iron and steelmaking slags). *Ironmak. Steelmak.* **2005**, *32*, 35–46. [CrossRef]
31. Tsakiridis, P.E.; Papadimitriou, G.D.; Tsivilis, S.; Koroneos, C. Utilization of steel slag for Portland cement clinker production. *J. Hazard. Mater.* **2008**, *152*, 805–811. [CrossRef]
32. Adolfsson, D.; Engström, F.; Robinson, R.; Björkman, B. Cementitious Phases in Ladle Slag. *Steel Res. Int.* **2011**, *82*, 398–403. [CrossRef]
33. Shi, C. Characteristics and cementitious properties of ladle slag fines from steel production. *Cem. Concr. Res.* **2002**, *32*, 459–462. [CrossRef]
34. Ortega-López, V.; Manso, J.M.; Cuesta, I.I.; González, J.J. The long-term accelerated expansion of various ladle-furnace basic slags and their soil-stabilization applications. *Constr. Build. Mater.* **2014**, *68*, 455–464. [CrossRef]
35. Manso, J.M.; Ortega-López, V.; Polanco, J.A.; Setién, J. The use of ladle furnace slag in soil stabilization. *Constr. Build. Mater.* **2013**, *40*, 126–134. [CrossRef]
36. Richardson, I.G.; Cabrera, J.G. The nature of C–S–H in model slag-cements. *Cem. Concr. Compos.* **2000**, *22*, 259–266. [CrossRef]
37. Akın Altun, İ.; Yılmaz, İ. Study on steel furnace slags with high MgO as additive in Portland cement. *Cem. Concr. Res.* **2002**, *32*, 1247–1249. [CrossRef]
38. Manso, J.M.; Rodriguez, Á.; Aragón, Á.; Gonzalez, J.J. The durability of masonry mortars made with ladle furnace slag. *Constr. Build. Mater.* **2011**, *25*, 3508–3519. [CrossRef]
39. Faraone, N.; Tonello, G.; Furlani, E.; Maschio, S. Steelmaking slag as aggregate for mortars: Effects of particle dimension on compression strength. *Chemosphere* **2009**, *77*, 1152–1156. [CrossRef]
40. Papayianni, I.; Anastasiou, E. Effect of granulometry on cementitious properties of ladle furnace slag. *Cem. Concr. Compos.* **2012**, *34*, 400–407. [CrossRef]
41. Rodriguez, Á.; Manso, J.M.; Aragón, Á.; Gonzalez, J.J. Strength and workability of masonry mortars manufactured with ladle furnace slag. *Resour. Conserv. Recycl.* **2009**, *53*, 645–651. [CrossRef]
42. Papayianni, I.; Anastasiou, E. Production of high-strength concrete using high volume of industrial by-products. *Constr. Build. Mater.* **2010**, *24*, 1412–1417. [CrossRef]
43. Montenegro, J.M.; Celemín-Matachana, M.; Cañizal, J.; Setién, J. Ladle Furnace Slag in the Construction of Embankments: Expansive Behavior. *J. Mater. Civ. Eng.* **2013**, *25*, 972–979. [CrossRef]
44. Sun, D.D.; Tay, J.H.; Cheong, H.K.; Leung, D.L.K.; Qian, G. Recovery of heavy metals and stabilization of spent hydrotreating catalyst using a glass–ceramic matrix. *J. Hazard. Mater.* **2001**, *87*, 213–223. [CrossRef]
45. Rađenović, A.; Malina, J.; Sofilić, T. Characterization of Ladle Furnace Slag from Carbon Steel Production as a Potential Adsorbent. *Adv. Mater. Sci. Eng.* **2013**, *2013*, 198240. [CrossRef]
46. Montenegro-Cooper, J.M.; Celemín-Matachana, M.; Cañizal, J.; González, J.J. Study of the expansive behavior of ladle furnace slag and its mixture with low quality natural soils. *Constr. Build. Mater.* **2019**, *203*, 201–209. [CrossRef]
47. Ortega-López, V.; Fuente-Alonso, J.A.; Santamaría, A.; San-José, J.T.; Aragón, Á. Durability studies on fiber-reinforced EAF slag concrete for pavements. *Constr. Build. Mater.* **2018**, *163*, 471–481. [CrossRef]

48. Pasetto, M.; Baldo, N. Recycling of waste aggregate in cement bound mixtures for road pavement bases and sub-bases. *Constr. Build. Mater.* **2016**, *108*, 112–118. [CrossRef]
49. Bocci, E. Use of ladle furnace slag as filler in hot asphalt mixtures. *Constr. Build. Mater.* **2018**, *161*, 156–164. [CrossRef]
50. Ziaee, S.A.; Behnia, K. Evaluating the effect of electric arc furnace steel slag on dynamic and static mechanical behavior of warm mix asphalt mixtures. *J. Clean. Prod.* **2020**, *274*, 123092. [CrossRef]
51. Skaf, M.; Ortega-López, V.; Fuente-Alonso, J.A.; Santamaría, A.; Manso, J.M. Ladle furnace slag in asphalt mixes. *Constr. Build. Mater.* **2016**, *122*, 488–495. [CrossRef]
52. Martínez-Echevarría, M.J.; Recasens, R.M.; del Carmen Rubio Gámez, M.; Ondina, A.M. In-laboratory compaction procedure for cold recycled mixes with bituminous emulsions. *Constr. Build. Mater.* **2012**, *36*, 918–924. [CrossRef]
53. Orden Circular 8/2001 Sobre RECICLADO DE FIRMES—Normativa de Carreteras. Available online: http://normativadecarreteras.com/listing/orden-circular-82001-sobre-reciclado-de-firmes/ (accessed on 29 September 2020).
54. UNE-EN 1430:2009 Bitumen and Bituminous Binders—Determination of Particle Polarity of Bituminous Emulsions. Available online: https://www.une.org/encuentra-tu-norma/busca-tu-norma/norma/?c=N0044069 (accessed on 29 September 2020).
55. UNE-EN 13075-1:2017 Bitumen and Bituminous Binders—Determination of Breaking Behaviour—Part 1: Determination of Breaking Value of Cationic Bituminous Emulsions, Mineral Filler Method. Available online: https://www.une.org/encuentra-tu-norma/busca-tu-norma/norma/?c=N0057840 (accessed on 29 September 2020).
56. UNE-EN 12846-1:2011 Bitumen and Bituminous Binders—Determination of Efflux Time by the Efflux Viscometer—Part 1: Bituminous Emulsions. Available online: https://www.une.org/encuentra-tu-norma/busca-tu-norma/norma/?c=N0047377 (accessed on 29 September 2020).
57. UNE-EN 12846:2003 Bitumen and Bituminous Binders. Determination of Efflux Time of Bitumen Emulsions by the Efflux Viscometer. Available online: https://www.une.org/encuentra-tu-norma/busca-tu-norma/norma/?c=N0029984 (accessed on 29 September 2020).
58. UNE-EN 1429:2013 Bitumen and bituminous binders—Determination of residue on sieving of bituminous emulsions, and determination of storage stability by sieving. Available online: https://www.une.org/encuentra-tu-norma/busca-tu-norma/norma?c=N0052189 (accessed on 29 September 2020).
59. UNE-EN 12847:2009 Bitumen and Bituminous binders—Determination of settling tendency of bituminous emulsions. Available online: https://www.une.org/encuentra-tu-norma/busca-tu-norma/norma/?c=N0044066 (accessed on 29 September 2020).
60. UNE-EN 13614:2011 Bitumen and bituminous binders—Determination of adhesivity of bituminous emulsions by water immersion test. Available online: https://www.une.org/encuentra-tu-norma/busca-tu-norma/norma?c=N0048094 (accessed on 29 September 2020).
61. UNE-EN 1431:2018 Bitumen and Bituminous Binders—Determination of Residual Binder and oil Distillate from Bitumen Emulsions by Distillation. Available online: https://www.une.org/encuentra-tu-norma/busca-tu-norma/norma?c=N0060676 (accessed on 29 September 2020).
62. UNE-EN 1426:2015 Bitumen and Bituminous Binders—Determination of Needle Penetration. Available online: https://www.une.org/encuentra-tu-norma/busca-tu-norma/norma/?c=N0055820 (accessed on 29 September 2020).
63. UNE-EN 1427:2015 Bitumen and Bituminous Binders—Determination of the Softening Point—Ring and Ball Method. Available online: https://www.une.org/encuentra-tu-norma/busca-tu-norma/norma/?c=N0055821 (accessed on 29 September 2020).
64. UNE-EN 13074-1:2019 Bitumen and Bituminous Binders—Recovery of Binder from Bituminous Emulsion or Cut-Back or Fluxed Bituminous Binders—Part 1: Recovery by Evaporation. Available online: https://www.une.org/encuentra-tu-norma/busca-tu-norma/norma?c=N0062153 (accessed on 29 September 2020).
65. UNE-EN 13074-2:2011 Bitumen and Bituminous Binders—Recovery of Binder from Bituminous Emulsion or Cut-Back or Fluxed Bituminous Binders—Part 2: Stabilisation after Recovery by Evaporation. Available online: https://www.une.org/encuentra-tu-norma/busca-tu-norma/norma/?c=N0047380 (accessed on 29 September 2020).

66. NLT-389/00 Fabricación y Curado de Mezclas de Materiales Áridos, reciclados o sin reciclar con emulsión bituminosa para ensayos de compresión y efecto del agua sobre su cohesión—Normativa de carreteras. Available online: http://normativadecarreteras.com/listing/nlt-38900-fabricacion-curado-mezclas-materiales-aridos-reciclados-sin-reciclar-emulsion-bituminosa-ensayos-compresion-efecto-del-agua-cohesion/ (accessed on 29 September 2020).
67. UNE 103501:1994 Geotechnic. Compactation Test. *Modified Proctor.* Available online: https://www.une.org/encuentra-tu-norma/busca-tu-norma/norma?c=N0007851 (accessed on 29 September 2020).
68. NLT-196/84 Envuelta y resistencia al desplazamiento por el agua de las emulsiones bituminosas—Normativa de carreteras. Available online: http://normativadecarreteras.com/listing/nlt-19684-envuelta-resistencia-al-desplaziamiento-agua-las-emulsiones-bituminosas/ (accessed on 29 September 2020).
69. NLT-161/98 Resistencia a compresión simple de mezclas bituminosas.—Normativa de carreteras. Available online: http://normativadecarreteras.com/listing/nlt-16100-resistencia-compresion-simple-mezclas-bituminosas/ (accessed on 29 September 2020).
70. NLT-162/00 Efecto del agua sobre la cohesión de las mezclas bituminosas compactadas. (ensayo de inmersión-compresión)—Normativa de carreteras. Available online: http://normativadecarreteras.com/listing/nlt-16200-efecto-del-agua-la-cohesion-las-mezclas-bituminosas-compactadas-ensayo-inmersion-compresion/ (accessed on 29 September 2020).
71. UNE-EN 933-1:2012 Tests for geometrical properties of aggregates—Part 1: Determination of particle size distribution—Sieving method. Available online: https://www.une.org/encuentra-tu-norma/busca-tu-norma/norma?c=N0049638 (accessed on 29 September 2020).
72. UNE-EN 12697-1:2013 Bituminous mixtures—Test methods for hot mix asphalt—Part 1: Soluble binder content. Available online: https://www.une.org/encuentra-tu-norma/busca-tu-norma/norma?c=N0050801 (accessed on 29 September 2020).
73. UNE-EN 1097-2:2010 Tests for Mechanical and Physical Properties of Aggregates—Part 2: Methods for the Determination of Resistance to Fragmentation. Available online: https://www.une.org/encuentra-tu-norma/busca-tu-norma/norma/?c=N0046026 (accessed on 16 September 2020).
74. UNE-EN 933-5:1999/A1:2005 Tests for geometrical properties of aggregates—Part 5: Determination of percentage of crushed and broken surfaces in coarse aggregate particles. Available online: https://www.une.org/encuentra-tu-norma/busca-tu-norma/norma/?c=N0034842 (accessed on 16 September 2020).
75. UNE-EN 933-3:2012 Tests for Geometrical Properties of Aggregates—Part 3: Determination of Particle Shape—Flakiness Index. Available online: https://www.une.org/encuentra-tu-norma/busca-tu-norma/norma?c=N0049063 (accessed on 16 September 2020).
76. UNE-EN 933-8:2012+A1:2015/1M: 2016 Tests for geometrical properties of aggregates—Part 8: Assessment of fines—Sand equivalent test. Available online: https://www.une.org/encuentra-tu-norma/busca-tu-norma/norma?c=N0056257 (accessed on 16 September 2020).
77. UNE 103103:1994 Determinación del Limite Liquido de un Suelo p. Available online: https://www.une.org/encuentra-tu-norma/busca-tu-norma/norma/?c=N0007830 (accessed on 29 September 2020).
78. UNE 103104:1993 Test for Plastic Limit of a Soil. Available online: https://www.une.org/encuentra-tu-norma/busca-tu-norma/norma?c=N0007831 (accessed on 29 September 2020).
79. UNE-EN 1097-7:2009 Tests for Mechanical and Physical Properties of Aggregates—Part 3: Determination of Loose Bulk Density and Voids. Available online: https://www.une.org/encuentra-tu-norma/busca-tu-norma/norma?c=N0042553 (accessed on 16 September 2020).
80. UNE-EN 1097-3:1999 Tests for Mechanical and Physical Properties of Aggregates—Part 3: Determination of Loose Bulk Density and Voids. Available online: https://www.une.org/encuentra-tu-norma/busca-tu-norma/norma/?c=N0009465 (accessed on 16 September 2020).
81. UNE-EN 12697-5:2020 Test Methods—Part 5: Determination of the Maximum Density. Available online: https://www.une.org/encuentra-tu-norma/busca-tu-norma/norma?c=N0063145 (accessed on 29 September 2020).

82. UNE-EN 12697-6:2012 Bituminous Mixtures—Test Methods for Hot Mix Asphalt—Part 6: Determination of Bulk Density of Bituminous Specimens. Available online: https://www.une.org/encuentra-tu-norma/busca-tu-norma/norma/?c=N0049868 (accessed on 29 September 2020).
83. UNE-EN 12697-8:2020 Bituminous Mixtures—Test methods—Part 8: Determination of Void Characteristics of Bituminous Specimens. Available online: https://www.une.org/encuentra-tu-norma/busca-tu-norma/norma/?c=N0063146 (accessed on 29 September 2020).

Publisher's Note: MDPI stays neutral with regard to jurisdictional claims in published maps and institutional affiliations.

© 2020 by the authors. Licensee MDPI, Basel, Switzerland. This article is an open access article distributed under the terms and conditions of the Creative Commons Attribution (CC BY) license (http://creativecommons.org/licenses/by/4.0/).

Article

The Influence of the Addition of Plant-Based Natural Fibers (Jute) on Biocemented Sand Using MICP Method

Md Al Imran [1,*], Sivakumar Gowthaman [1], Kazunori Nakashima [2] and Satoru Kawasaki [2]

1. Graduate School of Engineering, Hokkaido University, Sapporo 060-8628, Japan; gowtham1012@outlook.com
2. Faculty of Engineering, Hokkaido University, Sapporo 060-8628, Japan; k.naka@eng.hokudai.ac.jp (K.N.); kawasaki@geo-er.eng.hokudai.ac.jp (S.K.)
* Correspondence: imran@eis.hokudai.ac.jp; Tel.: +81-8011-706-6318

Received: 27 August 2020; Accepted: 18 September 2020; Published: 21 September 2020

Abstract: The microbial-induced carbonate precipitation (MICP) method has gained intense attention in recent years as a safe and sustainable alternative for soil improvement and for use in construction materials. In this study, the effects of the addition of plant-based natural jute fibers to MICP-treated sand and the corresponding microstructures were measured to investigate their subsequent impacts on the MICP-treated biocemented sand. The fibers used were at 0%, 0.5%, 1.5%, 3%, 5%, 10%, and 20% by weight of the sand, while the fiber lengths were 5, 15, and 25 mm. The microbial interactions with the fibers, the $CaCO_3$ precipitation trend, and the biocemented specimen (microstructure) were also evaluated based on the unconfined compressive strength (UCS) values, scanning electron microscopy (SEM), and fluorescence microscopy. The results of this study showed that the added jute fibers improved the engineering properties (ductility, toughness, and brittleness behavior) of the biocemented sand using MICP method. Furthermore, the fiber content more significantly affected the engineering properties of the MICP-treated sand than the fiber length. In this study, the optimal fiber content was 3%, whereas the optimal fiber length was s 15 mm. The SEM results indicated that the fiber facilitated the MICP process by bridging the pores in the calcareous sand, reduced the brittleness of the treated samples, and increased the mechanical properties of the biocemented sand. The results of this study could significantly contribute to further improvement of fiber-reinforced biocemented sand in geotechnical engineering field applications.

Keywords: jute; MICP; ureolytic bacteria; biocement; natural plant fiber

1. Introduction

Recently, significant interest in bio-mediated soil improvement has been highlighted as an innovative and effective approach for soil and ground improvement. Among the various bio-mediated soil development approaches, microbial-induced carbonate precipitation (MICP) has been recognized as a promising approach for soil improvement in recent years. The microbial urease hydrolyzes urea [$CO(NH_2)_2$] and produces ammonium and carbonate ions, and consequently increases the pH during the MICP, process resulting in an alkaline growing environment, which is favored for $CaCO_3$ precipitation. The $CaCO_3$ are the primary binding substances in between the sand particles, which lead to soil improvement.

Most of the previous studies [1–4] have focused on the impacts of various environmental factors on microorganism immobilization and strength improvement using several types of soil materials, including the capability of microorganisms to form $CaCO_3$ within sand particles and pores; the relationship between the precipitated $CaCO_3$ content and the strength of MICP-treated sand; the study of the engineering properties of MICP-treated sand, such as the volume, permeability, strength, and compressibility, which was assessed using introductory numerical simulations [5]. However, many recent experiments have demonstrated the mechanism of $CaCO_3$ deposition and improvement of soil strength after curing samples using the MICP method. Earlier research also demonstrated the non-uniformity of precipitation of $CaCO_3$ and brittle failure behavior of MICP-treated soil [6]. One study showed that the MICP-treated soil tended to fail at a low axial strain level in unconfined and triaxial compression tests. The axial stress of the samples also dropped rapidly after the peak stress, because most of the $CaCO_3$ precipitated non-uniformly close to the influent of the specimen column and hindered the biocementation process in the deeper location of the specimen [7].

To improve this shortcoming, several studies have been conducted to improve the ductility and toughness of sand after MICP curing. Although sand solidified by microorganisms can reach a high strength, brittle failure of the substances mainly occurs and the strength is lost immediately, leading to safety hazards and problems. Therefore, the development of new mechanisms to improve the toughness, ductility, and durability of MICP-treated soil is a critical research need. The toughness, ductility, and durability of MICP-treated soil can be improved by the addition of fibrous materials. Earlier studies showed that using fibrous materials increased bacterial interactions with the cementation solution and facilitated uniform $CaCO_3$ precipitation within the sand pores in MICP-treated soil [8]. Several studies have also revealed that the addition of fiber materials can substantially increase the engineering properties of MICP-treated soil (for example the shear strength, ductility, brittleness behavior, internal friction, and rigidity) [9]. However, to address these challenges, most of these previous studies focused on using synthetic-polymer-based fiber materials, such as non-woven geotextile fiber, steel fiber, polypropylene fiber, glass fiber, and carbon fiber, which are associated with relatively high costs, health concerns, and environmental hazards [10,11]. Moreover, to date, only a few experiments have concentrated on the improvement of the engineering properties of soils treated with the MICP method using fiber materials. Therefore, developing a new approach and identifying cheap, readily available fiber while considering safety, environmental considerations, and sustainability will require further investigations. A detailed analysis to improve the drawbacks of MICP-treated sand via the inclusion of fibers was inspired by the current deficiencies in the traditional MICP process.

This study considered the use of natural jute fiber (*Corchorus capsularis*), because jute fiber has excellent physical mechanical properties and high resistance capacity (temperature, pH, salinity) and is readily available, inexpensive, light weight, sustainable and eco-friendly, requiring low amounts of energy [12,13].

The primary objectives of this work were to investigate the effects of a natural fiber (jute) on the MICP-treated soil. To improve the engineering properties (ductility, toughness, and brittleness behavior) of the biocemented sand specimen in terms of the fiber length and content (ratio), carbonate precipitation patterns and interactions within the microorganisms were investigated in this study. The mechanical properties of the MICP-treated sand, the microstructure of the specimen, and the interactions between the fiber and microorganisms were investigated and analyzed using unconfined compressive strength (UCS), scanning electron microscopy (SEM), and fluorescence microscopy.

2. Materials and Methods

2.1. Fiber

Previous research studies have indicated that jute fibers of appropriate length and content can substantially increase a soil's engineering properties, and adding fibrous material can contribute to a significant reduction in construction costs [14,15] by improving the engineering properties of treated samples. Moreover, jute fibers have a high initial modulus, high consistency in terms of tenacity and tensile strength, high rigidity, and a lower percentage of elongation during breakage, leading to their wide use in soil improvement. The properties of jube fibers, as well as their availability, cost, and environmental friendliness, were the motivation for using jute fibers in this study. Locally available jute fibers (100% natural) were used in this study (Figure 1) without any chemical treatment. The jute fibers were purchased from DCM Homac Co., Ltd., Sapporo, Japan, and the fibers were collected by Hayase Industries, Ltd., Tsuyama, Japan. The microstructure of the jute fibers was examined through scanning electron microscope images. SEM images of clusters and single filaments of jute fibers are presented in Figure 1.

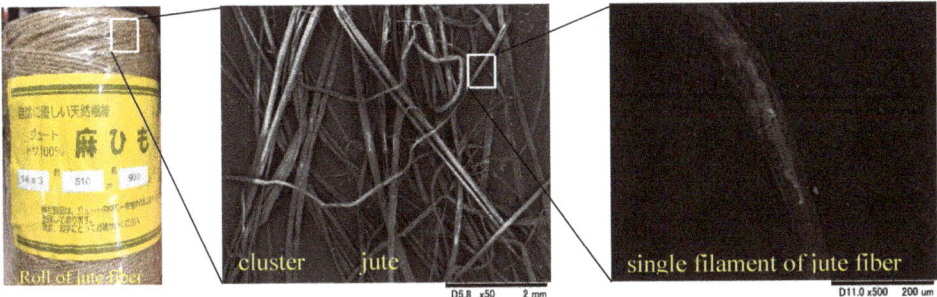

Figure 1. Appearance and microstructure images of jute fibers, obtained using SEM.

The key features of the jute fibers used in this study are shown in Table 1. It was shown that the efficiency of the fibers used to improve the soil highly depends on the properties of the fibers, such as the nature of fibers, the fiber length, and the fiber ratio [16]. The jute fibers were cut into three different lengths (5, 15, and 25 mm) and several percentages (content) by weight (0.5%, 1.5%, 3%, 5%, 10% and 20%) were used for mixture with "Mikawa" sand particles. The physical appearance of the prepared jute fiber samples is shown in Figure 2. All the jute fibers used in this study were placed in a dryer at 60 °C for 24 h before being mixed with the sand and receiving MICP treatment.

Table 1. The key features of the jute fibers used in this study.

Fiber Type	Thickness	Length (Total)	Weight	Type	Moisture Content	Colour
Jute	2 mm	510 m	900 g	Roll	3.4%	Golden-brown

Figure 2. The physical appearance of the jute fibers used in this study, which were cut into different length.

2.2. Microorganisms and Soil Properties

The bacteria used in this study was *Micrococcus yunnanensis* (hereafter denoted as G1), which was isolated from the coastal area of Porto Rafti, Greece [17]. The key features of this bacteria (Figure 3) are that it is known to exhibit comparatively high urease activity with salt-tolerant properties and can survive for extended periods of time in various temperature and pH conditions, as well as in nutrient-deficient conditions [18]. Liquid ZoBell2216 medium was used as the culture solution for the selected bacterial species. The culture medium was dissolved with hi-polypeptone (5.0 g/L), $FePO_4$ (0.1 g/L), and yeast extract (1.0 g/L). The components were mixed with artificial seawater and the pH was maintained at 7.6–7.8. The bacterial cells were precultured (using the ZoBell2216 medium) for 24 h at 30 °C in a shaker at 160 rpm. The precultured bacterial cells were then transferred (1 mL) into 100 mL of fresh ZoBell2216 medium and incubated at 30 °C at 160 rpm. The prepared bacterial culture solution was used for the MICP process. During the cultivation, the bacterial cell growth (OD_{600}) was determined and adjusted (by approximately 6) using a UV-Vis spectrophotometer (V-730, JASCO Corporation, Tokyo, Japan) and urease activity (1.5 ± 3 U/mL) was measured using the indophenol method [19].

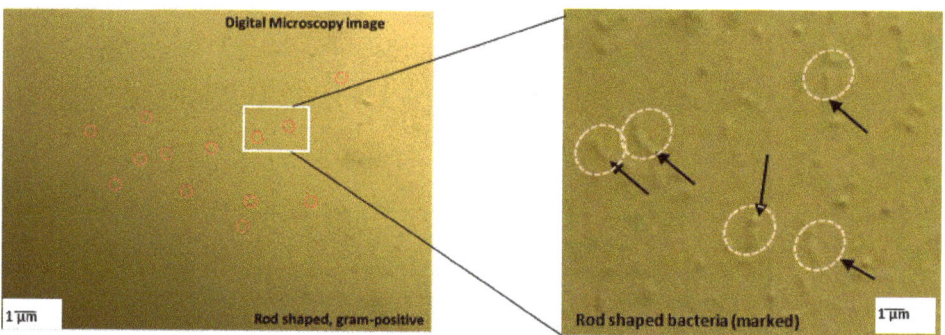

Figure 3. Digital microscopic image of the microorganism used in this study.

The soil used in this study was commercially available "Mikawa" sand. The maximum and minimum dry densities of the sand were 1.476 and 1.256 g/cm^3. The particle density and mean diameter were 2.66 g/cm^3 and 870 μm, respectively. The grain size distribution of "Mikawa" sand is presented in Figure 4. Before the MICP process, the sand was dried in an oven dryer at 110 °C for 24 h.

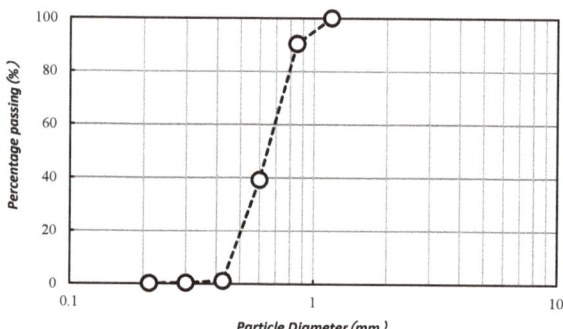

Figure 4. The grain size distribution of "Mikawa" sand.

2.3. Effects of Jute Fibers on the Process of $CaCO_3$ Precipitation

To investigate the interactions between jute fibers and $CaCO_3$, equal concentrations (0.5 mol/L) of $CaCl_2$ and urea solutions were used for the precipitation test in test tubes with and without fiber. The total volume of the mixers was adjusted up to 10 mL using distilled water, samples were kept in the shaker for 48 h at 30 °C, and the rpm was kept at 160. The testing conditions are shown in Tables 2 and 3, showing the fiber content and length, respectively. After 48 h, the resulting mixture was centrifuged to collect the crystal precipitate and the supernatant of solutions from the tube was removed separately using filter paper (Whatman filter paper, 11 μm (Global Life Sciences Technologies Ltd., Tokyo, Japan)). Both the filter papers and the tubes with the precipitate were oven-dried for 24 h at 110 °C, then subsequently the dry weights of the crystals were measured. The weight of the precipitated crystal was determined by contrasting the empty weight of the tube from the dry weight of the tube and the filter paper's dry weight. Using scanning electron microscopy (SEM; MiniscopeTM3000, Hitachi, Tokyo, Japan), the morphologies of crystals and jute fibers were analyzed. All the experiments were done in triplicate, and the mean value was plotted accordingly. Standard deviation was used to represent the error bars.

Table 2. Testing conditions for $CaCO_3$ precipitation with jute fibers (content).

Fiber Content [(%) mm]	$CaCl_2$ (M)	Urea (M)	Bacterial OD_{600}	Incubation Time (h)	Incubation Temperature (°C)
0	0.5	0.5	2	48	30
[(0.5) 15]	0.5	0.5	2	48	30
[(1.5) 15]	0.5	0.5	2	48	30
[(3) 15]	0.5	0.5	2	48	30
[(5) 15]	0.5	0.5	2	48	30
[(10) 15]	0.5	0.5	2	48	30
[(20) 15]	0.5	0.5	2	48	30

Table 3. Testing conditions for $CaCO_3$ precipitation with jute fibers (length).

Fiber Length [(mm) %]	$CaCl_2$ (M)	Urea (M)	Bacterial OD_{600}	Incubation Time (h)	Incubation Temperature (°C)
0	0.5	0.5	2	48	30
[(5) 3]	0.5	0.5	2	48	30
[(15) 3]	0.5	0.5	2	48	30
[(25) 3]	0.5	0.5	2	48	30

2.4. Sample Preparation

The designed materials and test setup for the MICP treatment are shown in Figure 5. The dried "Mikawa" sand (75 ± 5 g) was taken into a 50 mL standard syringe tube (diameter 3 cm, height 10 cm). In each case, oven-dried samples (as described earlier) were compacted into 3 layers by applying a hammer shock on each layer of the sand. A lab-grade filter paper was used to cover the bottom portion of each column. Each sand column was filled with consistently mixed jute fibers using an automatic mixer (kitchen aid 9KSM160 series), with different fiber lengths and contents used. To neutralize the electrostatic charge of fibers and sand grains, 10 mL of (DW) de-ionized water was added during the mixing process to ensure uniform distribution of fibers within the soil matrix.

Thereafter, 12 mL of bacterial culture solution (ZoBEll2216E) was injected from the top of the syringe and superfluous solutions were drained out at a controlled rate to achieve bacterial stabilization (approximately 2 h) within the soil matrix. At the later injection phase, 16 mL of cementation solution (30.0 g/L of urea, 55.0 g/L of $CaCl_2$, and 3.0 g/L of Bacto nutrient broth) was injected into the samples. The injected solution was kept at approximately 2 mL above the surface of the sand and maintained in a fully saturated condition. The prepared samples were kept in an incubator at 30 °C for 14 days.

The cementation solution was injected and drained every day for 14 days continuously. The pH values and Ca^{2+} concentrations from the outlet were measured every day. After 14 days of curing, the UCS of the samples were measured using an automated Instron 2511-308 load cell (Norwood, MA, USA) following the ASTM D7012 2014 standard. The axial strain rate was 0.036 mm/min until reaching the failure condition (critical stage). The testing conditions are presented in Tables 4 and 5, showing the fiber content and length, respectively.

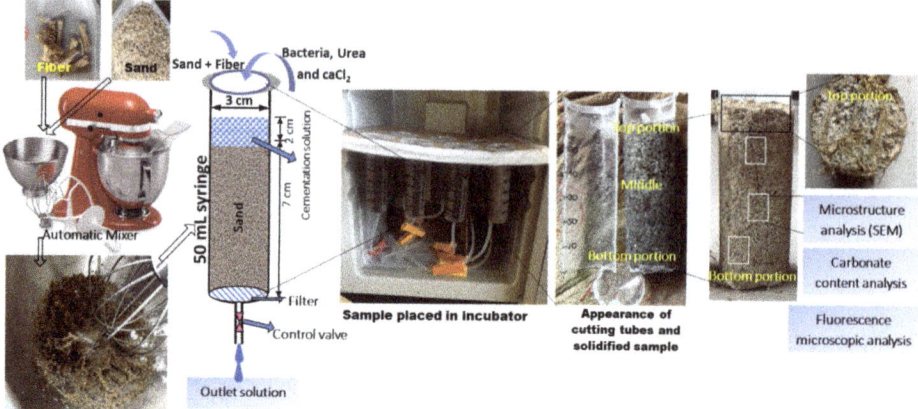

Figure 5. Sample preparation and test setup for the microbial-induced carbonate precipitation (MICP) process with the addition of jute fibers.

Table 4. Testing conditions for the sand solidification test (syringe) considering fiber content (%).

Cases	Fiber Content [(%) mm]	Cementation Solution Injection	Bacterial Injection	Bacterial OD_{600}	Curing Temperature (°C)	Curing Days
0	0	Everyday	Twice *	6	30	14
1	[(0.5) 15]	Everyday	Twice *	6	30	14
2	[(1.5) 15]	Everyday	Twice *	6	30	14
3	[(3) 15]	Everyday	Twice *	6	30	14
4	[(5) 15]	Everyday	Twice *	6	30	14
5	[(10) 15]	Everyday	Twice *	6	30	14
6	[(20) 15]	Everyday	Twice *	6	30	14

* Bacterial solution was injected at the beginning and after 7 days of the solidification test.

Table 5. Testing conditions for sand solidification test (syringe) considering fiber length.

Cases	Fiber Length [(mm) %]	Cementation Solution Injection	Bacterial Injection	Bacterial OD_{600}	Curing Temperature (°C)	Curing Days
0	0	Everyday	Twice *	6	30	14
1	[(5) 3]	Everyday	Twice *	6	30	14
2	[(15) 3]	Everyday	Twice *	6	30	14
3	[(25) 3]	Everyday	Twice *	6	30	14

* Bacterial solution was injected at the beginning and after 7 days of the solidification test.

The bacterial retention capacity (bacterial immobilization) during the MICP treatment was measured by the difference between the primary injected bacterial solution (OD_{600}) and the effluent solution (OD_{600}). To understand the cementation behavior of the treated samples, a pair of 0.5 MHz transducers with oscilloscope were also used to measure the primary and secondary shear wave velocities (Vp, Vs) of the treated samples. The transmitted signal was a 200 kHz square wave across the length of the cylindrical specimen (~6 cm). The velocities (Vp, Vs) of the treated specimens were calculated using SonicViewer-SX:5251 by measuring the time differences. The improvement ratio (IR) was calculated using the difference between the comparative values of the treated samples and the comparative values of the untreated samples. All the measurements were conducted after the samples were removed from the syringes and in dried condition.

The morphology, microstructure, and interactions of the precipitated $CaCO_3$, sand particles, and jute fibers were investigated using scanning electron microscopy (SEM). All the experiments were done in triplicate, and the mean value was plotted accordingly. The standard deviation was used to represent the error bars. The behavior of microorganisms and the effects of adding fibers to the bacterial culture solution were also investigated using an automatic fluorescence microscope (BZ-X800, KEYENCE Corporation, Osaka, Japan).

The precipitated $CaCO_3$ content of the MICP-treated specimens were measured using a simplified digital manometer device (Figure 6) under constant volume and temperature, followed by the ASTM standard method (ASTM D4373-14) [20].

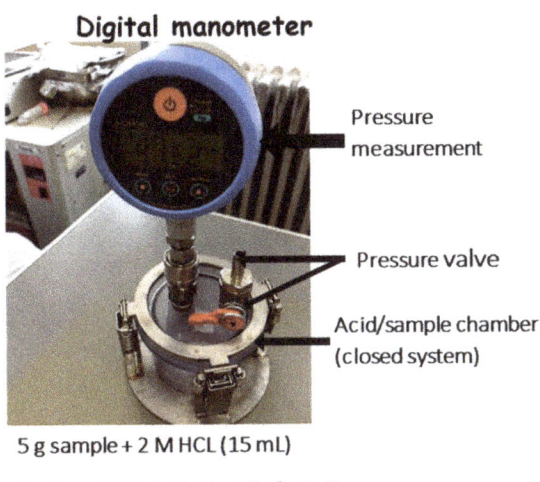

Figure 6. Digital manometer setup used to measure the $CaCO_3$ content.

3. Results and Discussion

3.1. Effects of Jute Fibers on the Process of $CaCO_3$ Precipitation

Studies have shown that the trend of $CaCO_3$ precipitation during the MICP process is greatly influenced by the fiber length and content (%) [21]. Therefore, it is essential to investigate the trends in $CaCO_3$ precipitation for individual bacterial species, because the $CaCO_3$ acts as the main binding material in between the substrate particles during the MICP process, leading to soil improvement. The effects of using the same fiber length and different fiber contents on $CaCO_3$ precipitation are presented in Figure 7. As can be shown, the amounts of precipitated $CaCO_3$ crystals varied significantly depending on the fiber content. By increasing the fiber content by 0.5%, the $CaCO_3$ precipitation was increased by approximately 29%, and reached 120% with the addition of 3% jute fibers compared to without fiber addition (Figure 7a). However, further addition of jute fibers decreased the amount of $CaCO_3$ (5% and 10% addition resulted in 84% and 23% improvements, respectively). With the addition of 20% jute fiber, the precipitation content dropped by 4%.

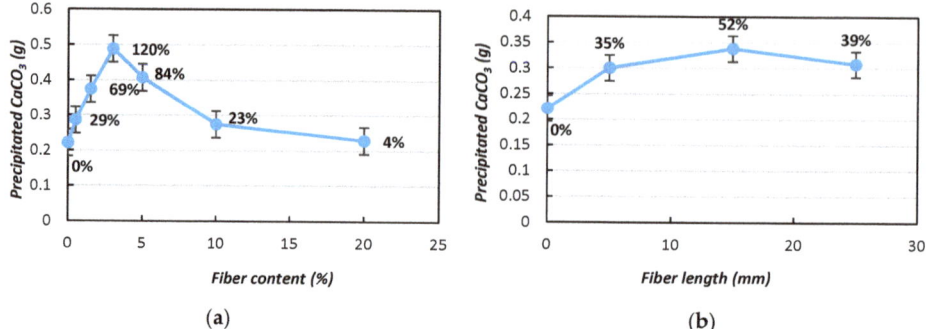

Figure 7. Effects of fiber content on CaCO$_3$ precipitation with the addition of jute fibers: (**a**) fiber content; (**b**) fiber length.

From this study, it was observed that the effect of CaCO$_3$ precipitation with 3% fiber is the best condition. From this study, it was also revealed that the higher fiber content intensely influenced the CaCO$_3$ precipitation. This study suggested that the optimum CaCO$_3$ content was obtained with an increase in fiber content up to 3%, while further increases of the fiber content led to decreased CaCO$_3$ precipitation compared to the other fiber content conditions. The reason for this phenomenon is that natural jute fibers have a non-regular cross sectional geometry and contain some chemical compounds such as pectin that act as accelerators (pectin breaks down into sugar and is used as the nutrient for the bacteria) [22] and inhibitors (excess amounts of sugar and other chemical compounds released from the jute fiber) of the microorganism growth (depending on the fiber content), which is also supported by previous studies [23].

Figure 7b shows the amounts of CaCO$_3$ precipitation with the same fiber content but variations in length (5, 15, and 25 mm, respectively). Figure 7b shows that the CaCO$_3$ precipitation increased to 35%, 52%, and 39% with 3% fiber content and lengths of 5, 15, and 25 mm, respectively, compared to without addition. The variation in the CaCO$_3$ precipitation amounts within the same set of samples was lower than when the fiber content was varied. The findings of this study indicate that the content (weight %) of the fiber addition is more important than the fiber length in the MICP treatment in order to promote the precipitation of CaCO$_3$, which is in good agreement with previous studies [24,25].

The microstructure analysis and interactions of precipitated CaCO$_3$ with jute fibers are shown in Figure 8 (with fiber and without fiber). The test results indicate that the precipitated crystals are irregular in shape and size and are coated around the jute fibers. The precipitated CaCO$_3$ also formed a CaCO$_3$ bridge, which could be very effective for binding and filling the void space in between the sand particles. Similar observations were also reported in previous studies [11,26]. The adsorption capacity of the microorganisms and CaCO$_3$ precipitation to the fibers was greatly influenced by the surface microstructures of the fiber. Different fiber and surface microstructures lead to different CaCO$_3$ precipitation patterns [27]. In addition, the reduction in the void space due to the CaCO$_3$ precipitation could be considered to be a primary strengthening factor [28,29] in the reduced brittleness behavior of the MICP-treated sand.

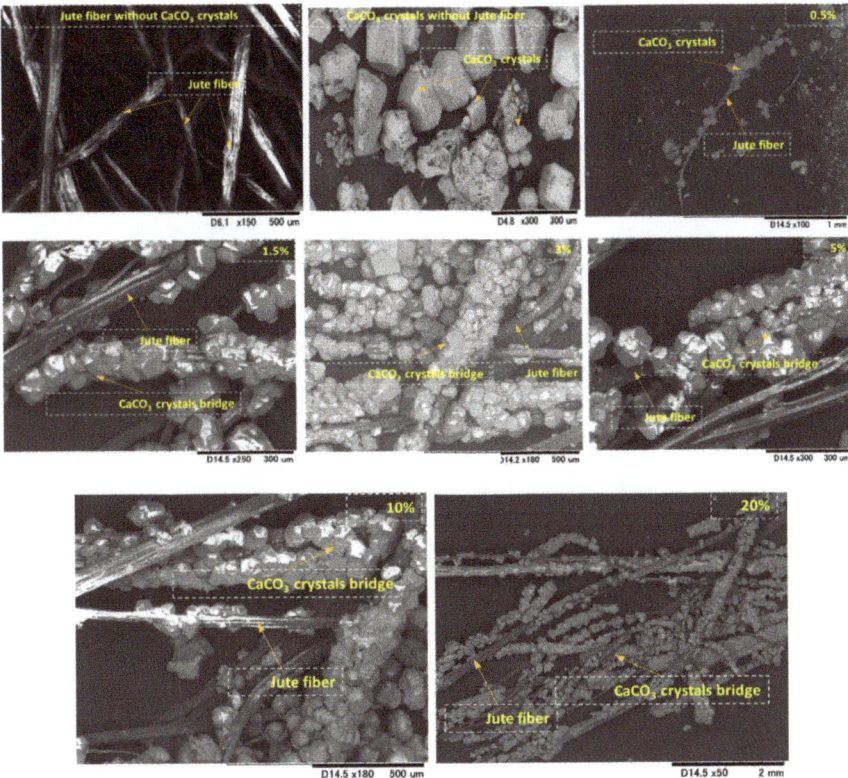

Figure 8. SEM images of precipitated CaCO$_3$ with and without jute fibers after the MICP process (without sand materials).

3.2. Effects of Fiber Inclusions on Microorganisms

The fluorescence microscopic images of the cultured bacterial cells (with and without fibers) are shown in Figure 9. The images show that the bacterial survival capacity persists for longer with the addition of jute fibers. The number of dead bacterial cells was also reduced after 10 days of cultivation with the addition of the fibers. The action behind this characteristic is the availability of biopolymers (cellulose, hemicellulose, lignin, pectin, and waxy substances) in the jute fibers [30]. When jute fibers are steeped in water, the water-soluble carbohydrate compounds (D-glucosidic bonds and hydroxyl groups) and biopolymers are broken down into simple sugars (galactose) by a bio-chemical mechanism [31], acting as a source of nutrient for the bacteria, which is essential for the bacteria to survive for a longer time than usual. However, a further multidisciplinary assessment could be conducted to quantify this mechanism.

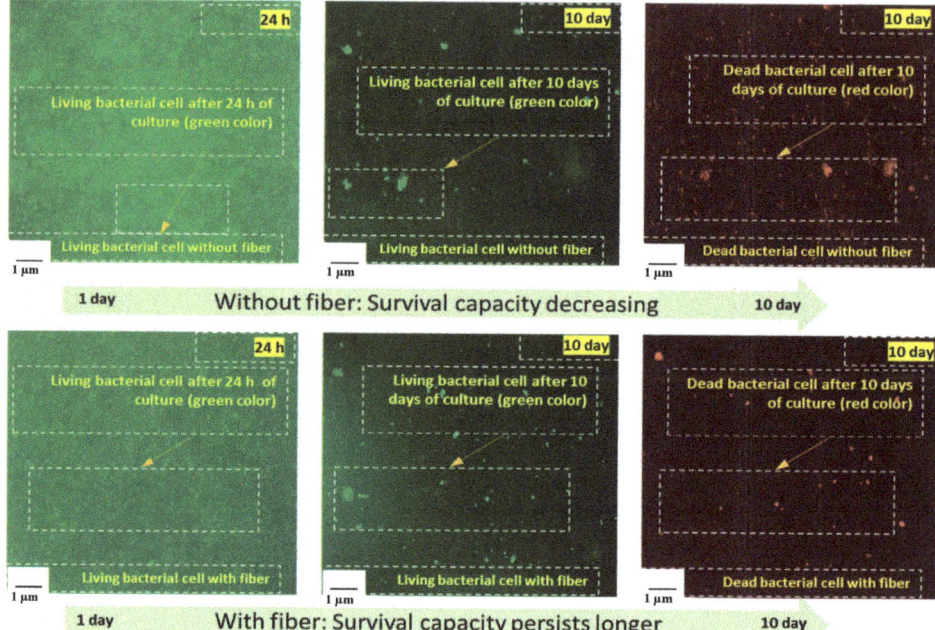

Figure 9. Fluorescence microscopic images living and dead bacterial cell (with and without fiber).

3.3. Variations in Strength after Fiber Inclusion in the MICP-Treated Sample

The stress–strain curves of the MICP-treated samples with different fiber lengths and contents are presented in Figure 10, which shows that the MICP-treated samples were significantly influenced by the addition of jute fibers (depending on the fiber content and length). The stress–strain curve of the MICP-treated biocemented sand without fiber was gradually compacted with increasing strain and stress, then failure occurred, which was considered as typical brittle failure. However, by increasing the fiber length (5, 15, and 25 mm), the stress on the biocemented sand reached the maximum strength and then entered the residual deformation stage. Failure occurred more slowly compared to the samples without fibers (Figure 10a). The slower rate of failure indicated improvement of the ductility behavior of the samples. In addition, as shown in Figure 10a, by increasing the fiber length, the strength (UCS) declined. The results of this study indicated that the addition of long fibers to the MICP-treated samples could meant they could be easily bent and the fibers eventually clustered within the sand, having a negative impact on the MICP treatment and leading to declines in the UCS. The results of this study were significant compared to a previous study [32].

Figure 10b shows that the unconfined compressive strength (UCS) of the biocemented sample initially increased (from 0.5% fiber content) and reached its maximum strength (UCS) with a fiber content of 3%. The unconfined compressive strength (UCS) interestingly decreased with further addition to the fiber content (up to a fiber content of 20%).

The reason for this was that the fiber was randomly distributed within the sand matrix by the interleaving mechanism and due to the cross-sectional geometry [33]. This mechanism led to several interlacing points forming between fibers, then a spatial distribution network and spatial stress area formed that were able to increase the bacterial retention and survival capacity (as mentioned earlier), meaning the sample was able to hold more bacteria (which resulted in increased $CaCO_3$) than samples without fibers. The stress area and the network controlled the deformation of the sand and increased the ductile behavior of the MICP-treated biocemented sand.

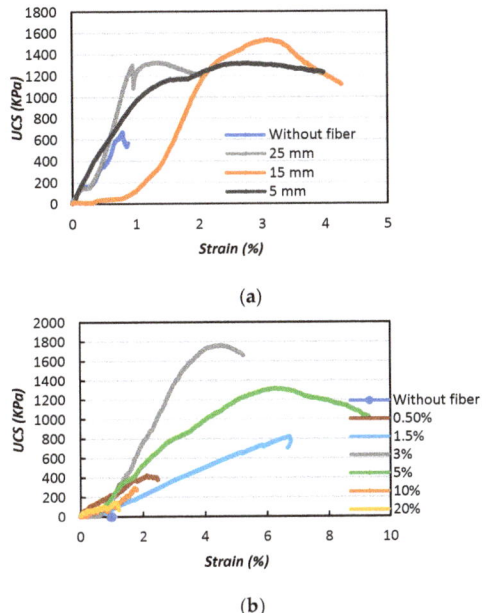

Figure 10. Effects of fiber on the unconfined compressive strength (UCS) of the biocemented sand with the addition of jute fibers: (**a**) fiber length; (**b**) fiber content.

The relationship between the estimated UCS and the $CaCO_3$ (%) from the results obtained in this study (considering the fiber length and content) is shown in Figure 11. By increasing the fiber length (5–15 mm), the $CaCO_3$ (%) increased and the UCS also increased. The maximum UCS was observed with a 15 mm fiber length (Figure 11a). By further increasing the fiber length (25 mm), the $CaCO_3$ (%) decreased and the UCS was dropped. In Figure 11b, it was also shown that the fiber content played a considerable role in the $CaCO_3$ (%) precipitation process and also improved the UCS. With the increases of the fiber content (0.5–3%), the $CaCO_3$ content also increased. As a result of the increasing $CaCO_3$, the UCS values of the treated samples also increased until reaching the maximum point (approximately 1.6 MPa). Further increases of the fiber content (5–20%) caused the $CaCO_3$ (%) precipitation and UCS of the treated samples to decrease [33,34]. The findings of this study clearly reveal that the fiber content (% weight) played a more significant role compared to the fiber length in terms of influencing the amount of $CaCO_3$ precipitation and improving the UCS in the treated samples.

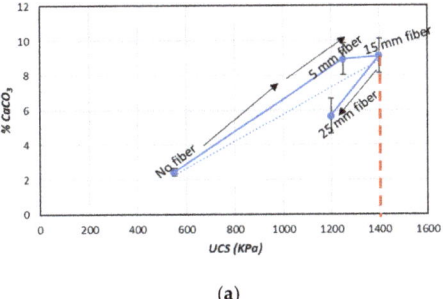

(**a**)

Figure 11. *Cont.*

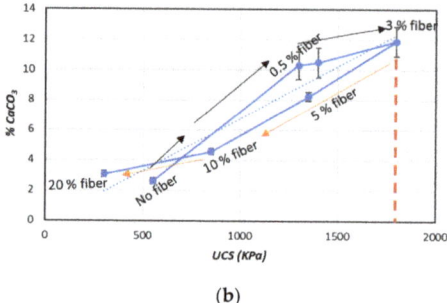

(b)

Figure 11. Average (%) CaCO$_3$ values for the biocemented sand sample after MICP treatment with the addition of jute fibers: (**a**) fiber length; (**b**) fiber content.

Figure 12 shows the effects of the fiber length on the soil's strength improvement (UCS). In the Figure 12a, it can be observed that the improvement ratio (IR) increased from 2.5 to 3 with appropriate increases of the fiber length (5–15 mm), however with further length increases the IR value dropped and a negative influence was evident. In Figure 12b, it can be observed that the IR value was indicates the significant influence of the fiber content in the improvement of the soil strength. A relatively greater improvements were achieved with the 0.5 and 3% additions of natural jute fibers, with improvement ratios of 2.5 and 3.1, respectively.

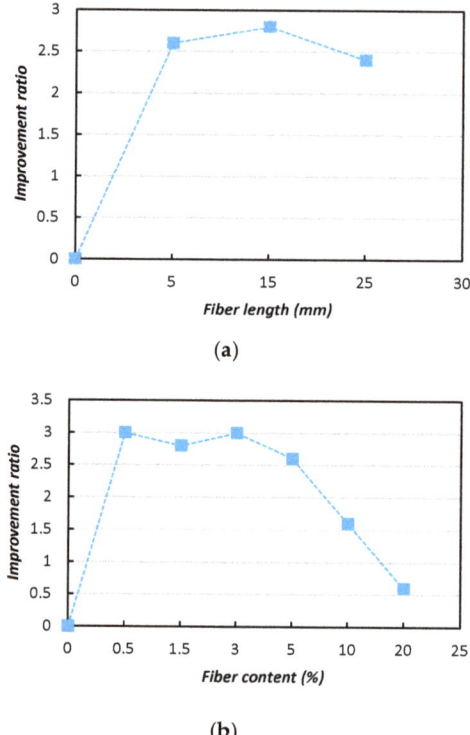

(a)

(b)

Figure 12. Improvement ratios (IRs) of the MICP-treated biocemented sample strength (UCS) with the addition of jute fibers: (**a**) fiber length; (**b**) fiber content.

In general, the results of this study suggest that the mixing of natural jute fibers with the MICP method could significantly improve biocemented sand, provided the appropriate length (i.e., 15 mm) and content (i.e., 3% by weight of sand) are used. The reason is that when jute fibers are mixed with sand, the cohesion, friction, and interface between the jute fibers and sand particles increase (depending on the fiber length and content). As a result, the frequent bacterial movement ensures and enhances the immobilization of the bacteria within the soil matrix (Figure 13a). The increased bacterial immobilization accelerates the uniform distribution of $CaCO_3$ within the sand matrix and consequently increases the effectiveness of the soil (in terms of the UCS, ductility, etc.). However, if the fiber content is too high (i.e., 20%), the bacterial movement is hindered, resulting in uneven distribution of the bacteria within the soil matrix. Similarly, the $CaCO_3$ precipitation occurs non-uniformly. In addition, increasing the fiber length by too much (i.e., 25 mm) reduces the efficiency of bacterial immobilization (Figure 13b) because of the uneven distribution of the fibers and the decreased retention capacity of the bacteria. As a result, the effectiveness of the soil's engineering properties is also decreased [34].

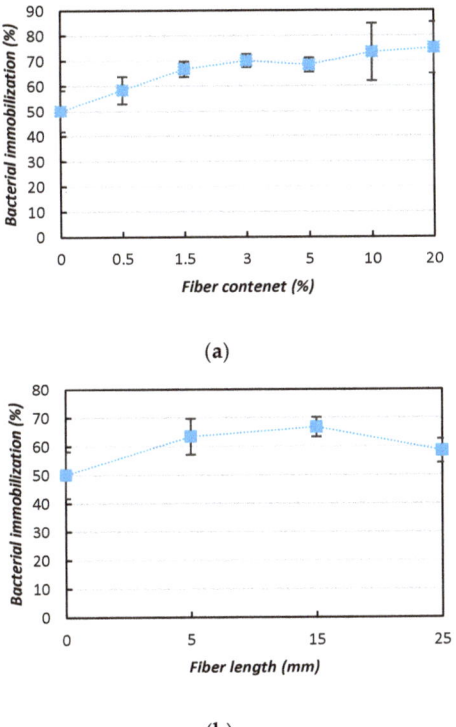

Figure 13. Influence of bacterial retention capacity with the addition of jute fibers: (**a**) fiber content; (**b**) fiber length.

3.4. Effects of Bacterial Immobilization on $CaCO_3$ Precipitation

Figure 14 summarizes the variations in the bacterial immobilization improvement ratios (in terms of retention capacity) with respect to different fiber inclusions. Figure 14 shows that for pure the MICP sample (without fiber addition), approximately 50% of the bacteria were flushed out and the bacterial retention capacity (immobilization) increased with the fiber content (Figure 14a), yielding average improvement ratio values of 1.2 to 1.6 (with fiber contents ranging from 0.5 to 20%). Moreover, increasing the fiber length (from 5 to 25 mm) decreased the bacterial immobilization (retention capacity)

(Figure 14b), giving an improvement ratio of 1.2 with the 15 mm fiber length. This study showed that the bacterial immobilization (retention capacity) was increased in the presence of fibers, as was also reported in previous studies [35,36].

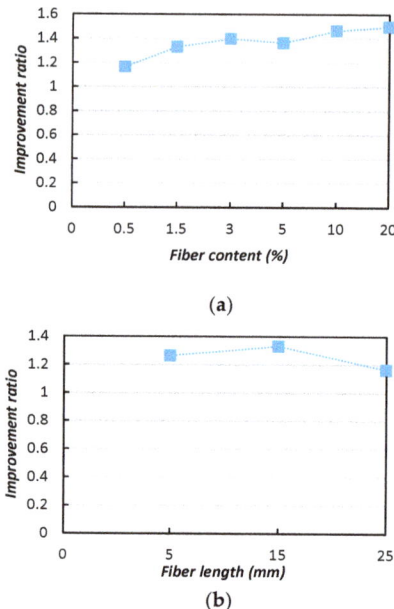

Figure 14. Improvement ratios for bacterial immobilization with the addition of jute fibers: (**a**) fiber content; (**b**) fiber length.

Figure 15 shows the effect of bacterial immobilization on the $CaCO_3$ content with the addition of jute fibers, considering both the fiber content (Figure 15a) and fiber length (Figure 15b). Figure 15a shows that the $CaCO_3$ content increased with increasing fiber content up to a certain amount. In this study, 3% fiber content result in the maximum amount of $CaCO_3$ (Figure 15a). With further increases of the fiber content, the bacterial movement [31] became stuck (due to the fiber structure, as described earlier), resulting in decreased $CaCO_3$ content within the soil matrix. A similar process was also observed when increasing the fiber length (Figure 15b).

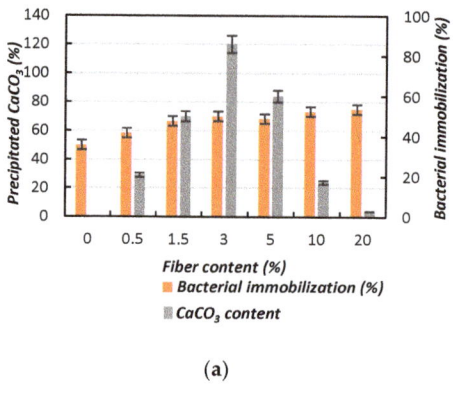

(**a**)

Figure 15. *Cont.*

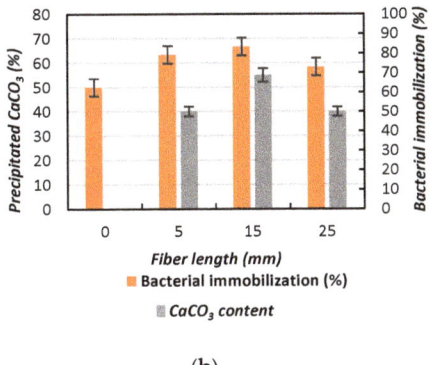

(b)

Figure 15. Effect of bacterial immobilization on the $CaCO_3$ content with the addition of jute fibers: (**a**) fiber content; (**b**) fiber length.

In this study, it was demonstrated that the production of $CaCO_3$ was greatly influenced by the bacterial immobilization capacity (retention ability), and the appropriate fiber content and length resulted in an increased $CaCO_3$ content.

3.5. Microstructure Analysis

The fracture morphologies of the biocemented samples with different fiber contents (0, 0.5, 1.5, 3, 5, 10, and 20%) and lengths (5, 15, and 25 mm) are also compared in Figure 16a,b, which shows the results of the unconfined compressive strength test. All the experiments were conducted in triplicate following the ASTM D2166 standard method.

In Figure 16 (considering both the fiber content and length), it can be seen that the addition of jute fibers in the MICP-treated sample significantly improved the unconfined compressive strength, and because of the lower biocementation level, the fractures generally started from the lower end (Figure 16a). Without the fibers, the fractures appeared throughout the whole sample from the bottom upward, suggested brittle failure of the samples. The results also showed that the fracture morphologies of the biocemented samples were closely interlinked with the different fiber contents (Figure 16a) and lengths (Figure 16b), due to the interaction and friction within the sand–fiber matrix. With increasing fiber content (i.e., 0.5–5%), the fiber formed a three-dimensional (3D) grid within the soil matrix, which restricted the development of the failure pattern and effectively improved the strength of the soil, while enhancing the brittleness delayed the overall damages of the MICP-treated sample [32]. However, further increasing the fiber content (i.e., 10–20%) resulted in "bulging" behavior. Regarding the fiber length, the use of short fibers (i.e, 5–15 mm) resulted in significant improvements for the biocemented specimen. The use of long fibers (i.e., 25 mm) led to the sudden fracture of the sample due to uneven distribution and bundles becoming entangled during the sample preparation. Therefore, for the actual engineering application, it is important to determine the optimum fiber content and length to be added to the soil in order to obtain the maximum effect.

Figure 17 shows the SEM images of MICP-treated biocemented samples with the addition of jute fibers (considering the length and content) and the distribution of $CaCO_3$ within the sand matrix. The images also show sand particles without MICP treatment, sand particles with fiber without MICP treatment, and biocemented sand particles with different fiber contents and lengths. From the microscopic images, it can be seen that the void spaces were dominant in both sand–fiber matrixes without MICP treatment. After the MICP treatment, the fibers were covered by $CaCO_3$ crystals (similar to a bridge). The $CaCO_3$ crystal bridge provided strong bonding between the sand particles and also filled the void space. Consequently, the cementation level of the treated sample increased significantly.

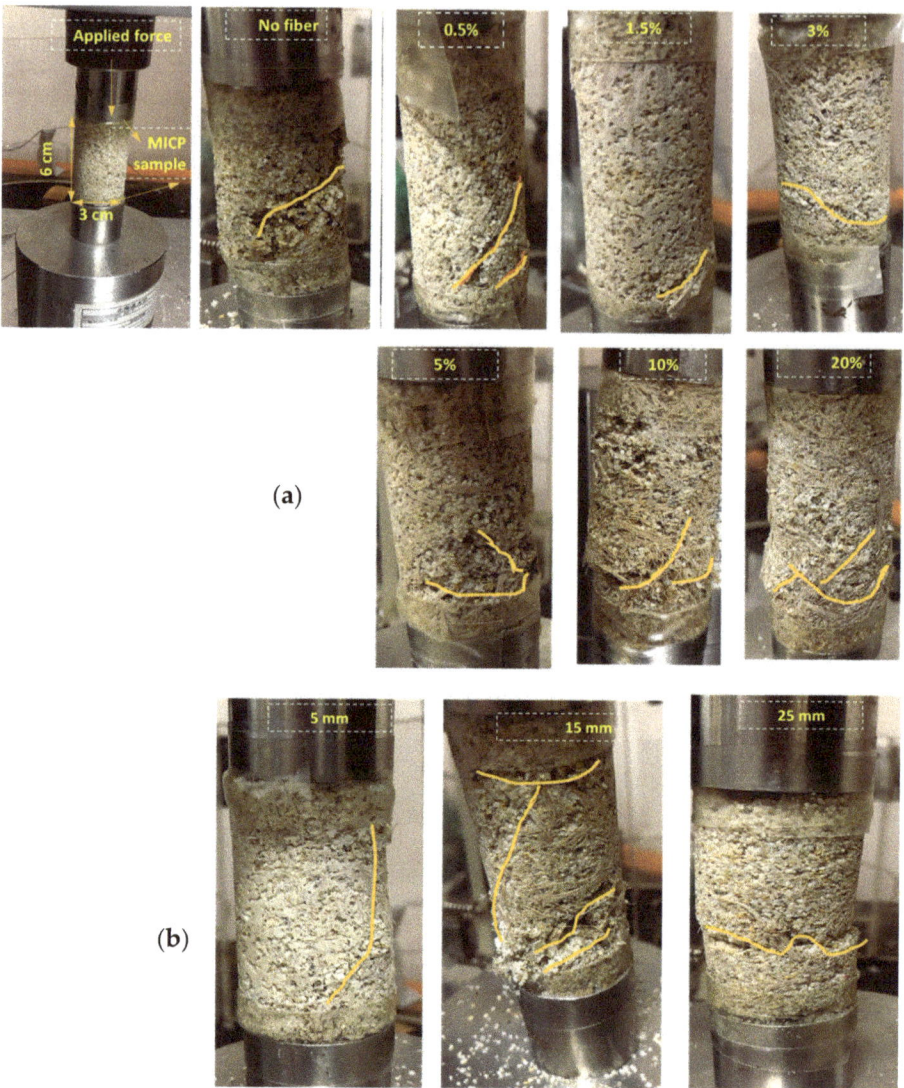

Figure 16. Failure behavior of the MICP-treated sample with addition of jute fibers: (**a**) fiber content; (**b**) fiber length.

The overall interactions and improvement of the MICP-treated sand are presented more clearly by a schematic diagram in Figure 18. By increasing the fiber content (up to a certain amount) and fiber length (up to a certain length), more $CaCO_3$ precipitation occurred in between the soil pore spaces and contact points, resulting in the soil having enhanced engineering properties. The results are also validated by the primary and secondary shear wave velocities (Vp, Vs), as shown in Table 6. A previous study also showed that an accelerated carbonation system enhanced the interface between the fibers and cementitious matrix in a sample, which improved the strong mechanical anchorage and interlocking effects [37] by filling the pores with calcite and fiber of the system. However, the findings of this study could play a significant role in improving the engineering properties of the soil using the MICP fiber matrix treatment.

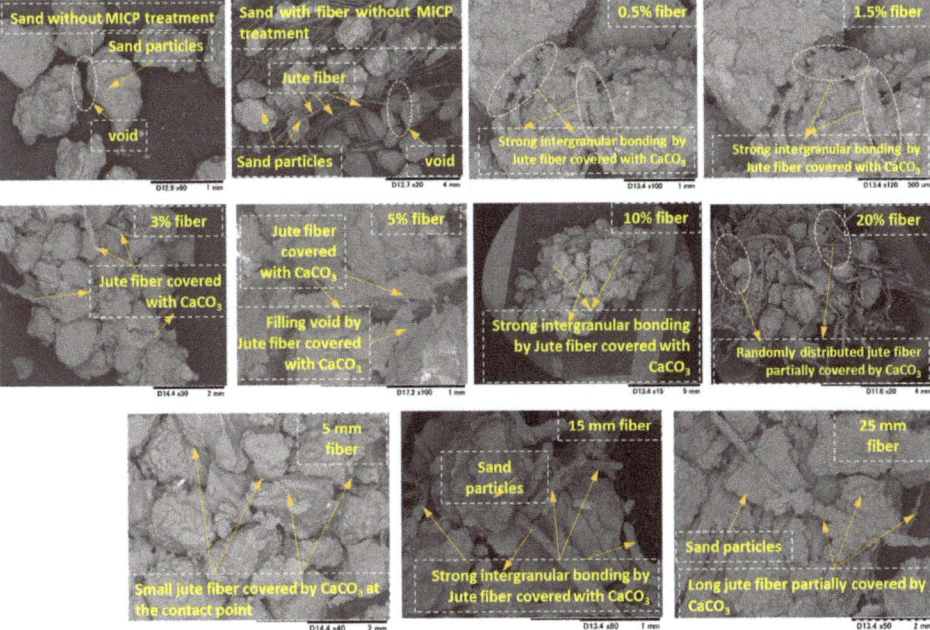

Figure 17. SEM images of MICP-treated biocemented samples with the addition of jute fibers (considering different lengths and contents) and distribution of $CaCO_3$ within the sand matrix.

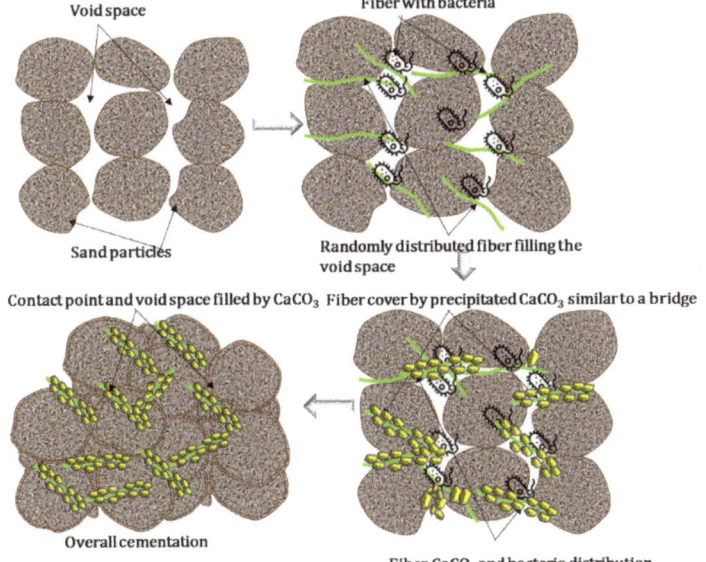

Figure 18. Schematic diagram of biocemented sample with the addition of jute fibers.

Table 6. Summary of test results for the biocemented sand after MICP treatment with jute fibers.

Fiber Content (%)	Unit Weight (g)	Vs (km/s)	Vp (km/s)	UCS (MPa)	Average CaCO$_3$ (%)
0	65.6	0.92	1.12	0.5	2.4
0.5	64.2	0.87	1.22	1.5	9.3
1.5	63.5	0.95	1.24	1.4	11.88
3	60.1	0.92	1.25	1.6	13.29
5	61.5	0.9	1.23	1.3	7.9
10	60.2	0.99	1.22	0.8	4.6
20	59.9	0.93	1.24	0.3	3.29
Fiber length (mm)					
5	62.4	0.92	1.28	0.5	2.4
15	61.9	0.87	1.12	1.3	8.4
25	63.6	0.95	1.27	1.4	9.7

4. Conclusions

This study was conducted to investigate the effects of the addition jute fibers to biocemented sand using MICP method. In this study, for the sand treatment using MICP method, the jute fiber contents were 0, 0.5, 1.5, 3, 5, 10, and 20% by sand weight and the lengths were 5, 15, and 25 mm. Based on the results of this study, the following conclusions could be outlined:

i. Jute fiber has significant effects on the microbial performance, CaCO$_3$ precipitation pattern, and solidification of sand. Using fluorescence microscopy, the survival capacity of the microorganisms was well demonstrated to be increased by the addition of jute fiber. The addition effectively improved not only the bacterial performance, but also the mechanical characteristics (UCS and ductility) of sand. The UCS of the sample increased with increasing fiber content; however, higher fiber addition past a point was found to decrease the UCS. From the results obtained in this study, the optimum jute fiber content was 3% and the optimum length was 15 mm;

ii. The CaCO$_3$ precipitation was positively correlated with the addition of jute fibers, which yielded significant improvement of the engineering properties of the soil. The SEM analysis suggested that the added jute fiber coupled well with CaCO$_3$ (i.e., CaCO$_3$ was attached on and along the surfaces of fibers), forming reliable bridges within the soil matrix, which tended to limit the development of failure planes within specimens. This process potentially increased the strength and toughness of the treated specimens compared to those of control biocemented specimens (without jute fibers);

iii. As the amount and length of jute fibers increased beyond the optimum level, the fibers tended to become entangled with each other during preparation of the samples, which hindered the entry of bacteria and reduced the space available for bacterial survival and CaCO$_3$ formation;

iv. In this study, natural jute fibers were used; however, the effects of chemically treated jute fibers and the roughness of jute fibers (surface roughness) have not been investigated in detail. In order to better understand the effects of fibers on soil stabilization (considering chemical pretreatment of the fiber, fiber roughness, etc.) using the MICP process, further studies are highly recommended.

Author Contributions: All the co-authors contributed equally to designing the experiments, analyzing the data, writing the manuscript, and completing the revisions and editing. All authors have read and agreed to the published version of the manuscript.

Funding: This work was partly supported by JSPS KAKENHI, grant number JP19H02229.

Conflicts of Interest: The authors declared no conflict of interest.

References

1. Mwandira, W.; Nakashima, K.; Kawasaki, S. Bioremediation of lead-contaminated mine waste by *Pararhodobacter* sp. based on the microbially induced calcium carbonate precipitation technique and its effects on strength of coarse and fine grained sand. *Ecol. Eng.* **2017**, *109*, 57–64. [CrossRef]
2. Amarakoon, G.G.N.N.; Kawasaki, S. Factors Affecting Sand Solidification Using MICP with *Pararhodobacter* sp. *Mater. Trans.* **2017**, *59*, 72–81. [CrossRef]
3. Al Imran, M.; Shinmura, M.; Nakashima, K.; Kawasaki, S. Effects of Various Factors on Carbonate Particle Growth Using Ureolytic Bacteria. *Mater. Trans.* **2018**, *59*, 1520–1527. [CrossRef]
4. Gowthaman, S.; Mitsuyama, S.; Nakashima, K.; Komatsu, M.; Kawasaki, S. Biogeotechnical approach for slope soil stabilization using locally isolated bacteria and inexpensive low-grade chemicals: A feasibility study on Hokkaido expressway soil, Japan. *Soils Found.* **2019**, *59*, 484–499. [CrossRef]
5. Omoregie, A.I.; Ngu, L.H.; Ong, D.E.L.; Nissom, P.M. Low-cost cultivation of *Sporosarcina pasteurii* strain in food-grade yeast extract medium for microbially induced carbonate precipitation (MICP) application. *Biocatal. Agric. Biotechnol.* **2019**, *17*, 247–255. [CrossRef]
6. Mortensen, B.M.; Haber, M.J.; Dejong, J.T.; Caslake, L.F.; Nelson, D.C. Effects of environmental factors on microbial induced calcium carbonate precipitation. *J. Appl. Microbiol.* **2011**, *111*, 338–349. [CrossRef] [PubMed]
7. Martinez, B.C.; DeJong, J.T.; Ginn, T.R.; Montoya, B.M.; Barkouki, T.H.; Hunt, C.; Tanyu, B.; Major, D. Experimental Optimization of Microbial-Induced Carbonate Precipitation for Soil Improvement. *J. Geotech. Geoenvironmental Eng.* **2013**, *139*, 587–598. [CrossRef]
8. Li, L.; Zhao, Q.; Zhang, H.; Amini, F.; Li, C. A Full Contact Flexible Mold for Preparing Samples Based on Microbial-Induced Calcite Precipitation Technology. *Geotech. Test. J.* **2014**, *37*, 917–921.
9. Qiu, R.; Tong, H.; Fang, X.; Liao, Y.; Li, Y. Analysis of strength characteristics of carbon fiber–reinforced microbial solidified sand. *Adv. Mech. Eng.* **2019**, *11*, 1–7. [CrossRef]
10. Chen, H.J.; Peng, C.F.; Tang, C.W.; Chen, Y.T. Self-healing concrete by biological substrate. *Materials* **2019**, *12*, 4099. [CrossRef]
11. Choi, S.-G.; Wang, K.; Chu, J. Properties of biocemented, fiber reinforced sand. *Constr. Build. Mater.* **2016**, *120*, 623–629. [CrossRef]
12. Gowthaman, S.; Nakashima, K.; Kawasaki, S. A state-of-the-art review on soil reinforcement technology using natural plant fiber materials: Past findings, present trends and future directions. *Materials* **2018**, *11*, 553. [CrossRef] [PubMed]
13. Wen, K.; Bu, C.; Liu, S.; Li, Y.; Li, L. Experimental investigation of flexure resistance performance of bio-beams reinforced with discrete randomly distributed fiber and bamboo. *Constr. Build. Mater.* **2018**, *176*, 241–249. [CrossRef]
14. Islam, M.S.; Ahmed, S.J. Influence of jute fiber on concrete properties. *Constr. Build. Mater.* **2018**, *189*, 768–776. [CrossRef]
15. Hejazi, S.M.; Sheikhzadeh, M.; Abtahi, S.M.; Zadhoush, A. A simple review of soil reinforcement by using natural and synthetic fibers. *Constr. Build. Mater.* **2012**, *30*, 100–116. [CrossRef]
16. Zhao, G.Z.; Li, J.; Qin, S.; Zhang, Y.Q.; Zhu, W.Y.; Jiang, C.L.; Xu, L.H.; Li, W.J. *Micrococcus yunnanensis* sp. nov., a novel actinobacterium isolated from surface-sterilized Polyspora axillaris roots. *Int. J. Syst. Evol. Microbiol.* **2009**, *59*, 2383–2387. [CrossRef]
17. Imran, M.; Kimura, S.; Nakashima, K.; Evelpidou, N.; Kawasaki, S. Feasibility Study of Native Ureolytic Bacteria for Biocementation Towards Coastal Erosion Protection by MICP Method. *Appl. Sci.* **2019**, *9*, 4462. [CrossRef]
18. Fang, X.; Yang, Y.; Chen, Z.; Liu, H.; Xiao, Y.; Shen, C. Influence of Fiber Content and Length on Engineering Properties of MICP-Treated Coral Sand. *Geomicrobiol. J.* **2020**, *37*, 582–594. [CrossRef]
19. Natarajan, K.R. Kinetic Study of the Enzyme Urease from Dolichos biflorus. *J. Chem. Educ.* **1995**, *72*, 556. [CrossRef]
20. American Society for Testing and Materials. *Standard Test Method for Rapid Determination of Carbonate Content of Soils*; ASTM D4373-14; ASTM International: West Conshohocken, PA, USA, 2014.

21. Gupta, M.K.; Srivastava, R.K.; Bisaria, H. Potential of Jute Fibre Reinforced Polymer Composites: A review ISSN 2277-7156 Review Article Potential of Jute Fibre Reinforced Polymer Composites: A Review. *Int. J. Fiber Text. Res.* **2015**, *5*, 30–38.
22. Choi, S.-G.; Hoang, T.; Alleman, E.J.; Chu, J. Splitting Tensile Strength of Fiber-Reinforced and Biocemented Sand. *J. Mater. Civ. Eng.* **2019**, *31*, 06019007. [CrossRef]
23. Li, M.; Li, L.; Ogbonnaya, U.; Wen, K.; Tian, A.; Amini, F. Influence of fiber addition on mechanical properties of MICP-treated sand. *J. Mater. Civ. Eng.* **2016**, *28*, 04015166. [CrossRef]
24. Al Qabany, A.; Soga, K.; Santamarina, C. Factors affecting efficiency of microbially induced calcite precipitation. *J. Geotech. Geoenvironmental Eng.* **2012**, *138*, 992–1001. [CrossRef]
25. Shao, W.; Cetin, B.; Li, Y.; Li, J.; Li, L. Experimental Investigation of Mechanical Properties of Sands Reinforced with Discrete Randomly Distributed Fiber. *Geotech. Geol. Eng.* **2014**, *32*, 901–910. [CrossRef]
26. Lei, X.; Lin, S.; Meng, Q.; Liao, X.; Xu, J. Influence of different fiber types on properties of biocemented calcareous sand. *Arab. J. Geosci.* **2020**, *13*, 317. [CrossRef]
27. DeJong, J.T.; Mortensen, B.M.; Martinez, B.C.; Nelson, D.C. Bio-mediated soil improvement. *Ecol. Eng.* **2010**, *36*, 197–210. [CrossRef]
28. Consoli, N.C.; Vendruscolo, M.A.; Fonini, A.; Rosa, F.D. Fiber reinforcement effects on sand considering a wide cementation range. *Geotext. Geomembr.* **2009**, *27*, 196–203. [CrossRef]
29. Munshi, T.K.; Chattoo, B.B. Bacterial population structure of the jute-retting environment. *Microb. Ecol.* **2008**, *56*, 270–282. [CrossRef]
30. Glöckner, F.O.; Fuchs, B.M.; Amann, R. Bacterioplankton compositions of lakes and oceans: A first comparison based on fluorescence in situ hybridization. *Appl. Environ. Microbiol.* **1999**, *65*, 3721–3726. [CrossRef]
31. Tang, C.; Shi, B.; Gao, W.; Chen, F.; Cai, Y. Strength and mechanical behavior of short polypropylene fiber reinforced and cement stabilized clayey soil. *Geotext. Geomembr.* **2007**, *25*, 194–202. [CrossRef]
32. DOS SANTOS, A.P.S.; CONSOLI, N.C.; BAUDET, B.A. The mechanics of fibre-reinforced sand. *Géotechnique* **2010**, *60*, 791–799. [CrossRef]
33. Teng, F.; Ouedraogo, C.; Sie, Y.C. Strength improvement of a silty clay with microbiologically induced process and coir fiber. *J. Geoengin.* **2020**, *15*, 79–88.
34. Zhao, Y.; Xiao, Z.; Fan, C.; Shen, W.; Wang, Q.; Liu, P. Comparative mechanical behaviors of four fiber-reinforced sand cemented by microbially induced carbonate precipitation. *Bull. Eng. Geol. Environ.* **2020**, *79*, 3075–3086. [CrossRef]
35. Nafisi, A.; Montoya, B.M.; Evans, T.M. Shear Strength Envelopes of Biocemented Sands with Varying Particle Size and Cementation Level. *J. Geotech. Geoenvironmental Eng.* **2020**, *146*, 04020002. [CrossRef]
36. Harkes, M.P.; van Paassen, L.A.; Booster, J.L.; Whiffin, V.S.; van Loosdrecht, M.C.M. Fixation and distribution of bacterial activity in sand to induce carbonate precipitation for ground reinforcement. *Ecol. Eng.* **2010**, *36*, 112–117. [CrossRef]
37. Urrea-Ceferino, G.E.; Rempe, N.; dos Santos, V.; Savastano Junior, H. Definition of optimal parameters for supercritical carbonation treatment of vegetable fiber-cement composites at a very early age. *Constr. Build. Mater.* **2017**, *152*, 424–433. [CrossRef]

© 2020 by the authors. Licensee MDPI, Basel, Switzerland. This article is an open access article distributed under the terms and conditions of the Creative Commons Attribution (CC BY) license (http://creativecommons.org/licenses/by/4.0/).

Article

Lime-Based Mortar Reinforced by Randomly Oriented Short Fibers for the Retrofitting of the Historical Masonry Structure

Michele Angiolilli, Amedeo Gregori * and Marco Vailati

Department of Civil, Building and Environmental Engineering, University of L'Aquila, 67100 L'Aquila, Italy; michele.angiolilli@graduate.univaq.it (M.A.); marco.vailati@univaq.it (M.V.)
* Correspondence: amedeo.gregori@univaq.it; Tel.: +39-0862-43-4141

Received: 3 June 2020; Accepted: 30 July 2020; Published: 6 August 2020

Abstract: Recent seismic events prompted research to develop innovative materials for strengthening and repair of both modern and historic masonry constructions (buildings, bridges, towers) and structural components (walls, arches and vaults, pillars, and columns). Strengthening solutions based on composite materials, such as the Fiber Reinforced Polymers (FRP) or the Fiber Reinforced Cementitious Matrix (FRCM), have been increasingly considered in the last two decades. Despite reinforcement made of short-fibers being a topic that has been studied for several years from different researchers, it is not yet fully considered for the restoration of the masonry construction. This work aims to experimentally investigate the enhancement of the mechanical properties of lime-based mortar reinforced by introducing short glass fibers in the mortar matrix with several contents and aspect ratios. Beams with dimensions of 160 mm × 40 mm × 40 mm with a central notch were tested in three-point bending configuration aiming to evaluate both the flexural strength and energy fracture of the composite material. Then, the end pieces of the broken beams were tested in Brazilian and compressive tests. All the tests were performed by a hydraulic displacement-controlled testing machine. Results highlight that the new composite material ensures excellent ductility capacity and it can be considered a promising alternative to the classic fiber-reinforcing systems.

Keywords: cultural heritage; durability; mechanical characterization; retrofitting; strengthening; quasi-brittle material; three-point bending test; energy fracture; NHL; composite material

1. Introduction

The use of masonry is very common in many historic constructions, both in architectural monuments and whole urban centers all over the world. This masonry is generally made of various and very poor materials, characterized by different typologies. The fragility of this heterogeneous material interferes with the ductility criteria based on energy dissipation, which nowadays constitutes the safety principles of structural design for the safeguarding of human lives [1]. The disasters generated by seismic actions discouraged the use of unreinforced masonry from ancient times until the more modern era. Suffice it to say that the adoption of retrofitting systems began with the primordial civilization, such as traditional earthquake-resistant timber frames [2,3].

The development of a fiber-based strengthening system began in the 1960s when the potential for adding steel fibers to enhance the ductility of concrete material was recognized. However, this technology has been commonly adopted for the reinforcement of masonry structures only in the last decade as an alternative to traditional systems, such as mortar injections, reinforced drilling, and reinforced concrete plaster. Indeed, because of the strict rules for the preservation of historic structures, conservation committees usually request structurally efficient but less intrusive techniques to protect the historical structures.

Among modern and innovative solutions of intervention on existing structures, composite materials, such as the Fiber Reinforced Polymers (FRP) or the Fiber Reinforced Cementitious Matrix (FRCM), have been increasingly considered for strengthening and repair of both modern and historic masonry constructions (buildings, bridges, towers) and structural components (walls, arches and vaults, pillars, and columns). These technologies consist of the use of composites material characterized by uni- or bi-directional long fibers.

These materials are proven to be effective in increasing the load-carrying capacity of masonry elements and improving their structural behavior through a reduction of critical brittle failure modes. Most importantly, the increase in strength is obtained with a lower increment of the structural weight, as compared to the traditional ones (e.g., reinforced concrete plaster).

The FRP technology consists of the application of laminates and rods. On one hand, the use of laminates involves the application of fiber sheets by manual lay-up to the surface of the masonry panels, which is previously prepared by sandblasting and puttying procedure. The fibers are impregnated by an epoxy resin, which after hardening enables the newly formed laminate to become an integral part of the strengthened member. On the other hand, the use of pultruded rods consists of placing them into grooves cut onto the surface of the member being strengthened. The groove is filled with an epoxy-based paste, and the rod is then placed into the groove and lightly pressed to force the paste to flow around the rod. The groove is then filled with more paste and the surface is leveled [4].

High tensile strength and stiffness-to-weight ratio, fatigue and corrosion resistance, easy in-situ feasibility and adaptability, and progressive reduction in production and distribution costs are the main characteristics that encouraged the diffusion of the FRPs [5]. The FRP is employed to improve the global behavior in the seismic zone (tying, connections among components, strengthening), to counteract specific incipient or developed damage (high compression, shear, and/or flexural conditions), and to repair very specific local weaknesses depending on the peculiar construction typology [4,6–8].

However, the FRPs' low fire resistance, high sensitivity to ultraviolet radiation when exposed to the open air, high toxicity, low vapor diffusion, and relatively short shelf life constitute obvious disadvantages for this retrofitting system [9]. As the executive phase is concerned, it is worth noting that the laying of the FRP materials must take place on completely dry surfaces, preventing the epoxy resin from coming into contact with moist parts; otherwise, the adhesion may be compromised.

Most of the drawbacks listed above are mainly related to the epoxy matrix used to embed and bond the fibers. That material is completely not compatible with the chemical property of the ancient mortars, leading to a severe breathable issue of the masonry walls. Thus, substituting the epoxy matrix with a mortar matrix appeared to be the most reasonable solution to improve the overall performance of externally-bonded composite systems. Furthermore, due to the nature of the FRP installation, the fracture may be caused in some areas where the masonry wall is not strengthened, particularly in the case of very brittle masonries, such as the irregular stone masonry.

Presently, the application of Fiber Reinforced Cementitious Matrix (FRCM) may overcome the disadvantages observed for the FRPs and represents the most favored choices in many projects [10]. Indeed, recent research [9–20] revealed the mechanical efficiency of the FRCM, its resistance to high temperatures and radiation, high vapor diffusion ability, and the possibility to perform installation even on a wet substratum.

In the FRCMs, the long fibers are embedded in a mortar matrix capable of ensuring the adhesion with the support. The function of the fibers is to carry tensile stresses, whereas the function of the matrix is to encapsulate and protect the fibers and transfer stresses from the mortar or masonry substrate to the fibers.

In FRCM composite systems, the fiber sheets or fabrics that are typically used in FRP are replaced with open fabric meshes in which the strands are spaced in both vertical and horizontal directions forming a bidirectional orthogonal grid. The behavior of masonry walls reinforced by FRCM tested under diagonal compression can be differentiated in three phases: (i) the load is carried mainly by the mortar matrix until cracking; (ii) the matrix undergoes a multi-cracking process resulting in the transfer of stresses from the matrix to the fibers; and (iii) the load is carried almost exclusively by the fabric [20].

Despite all advantages that this strengthening system can provide, it is characterized by a long application procedure that consists of three phases: (i) application of the first layer of mortar on the panel surfaces; (ii) application of the fiber grids on the panel surface lightly pressing them on the fresh mortar layer to have the fresh mortar passing through the grid openings; and (iii) application of a second finishing layer of mortar on the panel surfaces to cover the glass fabric while the previous mortar layer was still fresh. Even if this procedure can be considered easier than the ones concerned for the FRPs and concrete plaster, it still represents a limitation.

A common disadvantage of both the FRCMs and FRPs concerns the orientation of the fibers in specific directions: the FRCMs are characterized by fiber strands oriented in a bidirectional way; the FRPs are characterized by a prior defined fiber direction (usually along the diagonal and the edges of the wall). When the stress state is known, the proper use of such composites is expected to suitably orient the fibers in the direction of the maximum stress to optimize the efficiency of the material. Fibers activate their characteristics along their axial direction, whereas they have negligible properties in the other directions [5]. Hence, the composites with long fibers, such as the FRCM or FRP, are characterized by high resistance only in the direction of the fibers. However, stress may vary substantially in different load conditions. In particular, tensional states induced by seismic events do not act in a single and defined direction. In this case, the classic fiber-based systems may not be really efficient.

Therefore, it is necessary to consider the adoption of a diffuse reinforcement consisting of short fibers randomly oriented in the mortar matrix (discontinuous-fiber-reinforced composites) to ensure proper seismic capacity. Short and randomly distributed fibers can overcome the concern related to the material brittleness and poor resistance to crack initiation and growth [21].

The mechanical characterization of the lime-based mortar reinforced by randomly oriented short fibers is presented in this work, aiming to investigate the fiber content and the fiber type (different Aspect Ratio) on the flexural, tensile, and compressive strength as well as the energy fracture. Beams with dimensions of 160 mm × 40 mm × 40 mm with a central notch were prepared at the laboratory of the "Aquilaprem S.r.l." company (L'Aquila, Italy). Then, the specimens were tested at the LPMS (Laboratorio di Prove Materiali e Strutture) of the University of L'Aquila. First, the samples were tested in three-point bending configuration. Then, the end pieces of the broken beams with a size of 80 mm × 40 mm × 40 mm were tested in compression and Brazilian configurations. Such an experimental procedure was also employed in [22,23]. A final comparison between all the mechanical properties is proposed and analyzed.

2. Description of the Newly Short-Fibers-Based-Strengthening-System

In recent years, considerable interest has been aroused by different nature of short-fibers (steel, plastic, glass, cast iron, polypropylene, polyacrylonitrile, polyolefin, etc.) to enhance the mechanical properties of cementitious materials, characterized by brittle nature with a low tensile strength and strain capacity [24]. In particular, the use of the fibers is greatly increased especially for concrete structures (e.g., industrial concrete slabs, structural or nonstructural precast elements and tunnel coatings).

Incorporation of fibers into cementitious materials can produce materials with increased modulus, increased strength for high fiber content, decreased elongation at rupture, increased hardness even with relatively low fiber content, and improvements in cut, tear, puncture, and impact load resistance [25]. The enhancement is mainly ensured by preventing or controlling the initiation and propagation of cracks [26]. Another advantage concerns the easier execution of structural elements, as compared to the traditional technology (based on the use of reinforcing bars and/or welded mesh).

The performance of a fiber-reinforced material, although in part related to the elastic properties of the fibers (depending on their nature), depends on many factors, such as fiber geometry, fiber content, fiber dispersion, fiber orientation, and fiber–matrix adhesion. Among the several factors, the bond behavior at the fiber–matrix interface plays a role of primary importance. Indeed, the ultimate elongation of the fibers is about 2–3 orders of magnitude higher than the ultimate strain of the mortar matrix and, therefore, the failure of the mortar matrix takes place before fiber failure. In particular, the adherence property of the mortar usually increases for high mechanical properties of the mortar, namely compressive and tensile strengths.

The fibers provide the greatest benefits especially in the softening phase when the maximum resistance of the material is achieved. In that phase, fibers are arranged astride the damage allowing the transmission of forces through a "sewing effect" that prevents the brittle collapse of the material (as one would observe in the absence of fiber reinforcement). Hence, the aspect ratio of the fibers, as well as their shape (fibers with bent ends, hooked and wavy fibers, etc.), assumes considerable importance in the load-bearing capacity, when the cracks through the material occur, affecting the anchoring of the fiber from the matrix and, consequently, yield more efficient the effect of the fiber on the mechanical behavior of the fiber-reinforced material.

The sewing effect also depends on the number of fibers that are arranged astride the damage. Therefore, both the fiber content and the fiber distribution in the cement play an important role in the mechanical behavior of the composite material. Obviously, the higher the fiber content, the higher the fiber distribution in the cement. Therefore, the higher the fiber distribution in the cement, the higher the efficiency of the fiber reinforcement. However, it worth noting that high quantities of fibers also produce a reduction in the fluidity of the fresh product. This aspect should be taken into account in the mix-design phase.

Despite reinforcement made of short-fibers being a topic that is being studied for several years from different researchers, no commercial product made of lime-mortar reinforced with short-fibers is nowadays employed for the strengthening of the existing historical structures.

The idea of using short-fibers embedded in a lime-based mortar matrix for the strengthening of the walls of the historical stone masonries can be considered as a promising newly reinforcement system that may ensure an adequate safety level for seismic forces acting in any direction. The new material was conceived to be compatible with the old constituent material of existing historical masonry. Indeed, the low compatibility of the cement-based mortar of the classic strengthening system with the lime-based mortar of the masonry joints yet represents an issue. In several recent cases, extensive damage occurred to the ancient masonry due to the incompatibility of the cement-based mortars [27–29]. Current standards [30,31] define cement-based mortars to be inadequate for strengthening interventions of historical masonries. Natural Hydraulic Lime (NHL)-based mortars [29,32,33] are considered a promising alternative to cement materials when high chemical-physical compatibility with historical substrates is strictly required.

The new composite material presents high flexibility in its application methodologies to the historical masonry structures. Indeed, it can be used as a coating to the masonry surfaces or in the structural repointing technique. The latter consists of replacing the deteriorated mortar or filling the missing mortar in the joints by employing the new composite material, allowing both to enhance the shear capacity of walls and preserve the original aesthetic of the masonry texture. Indeed, when choosing a retrofit method, its impact on the aesthetics of the building being retrofitted needs to be evaluated [4]. Aesthetic considerations are fundamental for historic structures. Many unreinforced masonry buildings are part of the cultural heritage of a determined city or country. Thereby, the preservation of their aesthetic and architecture is of main importance and retrofit work should be carried out with the least possible irrevocable alteration to the building's appearance. It is recognized that the use of external reinforcing, such as the FRP or the FRCM, can alter the aesthetic of a masonry wall and resulted as unsatisfactory to retrofit churches and historical buildings after the last earthquakes in Europe. The use of the structural repointing by using the SFRLM is an alternative to strengthen masonry walls where aesthetics is an important issue.

3. Description of the Materials

Here, a description of the experimental investigation performed on the novel composite material for the retrofitting of masonry structures is presented. That material consisted of a lime-based mortar reinforced with short-fiber randomly oriented in the mortar matrix. In this study, zirconia-alkali-resistant glass fibers were employed. In particular, the effect of the aspect ratio of the fibers (i.e., the ratio between the length and the average diameter of the fibers) and the fiber content on both the tensile and compressive strengths as well as the workability of the product were investigated.

For the development of the product, the glass fibers were closely selected based on their length ℓ_f. From the literature, it is known that the higher the fiber length, the higher the mechanical properties of the fibrous-product. However, the long fiber creates a problem in the mixing phase, especially with the goal to spray the fresh fibrous product as a coating to the masonry surface (the main possible application of this product), with a consequent loss in the application easiness or the total inapplicability of the product. Hence, the limit of the fiber length was an important factor considered in the development of the new material.

The diameters d_f of the single yarn of both the glass fiber type were almost the same (ranging from 0.0135 to 0.014 mm). However, the fibers were originally impregnated with the matrix resin from manufactures. This has resulted in greater effective diameters of the glass fibers (about from 0.3 mm to 0.5 mm), whose values were not provided by manufacturers. Hence, a measurement of the effective diameters d_f^* of the glass fiber was performed by measuring their diameters by an electronic micrometer. Actually, the cross-section of the fibers was not perfectly circular and therefore d_f^* represents the diameter of the fiber with an ideal circle cross-section of the fiber.

Table 1 summarizes the geometry (fiber length ℓ_f, diameters of the single yarn d_f and the diameters of the strand d_f^*) and the mechanical properties (density ρ_f, Young's Module E_f, tensile strength $f_{t,f}$, ultimate strain $\varepsilon_{u,f}$ and moisture content MC) of the two glass short-fibers (F1 and F2) used in the experimental campaign (illustrated in Figure 1a,b).

Table 1. Geometrical and mechanical properties of the glass fiber used in the experiments.

Name	ℓ_f [mm]	d_f [mm]	d_f^* [mm]	ρ_{fib} [kg/m^3]	E_f [MPa]	$f_{t,f}$ [MPa]	$\varepsilon_{u,f}$ [%]	MC [%]
F1	24	0.0140	0.476	2680	72,000	1700	3.7	0.6
F2	13	0.0135	0.316	2680	72,000	1700	3.7	0.5

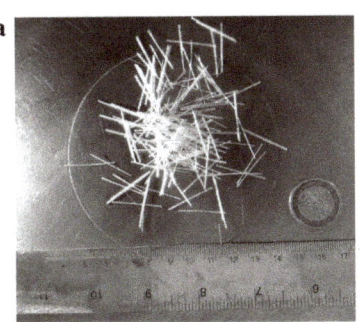
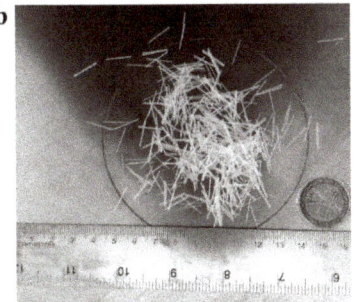

Figure 1. Two fiber types used in the experimental campaign: the F1 (**a**) and the F2 (**b**).

The following results were obtained by assuming the same values of content of lime (NHL 3.5 content equal to 30% of the total weight of the binder; the remaining 70% was a Portland cementitious binder), water content (equal to 80% of the binder weight), sieve curve (sand content equal to 65% of the total weight of the product with size ranging from 0.1 mm to 1.2 mm), and fluidizer content (equal to 0.2% of the total weight of the product). The natural hydraulic lime mortar was assumed in that content because it is intrinsically characterized by a higher variability of its mechanical properties [34], as compared to the cement. Hence, to better investigate the effect of the fiber type (the F1 and the F2) and the fiber content F on the mechanical properties of the composite material, authors decided to assume that lime content aimed to have results characterized by a lower dispersion of the data. Anyhow, that content was enough to obtain the so-called "lime-based" mortar instead of the "pure-lime mortar".

Furthermore, the choice to employ the same mix-design for the reinforced and unreinforced mortar specimens was also due to the intention to better investigate the effect of the fiber properties. Therefore, fresh products were characterized by different consistency. In particular, F was assumed equal to 1.5%, 2.0%, and 2.5% of the total weight of the product (corresponding almost to 1%, 1.3%, and 1.6% of the total volume of the product, respectively) It is worth noting that, even for the unreinforced samples, the mix design of the product (water content, sieve curve, and content of additives) was the same as the fiber-reinforced ones. Hence, the enhancement of the mechanical properties of the fibrous mortar, as compared to the unreinforced mortar, was merely due to the fiber type and fiber content.

For each batch of the product, immediately after the slump test (described in Section 4.1), three mortar samples measuring 160 mm in length, 40 mm in height, and 40 mm in thickness were cast according to the standard code EN 1015-11 [35]. Fibers were added during the mixing phase. The specimens were cast in molds and were kept moist for 48 h in the environmental chambers. Next, the samples were demolded and left in laboratory conditions (room temperature and ambient humidity of about 20° and 60%, respectively) for 26 days, for a total age of 28 days, before testing in a three points bending test (Section 4.2). After the 3PBTs, the end pieces of the broken beams were used to determine their tensile strength (Section 4.3) and compressive strength (Section 4.4).

In particular, a total of 21 mortar specimens were prepared, namely three unreinforced samples and 18 fibrous mortar samples (three samples for each of the three fiber contents adopted for both the two fiber types).

4. Methods

4.1. Characterization of the Consistency

The slump test is an empirical method that measures the workability of fresh mortar (or fresh concrete). More specifically, it measures the consistency of freshly made mortar in a specific batch of the product that can be subsequently used for the mechanical characterization tests, namely three points bending test, compressive test, and direct-indirect tensile test. It is a term that describes the state of fresh mortar. In particular, it is the relative mobility or ability of freshly mixed mortar to flow. It includes the entire range of fluidity from the driest to the wettest possible mixtures [36]. Consistency is a term very closely related to workability.

Practical evaluation of the consistency of the paste can be performed according to the procedure reported in EN 1015-3 [35] and ASTM C1437 [37]. In particular, the so-called slump test consists of measuring the mean diameter of the fresh mortar, previously cast in a specific steel mold, after 20 strokes of the flow table. The measure of the diameter value of the fresh mortar at the end of the test represents the "slump" of the product. The higher the slump value, the higher the workability of the fresh mortar.

4.2. Characterization of the Flexural Strength

To evaluate the enhancement of the innovative fibrous lime mortar material in terms of its flexural strength as well as its fracture energy, three points bending test (3PBT) was carried out on several specimens measuring 160 mm in length, 40 mm in height, and 40 mm in thickness. A 2 mm thick notch (t_n) was fabricated on the mortar samples (by using wet sawing) with a depth a of 6 mm, resulting with a notch to beam depth ratio a/d of 0.15.

The notches were fabricated on the mortar samples to minimize irreversible deformations outside the fracture zone, avoiding large parts with high stresses outside this zone. For that issue,

Hillerborg [38] suggested a depth notch of 0.3–0.4 times the beam depth. However, in the present research, the a/d ratio was chosen equal to 0.15, since the higher values of that ratio would have led to a larger dispersion of the results because of the presence of the short-fiber.

Indeed, the lower a/d ratio leads to an increase in the number of fibers passing through the unnotched ligament $(d - a)$. Hence, one can easily understand that, by increasing the number of fibers passing through the unnotched ligament, less scattering of the results can be obtained. Furthermore, the a/d suggested by Hillerborg referred to concrete material, in which the ratio between the size of the fracture plane and the size of the maximum aggregate is different from the mortar case. The final important reason in the choice of a a/d ratio equal to 0.15 was due to the nature of the material: lime-based mortar beams would have easily broken during handling in case of deeper notches.

The scheme of the notched beam used for the 3PBT is illustrated in Figure 2a. In particular, the nominal distance between the supports L was 100 mm, whereas both the width d and the thickness b were equal to 40 mm. The loading P was introduced at the midspan of the beam. The two rollers at the bottom allowed for free horizontal movement.

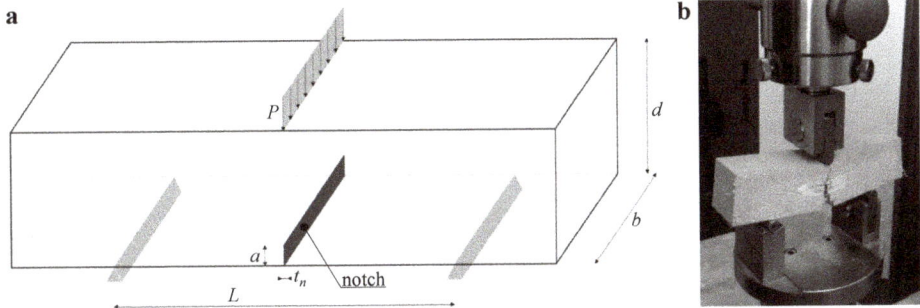

Figure 2. (a) notched beam adopted for the 3PBT; (b) picture of the 3PBT carried out at the LPMS of L'Aquila.

The three-point bending test was performed on notched beams to determine the maximal tensile stress σ_f as well as the fracture energy G_f. Indeed, as described by a Bazant's work [39], the fracture energy of quasi-brittle material is a basic material characteristic needed for a rational prediction of brittle failures of such structures.

Failure of quasi-brittle structures generally consists of numerous micro-cracks that might result in fracturing of the structures under loads. Thus, a micro-crack in quasi-brittle material may become a potential source of crack propagation leading to probable catastrophic failure. Definitely, the failure mechanism can be studied by quantifying the energy consumed in crack propagation and the formation of new crack surfaces.

In principle, the fracture energy as a material property should be a constant, and its value should be independent of the method of measurement, various test methods, specimen shapes, and sizes. However, these variables lead to very different results (e.g., [39–42]).

Despite the scientific interest in chaotic stone constructions, the test method for the determination of G_f and even its precise definition has been a subject of intense debate among researchers because it has been found to vary with the size and shape of the test specimen and with the test method used. The commonly used method for measuring the fracture energy is the work-of-fracture method recommended by RILEM [43,44], in which the total energy G_f is evaluated by dividing the total applied energy by the projected ligament area, as follows:

$$G_f = \frac{W_0 + m\, g\, \delta_0}{(d - a)\, b} \quad (1)$$

where W_0 is the area of the complete load-deflection curve, m is the weight of the beam between the supports, calculated as the beam weight multiplied by $1/L$, g is the acceleration due to gravity, δ_0 is the deformation at the final failure of the beam, d is the beam height, b is the beam width, and a is the notch depth. For stable test performance and to obtain reliable test data, the self-weight compensation [45] was used.

The flexural stress σ_f was calculated by using the following equation:

$$\sigma_f = \frac{3\,P\,L}{2\,b\,(d-a)^2} \tag{2}$$

For $P = P_{MAX}$, this equation gives the flexural strength.

The 3PBTs were performed in a displacement controlled hydraulic testing machine, by using the Zwick Roell test machine of the LPMS (Laboratorio di Prove Materiali e Strutture) of L'Aquila (Figure 2b). The specimens were loaded at a constant displacement rate of 0.5 mm/min. Both the force P and vertical mid-span displacement (or deflection) δ were directly recorded through the test machine.

4.3. Characterization of the Tensile Strength

The tensile strength of quasi-brittle material, such as mortar or concrete, can be determined from different types of tests, namely direct or indirect tensile tests. The direct test concerns the execution of direct pull tests, whereas the indirect tensile test concerns the execution of splitting tensile tests (also called the diametrical compression test, split-tension test, and Brazilian test (BT) among other names).

There are many technical difficulties in executing a true tensile strength test. A uniform stress distribution which makes it possible to calculate the true tensile strength is difficult to obtain. The method commonly used to determine tensile properties of quasi-brittle material is the flexural beam test by three-point loading on a beam over a span (the 3PBT described in Section 4.2). The flexural strength is computed from the bending moment at failure, assuming an ideal straight line stress distribution according to Hooke's law. However, the calculated flexural strength may be higher than the true tensile strength [46].

Many attempts have been made to find a substitute for the 3PBT and the splitting tensile test of a cylindrical specimen may be the solution to the problem [47]. The splitting tensile strength test method has many merits compared with the direct tensile test method; for example, it can be conducted much more easily and the scatterings of the test results are very narrow. The BT is straightforward and economic and can be used on cylindrical specimens (fabricated in molds or extracted concrete cores) or flat disk-shaped specimens as well as cubes or prisms [48]. In addition, the test can be performed with the same machine that is used to perform direct compression tests, and samples identical in shape and geometry as those used in direct compression can be employed. The BT is useful to experiment brittle or quasi-brittle materials that have a much greater compression strength than their tensile strength and that are susceptible to brittle ruptures. Researchers have indicated that, among the three testing methods (direct tensile, splitting tensile, and flexural tests), the splitting tensile test gives the most accurate measurement of the true tensile strength of mortar or concrete materials in a wide strain rate [48].

In the BT, the sample is compressed with load concentrated on a pair of antipodal points. In this way, tensile stress is induced in the direction perpendicular to the applied load, and it is proportional to the magnitude of the applied load.

When the induced stress exceeds the tensile strength, fracture initiates at the geometric center of the sample. In agreement with the Griffith criterion [49], the exact center of the sample is the only point at which the conditions for failure under tension are satisfied because, in this site, the tensile stress equals the uniaxial strength of the tested material. Indeed, the BT result is accepted if fracture initiates at the center of the sample, and in this case, the measured value is representative of the tensile strength of the tested material. In the BT, the specimen must fail along the vertical line between compression points; otherwise, the observed failure mode is considered invalid. In the case of unreinforced specimens, the test typically ends with a sudden failure of the specimen when it reaches the maximum load due to the propagation of an unstable crack. Since its invention, the BT has motivated a wide variety of studies. One can gain an idea of its impact if one considers that the use of concrete test specimens has been standardized into norms in various countries, such as UNI EN 12390-6, ASTM C-496. However, the BT is far from a universal test, and it is unknown whether a geometric configuration exists that favors effective, robust testing that is less sensitive to other experimental parameters [50].

In the present research, to obtain the tensile strength of the mortar specimens, instead of performing a direct tension test, which is of needless difficulty, Brazilian Tests (BT) (ASTM C496) were conducted using 40 mm × 40 mm × 80 mm prismatic mortar specimens obtained from the two-half specimens tested in 3PBT (see Figure 3a,b).

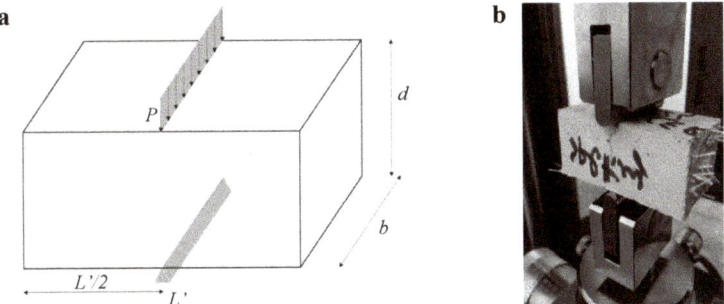

Figure 3. (a) scheme employed for the Brazilian test (BT); (b) picture of the BT carried out at the LPMS of L'Aquila.

The tensile stress f_t was calculated by using the following equation:

$$f_t = \frac{2P}{\pi b d} \quad (3)$$

By Equation (3), one can compute the tensile strength when $P = P_{max}$. The BTs were performed in a displacement controlled hydraulic testing machine, by using the Zwick Roell test machine of the LPMS of L'Aquila. The specimens were loaded at a constant displacement rate of 0.5 mm/min. Both the force P and vertical mid-span displacement (or deflection) δ were directly recorded through the test machine.

4.4. Characterization of the Compressive Strength

To have a complete overview of the mechanical behavior of the fibrous lime mortar material, the Compression Tests (CT) were also carried out to obtain the compressive strength of the specimens. In particular, the tests were performed on 40 mm × 40 mm × 80 mm prismatic mortar specimens obtained from the two-half-length specimens result from 3PBT after testing the full-length specimen (Figure 4).

The distributed load P was applied on a squared area of 40 mm, while the specimens were placed on a squared area of 40 mm. One can compute the compressive stress by using the following equation:

$$f_c = \frac{P}{b L''} \quad (4)$$

By Equation (4), one can compute the compressive strength when $P = P_{max}$. Moreover, the vertical strain ε_v is computed by the following equation:

$$\varepsilon_v = \frac{\delta}{d} \quad (5)$$

The CTs were performed in a displacement controlled hydraulic testing machine adopting a constant displacement rate of 1 mm/min. Even in this case, both P and δ were directly recorded by the test machine.

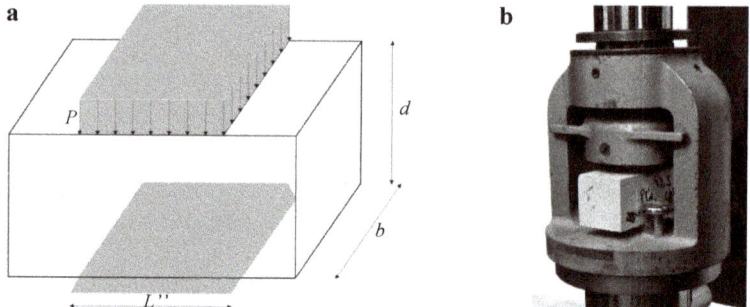

Figure 4. (a) scheme employed for the Compression Test (CT); (b) picture of the CT carried out at the LPMS of L'Aquila.

5. Results and Discussion

In Figure 5, one can observe the results of the slump tests carried out for the mortar specimens strengthened by the F1 and the F2 fibers. In that figure, the slump value is related to the fiber content F adopted in the mix-design of the fibrous mortar. It is worth noting that a linear variation of the slump (indicated by the dotted lines) was assumed from the case of the unreinforced mortar ($F = 0\%$) to the reinforced mortar with the content fiber of 1.5% as no tests were carried out by using fiber content in that range (0%–1.5%).

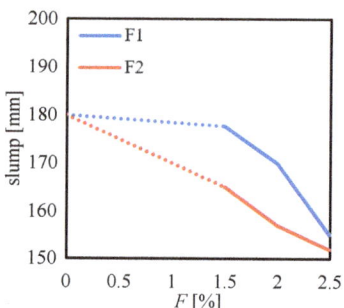

Figure 5. Slump measured for the F1 and the F2 fibers by varying the fiber content F.

As expected, in Figure 5, one can observe a decrease in the slump value by increasing the fiber content. This trend can be observed for both the fiber types. For the F1 case, one can see that no strong differences in terms of slump can be noted by comparing the unreinforced samples ($F = 0\%$) and the $F = 1.5\%$ case. This means that a significant compaction effect of the F1 fiber on the mortar matrix can be achieved only with fiber content higher than 1.5%. This is obviously related to the content of additives introduced in the product to increase its workability. Indeed, for lower additive content, one would observe a higher reduction of the slump value even for F lower than 1.5%. Moreover, in Figure 5, one can instead observe an almost linear decrease in the slump value for the F2 case by increasing the fiber content.

Results plotted in Figure 5 indicate that the F1 fiber leads to higher workability of the product (higher slump value), as compared to the F2 one. This trend can be observed for all the fiber content F.

The large difference in the results plotted in Figure 5 between the two fiber types may be mainly due to their different aspect ratio AR. Indeed, for the same F, a higher value of the AR leads to a lower number of fibers per unit volume of the product, and vice-versa.

In particular, the volumes of the single fiber strands were 4.269 mm^3 ($0.476^2 \pi/4 \times 24$) and 1.019 mm^3 ($0.316^2 \pi/4 \times 13$) for the F1 and F2 fibers, respectively. It is worth noting that the density of the two fibers was the same (2680 kg/m^3). Hence, at the same fiber content, the F2 fiber ensured almost four times the number of the fiber strands into the matrix of the mortar specimens, as compared to the F1. Therefore, due to the lower number of fibers, the F2 fiber allowed a higher compaction effect on the mortar matrix, as compared to the F1 fiber. This was reflected in the workability of the product: the lower the number of fibers, the higher the workability of the product.

Furthermore, the trend of Figure 5 may be influenced by the absorption level of the two types of fibers since they were produced by different manufacturers and different types of primers may have been employed for them. In particular, the moisture content (MC) declared from the two manufactures is equal to 0.6% and 0.5% for the F1 and F2 fibers, respectively, where MC is defined as the weight of water in a material express as a percentage of the total weight of the material (for more details on the effect of moisture content on the mechanical properties of glass fiber, the reader is referred to a recent study [51]). Hence, F1 fiber tends to absorb slightly more water than the F2 fiber. Despite this difference, the workability of the fresh mortar with F1 fiber is higher than the one with F2. Definitely, the higher workability obtained for the F1 fiber is mainly related to other factors, such as the number of the fibers into the mortar (depending on the fiber geometry).

Figure 6a shows the relation measured between the flexural stress σ_f and the deflection δ obtained for the unreinforced mortar specimens under three-point bending tests (3PBT). Plots are referred to three tests. The mean value of the maximum flexural stress is equal to 2.9 MPa. Moreover, one can observe an almost perfect brittle behavior after the achievement of the maximum flexural stress. Indeed, the σ_f suddenly drops after that point.

Figure 6b–d show the σ_f–δ plots obtained for the mortar specimens reinforced by the two fiber types (F1 and F2) tested under 3PBT by assuming different fiber content F (1.5%, 2.0%, and 2.5%). Plots are referred to a number of three tests for each fiber type and fiber content. One can see that, for $F = 1.5\%$ and $F = 2.0\%$, no strong differences can be noted between the two different fiber types (F1 and F2) in terms of flexural strength. On the contrary, for the higher fiber content ($F = 2.5\%$), one can see that the F1 fiber leads to higher flexural strength, as compared to the F2 fiber. From those figures, it is clear that the F1 fiber leads to higher energy fracture for all the fiber content, as compared to the F2 fiber.

For a better interpretation of the effect of both the fiber content and fiber type on the mechanical properties of the composite material, one can see Figure 7a,b. In particular, Figure 7a shows the variation of the flexural strength σ_f, computed by using the equations Equation (2), as a function of the fiber content F. One can see that, for lower fiber content ($F = 1.5\%$), no differences can be noted between F1 and F2. Then, for higher values of fiber content ($F = 2.0\%$ and $F = 2.5\%$), one can see different results by comparing the two fiber types. In particular, for the F1 case, one can observe a gradual increase in the flexural strength up to achieve almost 15 MPa (when $F = 2.5\%$). Instead, for the F2 case, a high increase in the flexural strength is observed from $F = 1.5\%$ to $F = 2.0\%$, whereas one can observe a slight difference by comparing the case of $F = 2.0\%$ and $F = 2.5\%$.

In general, from Figure 7a, one can see the benefit of the fiber content on the flexural strength albeit the variability of the results may be affected by many factors (i.e., specimens properties and failure mode).

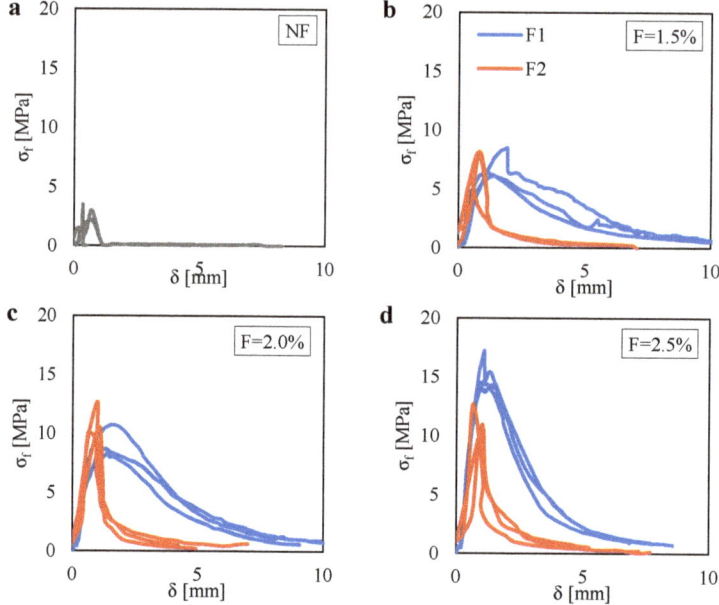

Figure 6. Flexural stress σ_f related to the vertical displacement δ measured in the 3PBT for: (**a**) the unreinforced case (NF). Reinforced cases by assuming two fiber types (F1 and F2) with fiber contents equal to (**b**) 1.5%; (**c**) 2.0% and (**d**) 2.5%.

Figure 7b shows the variation of the fracture energy G_f, computed by using the Equation (1), as a function of the fiber content. Respect to the results noted in terms of σ_f, one can observe a clearer trend in terms of G_f. For all the fiber content, one can see that the fracture energy computed for the F1 is more than two times the one computed for the F2. This is due to the higher contact surface of the F1 fiber with the mortar matrix. Indeed, the higher fiber length as well as the higher diameter of the F1 ensured a higher bond behavior (adherence level) at the fiber–mortar interface, as compared to the F2 fiber.

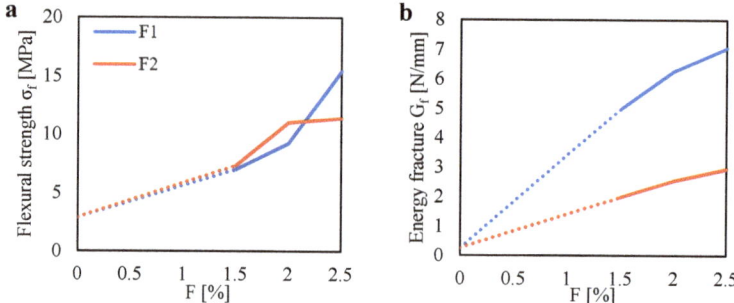

Figure 7. (**a**) flexural strength σ_f and (**b**) fracture energy G_f measured in the 3PBT for two fiber type (F1 and F2) with different fiber content.

Figure 8a shows the relation measured between the tensile stress f_t and the deflection δ obtained for the unreinforced mortar specimens under Brazilian Tests (BT). Plots are referred to two tests (one specimen was broken before testing). The mean value of the tensile strength is equal to 2.0 MPa. It is worth noting that the flexural strength computed in the 3PBT was equal to 2.9 MPa, which is 1.45 times higher than the tensile strength.

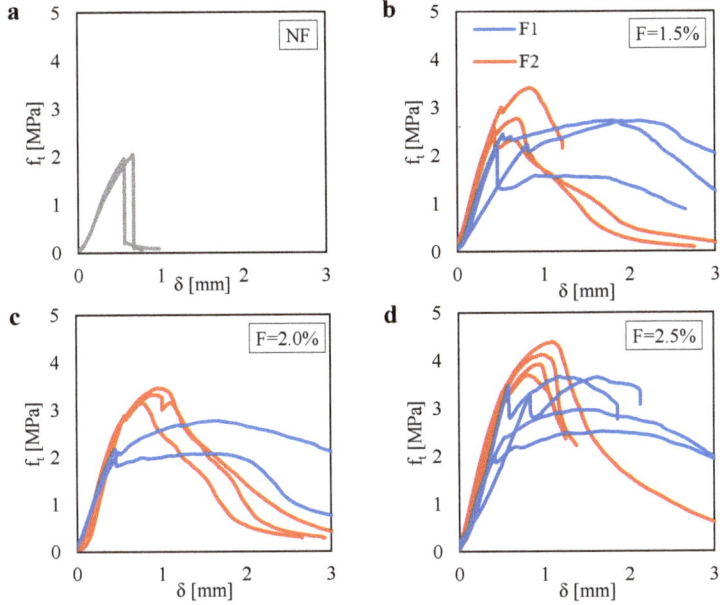

Figure 8. Tensile stress f_t related to the vertical displacement δ measured in the BTs for: (**a**) the unreinforced case (NF). Reinforced cases by assuming two fiber types (F1 and F2) with fiber contents equal to (**b**) 1.5%; (**c**) 2.0% and (**d**) 2.5%.

As already observed for the unreinforced specimens tested in 3PBT, one can observe an almost perfect brittle behavior. Indeed, f_t suddenly drops after the achievement of the maximum tensile strength.

Figure 8b–d show the f_t–δ plots obtained for the fiber-reinforced mortar specimens tested under Brazilian Test (BT) by varying the fiber content therein the mortar matrix. Plots are referred to three tests (except the case of $F = 2.0\%$ with the F1 fiber, for which two tests are presented).

For a better interpretation of the effect of both the fiber content and fiber type on the tensile behavior of the composite material, one can see Figure 9. In particular, that figure shows the variation of the tensile strength f_t, computed by using the Equation (3) for $P = P_{max}$, as a function of the fiber content.

One can see that higher values of tensile strength can be noted for the F2, as compared to the F1, for all the fiber content F. In particular, for the F2 case, one can observe a gradual increase in the tensile strength up to achieve almost 4 MPa for $F = 2.5\%$. It is worth noting that, for the same fiber content, a flexural strength equal to 11.4 MPa (2.85 times the f_t value) was computed for the 3PBT. Instead, for the F1 case, one can observe no differences in the tensile strength between the $F = 1.5\%$ and the $F = 2.0\%$ cases, in which f_t is equal to 2.5 MPa. Then, an increase in f_t is observed for $F = 2.5\%$, in which f_t is equal to 3.2 MPa. By comparing that value with the flexural strength computed for the same fiber content ($F = 2.5\%$), one can see a high difference (σ_f is 4.8 times the f_t value).

In general, from Figure 9, one can see the benefit of the fiber content on the tensile strength. It is worth noting that the F2 fiber type leads to the higher tensile strength though it was characterized by the lower fiber length and lower AR, as compared to the F1 fiber. This may be due to the number of fibers that the F2 fiber type ensured respect to the F1 one, at the same fiber content. Indeed, a higher number of fibers lead to higher compaction of the mortar matrix (as discussed for the results of Figure 5) as well as a higher number of the fiber over the entire projected ligament area, where cracks develop.

Figure 10a shows the relation measured between the compressive stress f_c and the vertical strain ε_v obtained for the unreinforced mortar specimens under Compression Test (CT). Plots are referred to four tests. Each test was interrupted at 40% of the maximum load in the post-peak behavior. One can compute a mean value of the compressive strength equal to 18.1 MPa. Moreover, one can observe a typical softening curve of quasi-brittle materials. Indeed, for all the curves, one can see that the compressive stress suddenly drops after the achievement of the maximum stress.

Figure 10b–d show the f_c–ε_v plots obtained for the fiber-reinforced mortar specimens tested under Compression Test (CT) for different fiber content F. All the plots are referred to three tests. In general, from these results, one can see that the enhancement of the compressive strength due to both the fiber type and fiber content is more and more limited, as compared to the contribution offered by the fibers in the other mechanical properties. Furthermore, one can observe a large scattering of the results. In particular, an increase in the scattering of the results can be observed by increasing the fiber content. This trend was in line with what was expected. Indeed, the higher fiber content leads to lower homogeneity of the product with a consequent increase of the scatter in the results.

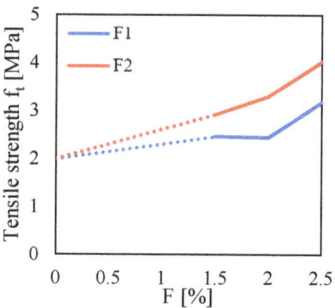

Figure 9. Tensile strength f_t measured in the BTs for two fiber type (F1 and F2) with different fiber content.

For a better interpretation of the effect of both the fiber content and fiber type on the compressive behavior of the composite material, one can see Figure 11. In particular, this figure shows the variation of the compressive strength f_c, computed by using Equation (4) for $P = P_{max}$, as a function of the fiber content.

As compared to the other results obtained for the 3PBT and the BT, the trends of the results obtained for the CT are less clear. Indeed, for the F1 case, one can see that, by increasing the fiber content from $F = 1.5\%$ to $F = 2.0\%$, a reduction of the mean compressive strength is observed. The same unusual trend was also observed for the F2 case by increasing the fiber content from $F = 2.0\%$ to $F = 2.5\%$. However, from a general point of view, one can see a clear increase in the compressive strength of the fiber-reinforced mortar as compared to the unreinforced case.

A final comparison of the results obtained by the three-point bending test (3PBT), Brazilian test (BT), and compression test (CT) is proposed in Figure 12a–d. In particular, this figure shows the increase (in percentage) in the flexural strength σ_f, fracture energy G_f, tensile strength f_t and compressive strength f_c, as compared to the unreinforced case. One can see that the best benefit in introducing fiber therein the mortar matrix is observed in terms of the fracture energy (up to almost 2500% for the F1 case and 1000% for the F2 case). Excellent increase in the flexural strength can also be noted (up to almost 450% for the F1 case and 300% for the F2 case). As compared to the other mechanical properties, a lower increase in the tensile strength and especially in the compressive strength can be observed (less than 100% for all the cases).

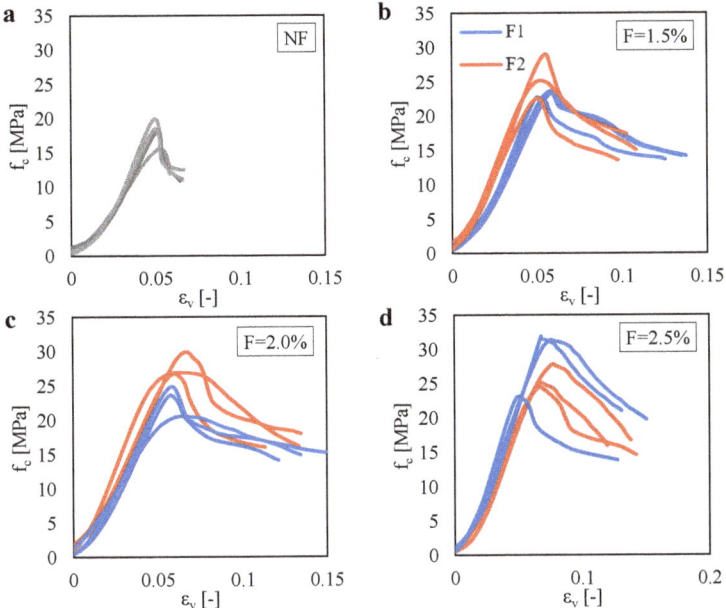

Figure 10. Compressive stress f_c related to the vertical deformation ε_v measured in the CTs for: (**a**) the unreinforced case (NF). Reinforced cases by assuming two fiber types (F1 and F2) with fiber content equal to (**b**) 1.5%; (**c**) 2.0%, and (**d**) 2.5%.

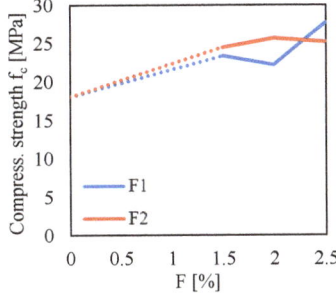

Figure 11. Compressive strength f_c measured in the CTs for two fiber type (F1 and F2) with different fiber content.

It is worth noting that the variability of the results may be caused by many factors: specimen properties irregularity of the cross-section, the presence of micro-cracks, misalignment of the samples with respect to their mid-thickness, different thickness of the specimens, the randomness of the fibers, and last but not the least the spatial randomness of material properties. Hence, it was normal to observe a large dispersion of the results (that is obviously higher for the fibrous materials).

Definitively, despite the F2 fiber ensured, in most cases, the higher flexural, tensile, and compressive strengths, its energy fracture was clearly smaller, as compared to the F1 fiber. This result highlighted the importance to investigate the softening behavior of fibrous mortar specimens. Hence, the F1 fiber is more recommendable for the retrofitting system as compared to the F2 fiber.

Table 2 summarizes the mean and standard deviation values of the flexural strength, fracture energy, tensile strength, and compressive strength computed for the new composite material by assuming different fiber type (F1 and F2) and different fiber content ($F = 0\%$, 1.5%, 2.0%, and 2.5%).

Table 2. Mechanical properties of the lime-based mortar reinforced by two types of short-fibers (F1 and F2) with different fiber content F.

Name	F [%]	σ_f [MPa]	G_f [N/mm]	f_t [MPa]	f_c [MPa]
NF	0	2.9 ± 0.71	0.32 ± 0.3	2.0 ± 0.1	18.1 ± 2.4
F1	1.5	7.0 ± 0.9	5.0 ± 1.0	2.5 ± 0.3	23.4 ± 0.5
F1	2.0	9.2 ± 0.9	6.3 ± 1.0	2.5 ± 0.3	22.2 ± 2.4
F1	2.5	15.4 ± 0.9	7.1 ± 0.8	3.2 ± 0.7	27.7 ± 4.5
F2	1.5	7.3 ± 1.5	1.9 ± 0.6	2.9 ± 0.5	24.5 ± 3.5
F2	2.0	11.0 ± 1.1	2.7 ± 0.4	3.3 ± 0.2	25.7 ± 6.4
F2	2.5	11.4 ± 0.8	2.9 ± 0.3	4.0 ± 0.3	25.2 ± 1.9

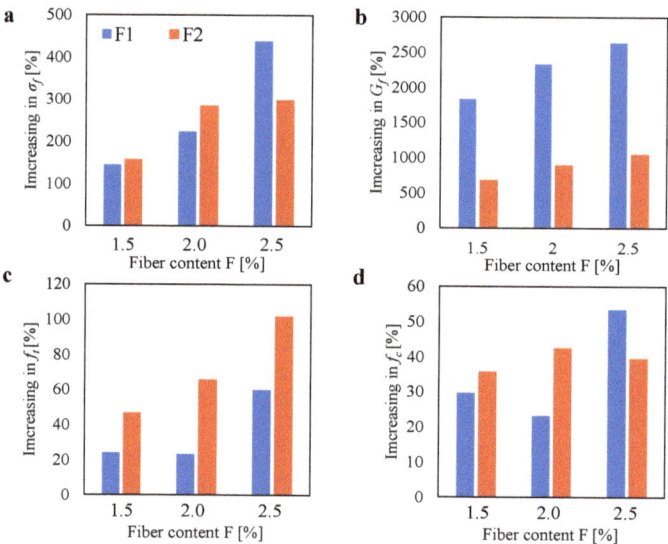

Figure 12. Increase in the mechanical properties (flexural strength σ_f, fracture energy G_f, tensile strength f_t and compressive strength f_c) with respect to the unreinforced case by varying the fiber content. Figure also shows the comparison between the two fiber types (the F1 and the F2).

6. Conclusions

Fiber Reinforced Mortar has proven on more than on occasion that it can increase the strength, ductility, and toughness of plain mortar. This study was not presented just to prove this point. The ultimate goal was to understand how various types of fibers and their content affect the same mortar type and to consider these fibers in binders more closely related to material compositions of historical mortars, namely the lime-based mortar. In particular, three fiber contents were investigated: 1.5%, 2.0%, and 2.5% of the total weight of the product (corresponding almost to 1%, 1.3%, and 1.6% of the total volume of the product, respectively). Experimental results showed that the contribution of diffuse short fibers greatly increased (almost proportionally to their content) the mechanical properties of the lime-mortar. In particular, the fracture energy was the mechanical property that most benefited from the use of short fibers up to about 2500% (for fiber content equal to 2.5%), as compared to the unreinforced case. This value was computed for the fiber characterized by the longer length (24 mm). For the lower fiber length (13 mm), it was computed an energy fracture of about 2–3 times less. Then, for the highest fiber content, the maximum increase in the flexural and tensile strengths was about 450% and 100%, respectively. The mechanical property less affected by the fiber contribution was the compressive strength as its maximum increase was only about 50%.

It is worth noting that the long fiber strands used in the classic fiber-based strengthening system (namely, the FRCMs and the FRPs) have mainly the function of carrying tensile stresses. Hence, the results obtained for the new composite material are important as it can significantly increase both the strength and ductility of the mortar material. Definitely, the new material can ensure physical-chemical compatibility with the special characteristics of the historical masonries and also increasing their strength and ductility, by standing as a promising alternative to the classic fiber-reinforcing systems.

The overriding wish of the authors is to boost the use of sustainable materials and reduce the impact of the intervention in the strengthening of historical masonries. To do so, they will focus their next research on the mortars made with vegetable resins, natural yarns such as sisal, hempen, and similar, also using recycled materials.

Author Contributions: Conceptualization, M.A. and A.G.; methodology, M.A., M.V., and A.G.; software, M.A.; validation, A.G.; formal analysis, M.A.; investigation, M.A. and M.V.; resources, A.G.; data curation, M.A.; writing—original draft preparation, M.A.; writing—review and editing, M.V. and A.G.; visualization, M.A.; supervision, A.G.; project administration, A.G.; funding acquisition, A.G. All authors have read and agreed to the published version of the manuscript.

Funding: The Ph.D. scholarship of the first author was co-financed by the Project "2014-2020 PON" (CCI 2014EN16M2OP005).

Acknowledgments: The authors gratefully acknowledge the support of the "Aquilaprem S.r.l." company (L'Aquila, Italy) for the concession of materials as well as Paolo Dacci and Eng. Daniele Martini for the development of the mix-design of the product.

Conflicts of Interest: The authors declare no conflict of interest. The funders had no role in the design of the study; in the collection, analyses, or interpretation of data; in the writing of the manuscript; nor in the decision to publish the results.

Abbreviations

The following abbreviations are used in this manuscript:

3PBT	Three-Point Bending Test
BT	Brazilian Test
CT	Compressive Test
F	Fiber Content
F1	glass Fiber with 24 mm in length
F2	glass Fiber with 13 mm in length

References

1. Angiolilli, M.; Gregori, A. Triplet test on rubble stone masonry: Numerical assessment of the shear mechanical parameters. *Buildings* **2020**, *10*, 49. [CrossRef]
2. Aloisio, A.; Fragiacomo, M.; D'Alò, G. The 18th-century baraccato of l'aquila. *Int. J. Archit. Herit.* **2019**, 1–15. [CrossRef]
3. Aloisio, A.; Fragiacomo, M.; D'Alò, G. Traditional TF Masonries in the City Centre of L'Aquila—The Baraccato Aquilano. *Int. J. Archit. Herit.* **2019**, 1–18. [CrossRef]
4. Tumialan, J.G.; Micelli, F.; Nanni, A. Strengthening of masonry structures with FRP composites. In *Structures 2001: A Structural Engineering Odyssey*; University of Missouri-Rolla: Rolla, MO, USA, 2001; pp. 1–8. [CrossRef]
5. Valluzzi, M. Strengthening of masonry structures with fibre reinforced plastics: From modern conception to historical building preservation. In *Structural Analysis of Historic Construction: Preserving Safety and Significance, Two Volume Set*; CRC Press: Boca Raton, FL, USA, 2008; pp. 53–66.

6. Schwegler, G. Strengthening of Masonry in Seismic Regions Using FRP. In *10th European Conference on Earthquake Engineering*; Empa Report N; Balkema: Rotterdam, The Netherlands, 1995.
7. Velazquez-Dimas, J.I.; Ehsani, M. Modeling Out-of-Plane Cyclic Behavior of URM Walls Retrofitted with Fiber Composites. *J. Compos. Constr.* **2000**, *4*, 172–181. [CrossRef]
8. Korany, Y.; Drysdale, R. Rehabilitation of masonry walls using unobtrusive FRP techniques for enhanced out-of-plane seismic resistance. *J. Compos. Constr.* **2006**, *10*, 213–222. [CrossRef]
9. Túrk, A.M. Seismic response analysis of masonry minaret and possible strengthening by fiber reinforced cementitious matrix (FRCM) materials. *Adv. Mater. Sci. Eng.* **2013**. [CrossRef]
10. Carozzi, F.G.; Bellini, A.; D'Antino, T.; de Felice, G.; Focacci, F.; Hojdys, Ł.; Laghi, L.; Lanoye, E.; Micelli, F.; Panizza, M.; et al. Experimental investigation of tensile and bond properties of Carbon-FRCM composites for strengthening masonry elements. *Compos. Part B Eng.* **2017**, *128*, 100–119. [CrossRef]
11. Papanicolaou, C.; Triantafillou, T.; Lekka, M. Externally bonded grids as strengthening and seismic retrofitting materials of masonry panels. *Constr. Build. Mater.* **2011**, *25*, 504–514. [CrossRef]
12. Tomaževič, M.; Gams, M.; Berset, T. Seismic strengthening of stone masonry walls with polymer coating. In Proceedings of the 15th World Conference on Earthquake Engineering (15 WCEE), Lisbon, Portugal, 24–28 September 2012; pp. 24–28.
13. Babaeidarabad, S.; Arboleda, D.; Loreto, G.; Nanni, A. Shear strengthening of un-reinforced concrete masonry walls with fabric-reinforced-cementitious-matrix. *Constr. Build. Mater.* **2014**, *65*, 243–253. [CrossRef]
14. De Luca, A. FRCM Systems. *Structure* **2014**, *23*, 22–24.
15. Gattesco, N.; Boem, I.; Dudine, A. Diagonal compression tests on masonry walls strengthened with a GFRP mesh reinforced mortar coating. *Bull. Earthq. Eng.* **2015**, *13*, 1703–1726. [CrossRef]
16. Pereira, M.V.; Fujiyama, R.; Darwish, F.; Alves, G.T. On the strengthening of cement mortar by natural fibers. *Mater. Res.* **2015**, *18*, 177–183. [CrossRef]
17. Marcari, G.; Basili, M.; Vestroni, F. Experimental investigation of tuff masonry panels reinforced with surface bonded basalt textile-reinforced mortar. *Compos. Part B Eng.* **2017**, *108*, 131–142. [CrossRef]
18. Gregori, A.; Marchini, G.; Martini, D.; Angiolilli, M. Experimental Characterization of FRCM Systems for Conservative and Strengthening Intervention on Monumental Real Estate Heritage (in italian). In *Proceedings of the 18th Italian National Association of Earthquake Engineering-ANIDIS*; Pisa University Press: Pisa, Italy, 2017; pp. 116–126.
19. Donnini, J.; Lancioni, G.; Corinaldesi, V. Failure modes in FRCM systems with dry and pre-impregnated carbon yarns: Experiments and modeling. *Compos. Part B Eng.* **2018**, *140*, 57–67. [CrossRef]
20. Angiolilli, M.; Gregori, A.; Pathirage, M.; Cusatis, G. Fiber Reinforced Cementitious Matrix (FRCM) for strengthening historical stone masonry structures: Experiments and computations. *Eng. Struct.* **2020**, accept for publication.
21. Bentur, A.; Mindess, S. *Fibre Reinforced Cementitious Composites*; CRC Press: Boca Raton, FL, USA, 2006.
22. Skłodowski, M. *Compact Diagnostic Test: Outline of Historical Monuments Testing Procedure*; Instytut Podstawowych Problemów Techniki PAN: Warsaw, Poland, 2006.
23. Skłodowski, M. Minor Destructive Testing of XVIII century brick wall using Compact Diagnostic Test CoDiT. *Adv. Mater. Res.* **2010**, *133–134*, 223–228. [CrossRef]
24. Pakravan, H.; Latifi, M.; Jamshidi, M. Hybrid short fiber reinforcement system in concrete: A review. *Constr. Build. Mater.* **2017**, *142*, 280–294. [CrossRef]
25. Lee, D.; Ryu, S. *The Influence of Fiber Aspect Ratio on The Tensile and Tear Properties of Short-Fiber Reinforced Rubber*; ICCM12: Paris, France, 1999.
26. Yao, W.; Li, J.; Wu, K. Mechanical properties of hybrid fiber-reinforced concrete at low fiber volume fraction. *Cem. Concr. Res.* **2003**, *33*, 27–30. [CrossRef]
27. Degryse, P.; Elsen, J.; Waelkens, M. Study of ancient mortars from Sagalassos (Turkey) in view of their conservation. *Cem. Concr. Res.* **2002**, *32*, 1457–1463. [CrossRef]
28. Moropoulou, A.; Cakmak, A.; Biscontin, G.; Bakolas, A.; Zendri, E. Advanced Byzantine cement based composites resisting earthquake stresses: The crushed brick/lime mortars of Justinian's Hagia Sophia. *Constr. Build. Mater.* **2002**, *16*, 543–552. [CrossRef]
29. Lanas, J.; Alvarez-Galindo, J.I. Masonry repair lime-based mortars: Factors affecting the mechanical behavior. *Cem. Concr. Res.* **2003**, *33*, 1867–1876. [CrossRef]

30. Council, N.R. *R1: Istruzioni per la Progettazione, l'Esecuzione ed il Controllo di Interventi di Consolidamento Statico Mediante L'utilizzo di Compositi Fibrorinforzati*; CNR-DT 200; CNR: Rome, Italy, 2013.
31. Ministero per i Beni e le Attivita' Culturali. *Linee Guida per la Valutazione e Riduzione del Rischio Sismico del Patrimonio Culturale Allineate Alle Nuove Norme Tecniche per le Costruzioni (d.m. 14 gennaio 2008)*; Gangemi Editore spa: Rome, Italy, 2011.
32. Alecci, V.; De Stefano, M.; Focacci, F.; Luciano, R.; Rovero, L.; Stipo, G. Strengthening masonry arches with lime-based mortar composite. *Buildings* **2017**, *7*, 49. [CrossRef]
33. Angiolilli, M.; Gregori, A.; Martini, D. Mechanical characterization of the cyclic behavior of historic masonry panels reinforced by FRCM system subject to diagonal compression tests (in italian). In *Proceedings of the 18th Italian National Association of Earthquake Engineering-ANIDIS*; Pisa University Press: Ascoli Piceno, Italy, 2019; pp. 116–126. [CrossRef]
34. Drougkas, A.; Roca, P.; Molins, C. Compressive strength and elasticity of pure lime mortar masonry. *Mater. Struct.* **2016**, *49*, 983–999. [CrossRef]
35. EN 1015-11. *Methods of Test for Mortar for Masonry—Part 11*; European Committee for Standardization: Brussels, Belgium, 2007.
36. Mather, B.; Ozyildirim, H.C. *Concrete Primer*; American Concrete Institute: Indianapolis, IN, USA, 2002.
37. Standard, A. C1437: Standard Test Method for Flow of Hydraulic Cement Mortar. In *Annual Book of ASTM Standards*, ASTM: West Conshohocken; PA, USA, 2007.
38. Hillerborg, A. *Concrete Fracture Energy Tests Performed by 9 Laboratories According to a Draft RILEM Recommendation*; Report to RILEM TC50-FMC, Report TVBM-3015; Division of Building Materials: Lund, Sweden, 1983.
39. Bazant, Z.P.; Pfeiffer, P.A. Determination of fracture energy from size effect and brittleness number. *ACI Mater. J.* **1987**, *84*, 463–480.
40. Bažant, Z.P.; Lin, F.B. Nonlocal smeared cracking model for concrete fracture. *J. Struct. Eng.* **1988**, *114*, 2493–2510. [CrossRef]
41. Hillerborg, A. Results of three comparative test series for determining the fracture energyG F of concrete. *Mater. Struct.* **1985**, *18*, 407–413. [CrossRef]
42. Duan, K.; Hu, X.; Wittmann, F.H. Boundary effect on concrete fracture and non-constant fracture energy distribution. *Eng. Fract. Mech.* **2003**, *70*, 2257–2268. [CrossRef]
43. Recommendation, R.D. Determination of the fracture energy of mortar and concrete by means of three-point bend tests on notched beams. *Mater. Struct.* **1985**, *18*, 285–290.
44. Planas, J. (Ed.) *Report 39: Experimental Determination of the Stress-Crack Opening Curve for Concrete in Tension-Final Report of RILEM Technical Committee TC 187-SOC*; RILEM Publication: Bagneux, France, 2007; Volume 39.
45. Bazant, Z.P.; Planas, J. *Fracture and Size Effect in Concrete and Other Quasibrittle Materials*; CRC Press: Boca Raton, FL, USA, 1997; Volume 16.
46. Min, F.; Yao, Z.; Jiang, T. Experimental and numerical study on tensile strength of concrete under different strain rates. *Sci. World J.* **2014**, *2014*. [CrossRef]
47. Martin, J.; Stanton, J.; Mitra, N.; Lowes, L.N. Experimental testing to determine concrete fracture energy using simple laboratory test setup. *ACI Mater. J.* **2007**, *104*, 575.
48. Rocco, C.; Guinea, G.; Planas, J.; Elices, M. Review of the splitting-test standards from a fracture mechanics point of view. *Cem. Concr. Res.* **2001**, *31*, 73–82. [CrossRef]
49. Griffith, A. The Phenomena of Rupture and Flow in Solids (Philosophical Transactions/Royal SOCIETY of London Ser. A, v. 221). *Royal Soc.* **1920**. [CrossRef]
50. García, V.J.; Márquez, C.O.; Zúñiga-Suárez, A.R.; Zuñiga-Torres, B.C.; Villalta-Granda, L.J. Brazilian test of concrete specimens subjected to different loading geometries: Review and new insights. *Int. J. Concr. Struct. Mater.* **2017**, *11*, 343–363. [CrossRef]
51. Chaichanawong, J.; Thongchuea, C.; Areerat, S. Effect of moisture on the mechanical properties of glass fiber reinforced polyamide composites. *Adv. Powder Technol.* **2016**, *27*, 898–902. [CrossRef]

© 2020 by the authors. Licensee MDPI, Basel, Switzerland. This article is an open access article distributed under the terms and conditions of the Creative Commons Attribution (CC BY) license (http://creativecommons.org/licenses/by/4.0/).

Article

Effects of a New Type of Shrinkage-Reducing Agent on Concrete Properties

Mari Masanaga [1], Tsuyoshi Hirata [1], Hirokatsu Kawakami [1,*], Yuka Morinaga [2], Toyoharu Nawa [2] and Yogarajah Elakneswaran [2]

[1] Ethylene Oxide (EO) Research and Development Department, NIPPON SHOKUBAI, Osaka 564-0034, Japan; mari_masanaga@shokubai.co.jp (M.M.); t_hirata56@me.com (T.H.)
[2] Division of Sustainable Resources Engineering, Faculty of Engineering, Hokkaido University, Kita 13, Nishi 8, Kita-ku, Sapporo 060-8628, Japan; morinaga@eng.hokudai.ac.jp (Y.M.); nawa@eng.hokudai.ac.jp (T.N.); elakneswaran@eng.hokudai.ac.jp (Y.E.)
* Correspondence: hirokatsu_kawakami@shokubai.co.jp

Received: 5 June 2020; Accepted: 2 July 2020; Published: 6 July 2020

Abstract: Shrinkage-reducing agents have been developed to mitigate shrinkage and to control cracks in concrete. This study aims to evaluate the impact of a newly developed shrinkage-reducing agent (N-SRA) on concrete properties and to compare its properties with a conventional shrinkage-reducing agent (C-SRA). The hydration rate, compressive strength, splitting tensile strength, shrinkage, occurrence of cracking, and freezing and thawing were investigated. N-SRA showed higher surface tension than C-SRA and reduced shrinkage to the same degree as C-SRA with half the dosage of C-SRA. The addition of N-SRA or C-SRA did not influence the early compressive strength but slightly reduced splitting tensile strength at seven days. Concrete with N-SRA showed higher compressive strength at 28 days than those of concrete with C-SRA or without SRA. Furthermore, concrete with N-SRA extended the period for the occurrence of shrinkage cracking under restrained conditions. It was found that N-SRA provided excellent freezing and thawing resistance because of the formation of good air voids, while C-SRA demonstrated inefficient behaviour in such an environment.

Keywords: shrinkage-reducing agent; compressive strength; splitting tensile strength; freezing and thawing; spacing factor

1. Introduction

Shrinkage is a common problem in concrete, when it is not properly handled. Shrinkage causes cracking, accelerates ingress of deleterious ions (chloride, sulphate, etc.), and eventually leads to failure, thereby shortening the service life of concrete structures. A possible way to mitigate shrinkage-induced cracking in concrete is to add a special type of organic chemical admixture called shrinkage-reducing agent (SRA) to the concrete mixture. The practice of using SRA, a class of surfactants, was initiated in Japan in the 1980s and is now available worldwide [1]. Conventional SRA is a polyoxyalkylene with or without a hydrophobic group at the end (e.g., a blend of propylene glycol derivatives [2], 2-butoxy ethanol with four ethylene oxide adduct [3]), and forms micelles in aqueous solution due to amphiphilic molecules [4,5]. Various studies have been conducted on the performance of conventional SRAs for shrinkage reduction [2–13].

When the water in hardened concrete evaporates, a meniscus is formed at the air–solution interface of the capillary pore when the surrounding humidity is lower. Consequently, the surface tension in the meniscus pulls the pore walls inward causing shrinkage in the concrete. However, the addition of conventional SRA to the concrete mixing water lowers the surface tension (to approximately 40 dynes/cm) of the pore solution in the hardened concrete, resulting in a reduction in the drying shrinkage. The conventional SRA remains in the pore system even after the concrete hardens, and it continues to reduce the surface tension effect that contributes to drying shrinkage. It has been reported that, in addition to reducing the pore solution surface tension, the conventional SRA has several other beneficial effects [5,14–16]. Conventional SRA helps to reduce the capillary stress formed by autogenous shrinkage and the associated restrained shrinkage cracking in high-strength concrete with low water-to-cement ratio [5]. The conventional SRA shows a lower weight loss in concrete, indicating the prevention of water evaporation. Moreover, conventional SRA exhibits the absence of autogenous shrinkage [6] and resistance to restrained shrinkage cracking, through extended time to cracking [2] and shallower crack width [8,17], when compared to concrete without SRA. It has also been reported that SRA significantly improves durability by reducing sorptivity and wetting moisture diffusivity [16] and increasing the pore solution viscosity, which can reduce the diffusion coefficient of ions [15]. Nevertheless, conventional SRA also shows some negative effects, such as a slight decrease in the early age compressive and flexural strengths [5–8], a delay in setting [5,6,17,18], and freezing and thawing damage depending on the air void system [19].

Although most studies suggest that lowering the surface tension of the pore solution is the main mechanism for the shrinkage reduction by conventional SRA [2–9], recent studies have shown that the shrinkage-reduction mechanism of SRA is very complicated, and that lowering the surface tension is not the only reason for shrinkage reduction [9,16,20]. An increasing number of studies have focused on the development of new types of SRA to overcome the negative effect of conventional SRA and to enhance its performance and economic efficiency. Therefore, in this study, a newly developed shrinkage-reducing agent (N-SRA) was introduced, and its properties and performance were compared with those of a conventional SRA. Finally, the acting mechanism of N-SRA is discussed.

2. Experimental

2.1. Materials

Ordinary Portland cement (OPC) manufactured by Taiheiyo cement corporation was used in this study. The chemical composition and physical properties of OPC are given in Table 1.

Table 1. Chemical composition and physical property of cement.

Mineral Composition	(%)
C_3S	56
C_2S	18
C_3A	9
C_4AF	9
Loss on ignition	1.91
Specific surface area	3300 cm^2/g
Density	3.16 g/cm^3

N-SRA is a newly developed, completely water-soluble polymer consisting of hydrophilic monomers with an optimal molecular weight. The design concept of N-SRA itself does not form a micelle in the pore solution of hydrated cement. A commercial polyoxyalkylene-type SRA called C-SRA was used for comparison. C-SRA is five-molar ethylene oxide and one-molar propylene oxide adduct to a methyl alcohol based on ^1H NMR analysis. Generally, the structure of C-SRA, which has hydrophilic group as well as hydrophobic group, is called amphiphilicity, which is widely used as a surfactant in industry.

The hydrodynamic radius (R_H) was determined by dynamic light scattering (DLS) method using a Zetasizer Nano ZSP (Malvern Instruments Ltd., Malvern, UK). The hydrodynamic radius was calculated by the Stokes-Einstein equation, which defines the relationship between the hydrodynamic radius of a sample and its diffusion rate due to Brownian motion. The sample concentration was 0.5 wt.% in the solution, and the measurement temperature was set to 25 °C. The pore solution of hydrated cement was synthesized by dissolving the following amount of chemicals: 1.72 g/L $CaSO_4$: $2H_2O$, 6.959 g/L Na_2SO_4, 4.757 g/L K_2SO_4, and 7.12 g/L KOH (pH = 13.1). The determined R_H in the simulated pore solution was 9 nm and 459 nm for N-SRA and C-SRA, respectively. The results suggest that N-SRA is sufficiently dissolved. On the other hand, C-SRA is agglomerated and formed micelles due to hydrophobic association via the salting-out effect.

2.2. Casting and Testing of Concrete

The concrete mixture proportions and properties of the fresh concrete are summarised in Table 2. The starting materials were selected in accordance with JIS A 5308, and the concrete formulations were determined as reference to JIS A 6204. Figure 1 shows the aggregate grading curves used in the preparation of concrete. Concrete made with C-SRA and N-SRA are denoted as Concrete-C and Concrete-N, respectively. Concrete without SRA is used as the reference. The SRA dosage was 2% and 1% by weight of cement (% by weight of cement (BWOC)) for C-SRA and N-SRA, respectively. The concrete was cast with a dual-axle revolving-paddle mixer with a capacity of 50 L. First, cement, fine aggregates, and coarse aggregates were mixed without water for 10s. Second, water with superplasiticiser (SP), Air-entraining (AE) agent, defoamer, and N-SRA or C-SRA were added, and mixed for 1 min. Mixing was then stopped to scrape the mortar off the mixer wall, followed by another mixing for 1 min. A polycarboxylate ether type superplasticizer (MasterGlenium SP8SV, BASF, Tokyo, Japan) was used as the SP, and its dosage was adjusted to obtain the target slump value of 18 ± 2 cm. It should be noted that the SP dosage for the concrete with N-SRA was lower than that in the concrete with C-SRA because N-SRA has a slight water-reducing effect. The air content was controlled to the range of 4.5 to 5.5 vol.% by using both AE agent and defoamer. Both coarse and fine aggregates were used in the saturated surface-dry condition. The amounts of SP, SRA, AE agent, and defoamer were subtracted from the amount of mixing water.

Table 2. Proportions, constituents, and properties of concrete mixtures.

	Reference	Concrete-N	Concrete-C
Coarse aggregate (kg/m^3)	939	939	939
Fine aggregate (kg/m^3)	853	853	853
Cement (kg/m^3)	309	309	309
Water (kg/m^3)	170	170	170
Superplasticizer (% BWOC)	0.80	0.30	0.80
N-SRA (% BWOC)	0	1.0	0
C-SRA (% BWOC)	0	0	2.0
w/c	0.55	0.55	0.55
Slump (cm)	19.0	18.5	19.0
Slump flow (cm)	290	285	295
Air (%)	5.5	5.4	4.9
Spacing factor (μm)	316	283	439
Setting time: Initial (h)	6:00	6:45	6:40
Final (h)	8:15	9:15	9:15

The workability of concrete and the air void content in concrete were determined in accordance with Japanese Industrial Standard JIS A 1101 (the slump test) and JIS A 1118 (air content: a pressure method), respectively. The air void characteristics of concrete were also identified using an air void analyser (AVA, Germann Instrument, Copenhagen, Denmark) [21]. The size distribution of air bubbles can be measured through the buoyancy changes of the air bubbles by Stokes' law. Initial and final setting times were determined in accordance with JIS A 1147. The fresh concrete was cast into the formworks, covered by plastic sheeting, and stored at 20 °C until further experiments.

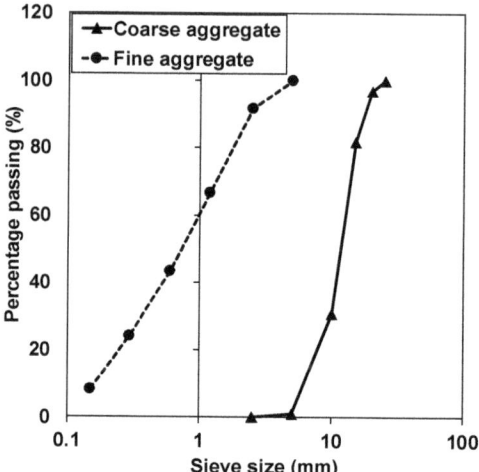

Figure 1. Grading curves for fine and coarse aggregates.

2.3. Measurement of Hardened Concrete Properties

Standard test methods were adopted to determine the hardened concrete properties such as free shrinkage strain (JCI-TC083A, formwork: JIS A 1132), compressive strength (JIS A 1108, formwork: JIS A 1132), splitting tensile strength (JIS A 1113, formwork: JIS A 1132), and freeze–thaw resistance (JIS A 1148, formwork JIS A 1132). To measure the free shrinkage strain, a strain gauge (PLT-60-5LT, Tokyo Sokki Kenkyujo Co., Ltd. Tokyo, Japan) was placed inside the formwork. Restrained shrinkage cracking test was conducted according to the shrinkage crack evaluation test method for concrete, which was proposed by the Japan Concrete Institute (Figure 2).

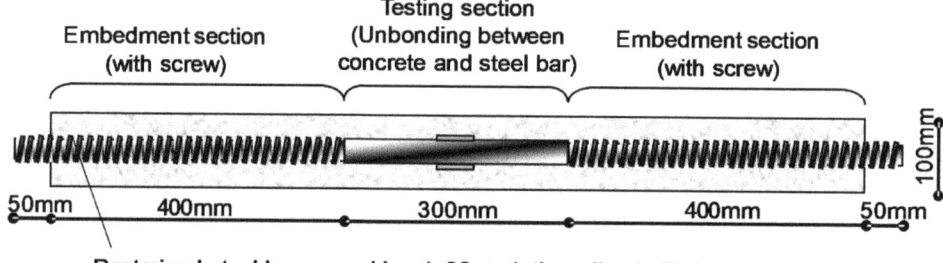

Figure 2. Schematic diagram of specimen for restrained shrinkage cracking test.

2.4. Cement Hydration

The heat generated during cement hydration was quantified using a conduction calorimeter (TAM Air, TA Instruments, New Castle, UK) at 20 °C. A cement paste with a water-to-cement ratio (w/c) of 0.5 and N-SRA dosage of 1% BWOC was used for the measurement.

For the XRD-Rietveld analysis, cement paste with w/c ratio of 0.5 was prepared using Hobart mixer N50. Initially, water was added to cement, and mixed for 30 s. Mixing was then stopped to scrape cement paste off the wall, followed by another mixing for 3 min. N-SRA was added to the mixing water and its amount was subtracted from the amount of mixing water. The prepared cement slurry was stored in a plastic bottle and agitated with a spatula at intervals of 1 h for 7 h to prevent bleeding. The cement slurry was poured into a settled formwork (φ 50 mm × height 10 mm) and sealed with a plastic sheet and then stored for 24 h at 20 °C. Afterwards, it was demoulded and cured in water. Specimens were taken after three and nine weeks of hydration and were broken into pieces using a hammer. Five grams of the obtained powder sample was mixed with 50 g acetone for 24 h to stop the hydration. After acetone was completely removed by suction filtration, the samples were stored at 40 °C for 6 h before the hydration rate was measured.

The XRD patterns were analysed using a Rigaku MultiFlex X-ray generator (Tokyo, Japan) with CuKα radiation ranging from 5° to 70° (2θ) at a scanning rate of 6.5° (2θ) per min and a step size of 0.02° (2θ). In addition to the qualitative phase analysis by XRD, the Siroquant-Rietveld program was used to quantify the phases generated in the solids. The quantitative phase analysis by XRD is abbreviated as Q-XRD. Based on Q-XRD and the measurements of weight loss on ignition, the weight of HCP in percentage can be converted to mass in grams. Loss on ignition was conducted by heating reacted solids to 950 °C for 2 h, and the weight loss of the solids due to heating was calculated. Hydration degree of hydrated cement by XRD was calculated according to [22].

3. Results

3.1. Surface Tension of SRA

The hydrophile-lipophile balance (HLB) value was calculated according to Griffin's method [23]:

$$HLB = 20 \times M_h/M \qquad (1)$$

where M_h is the molecular mass of the hydrophilic portion of the molecule and M is the molecular mass of the total molecule. The calculated HLB values for N-SRA and C-SRA were 18.9 and 14.2, respectively. Therefore, N-SRA is more hydrophilic, which does not lower the surface tension in the aqueous solution. To confirm this, the surface tension of the liquid phase was determined using a Lauda tensiometer. The measurements were conducted at 0.01, 0.05, 0.5, 1.0 and 5.0 wt.% aqueous solutions of SRAs at 20 ± 3 °C by the du Nouy method using a platinum ring. The experimental results are shown in Figure 3. The obtained results for C-SRA are consistent with the results reported in the literature [5,9]. Furthermore, the surface tension of N-SRA is higher than that of C-SRA, which agrees with the theoretical prediction.

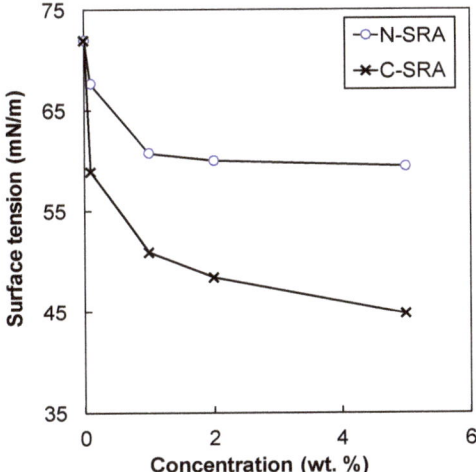

Figure 3. Surface tension of newly developed (N-SRA) and conventional (C-SRA) shrinkage-reducing agent in deionic water.

3.2. Concrete Properties

The measured setting times are tabulated in Table 2. There was no significant difference between N-SRA and C-SRA, indicating that both SRAs slightly delay the setting time by 45 min to 1 h (see Table 2). The initial heat evolution of cement hydration with N-SRA is slower than that of the reference (see Figure 4), which is attributed to the delay in setting time of N-SRA. The compressive strength and spilling tensile strength of concrete consisting of SRA were compared with those of the reference concrete in Figure 5. This demonstrates that the strength continues to develop with hydration. The results showed that the addition of SRA does not affect the early compressive strength of concrete, but it slightly reduces early splitting tensile strength. Furthermore, concrete with N-SRA shows somewhat higher compressive strength of concrete at 28 days in comparison to the concrete with C-SRA or reference concrete.

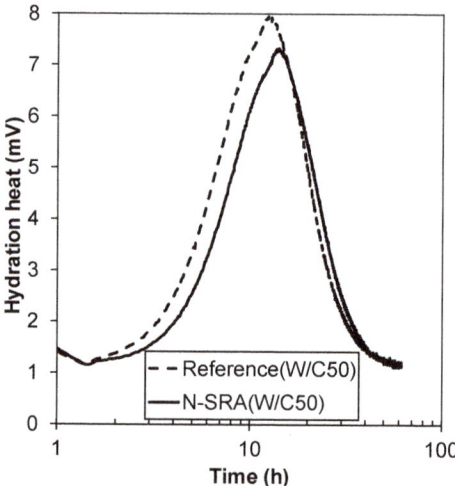

Figure 4. Calorimetric curve during the initial cement hydration.

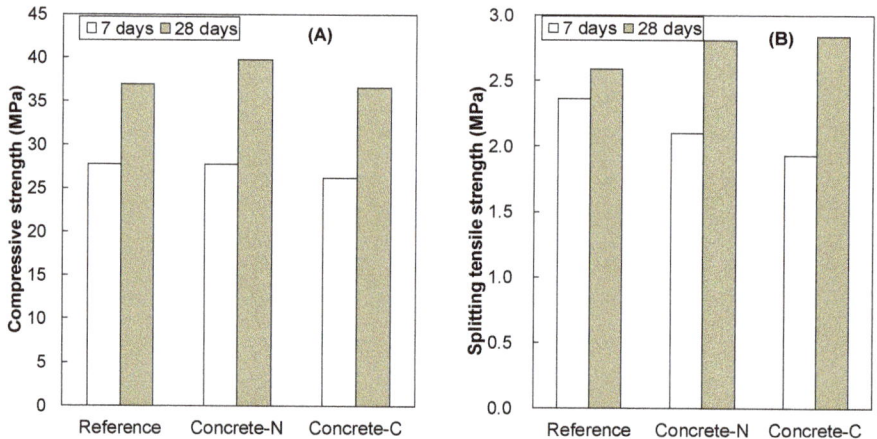

Figure 5. (**A**) Compressive strength and (**B**) splitting tensile strength of concrete with N-SRA and C-SRA compared with reference.

In order to understand the strength development behaviours of the samples, the total hydration degree of cement in the cement pastes with N-SRA and without SRA at curing periods of three and nine weeks was determined by XRD analysis. The results are shown in Figure 6. The hydration degree of cement for the concrete with N-SRA was almost the same as that of the reference until nine weeks. This suggests concrete with N-SRA achieved approximately the same later-age concrete strength (even after 28 days) as the concrete without SRA.

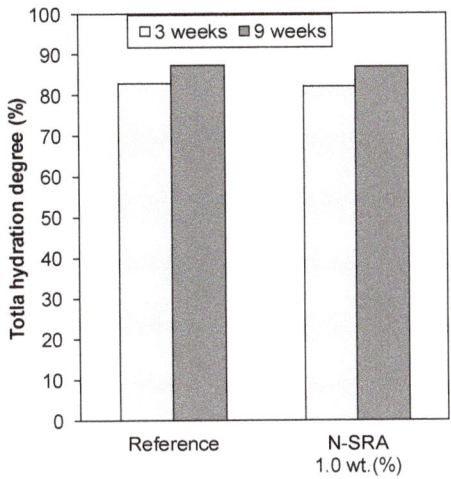

Figure 6. Total hydration degree of cement paste hydrated for three and nine weeks with and without N-SRA.

3.3. Concrete Shrinkage

The free shrinkage strain of concrete with N-SRA (Concrete-N) and C-SRA (Concrete-C) as a function of time are shown in Figure 7. Inclusion of SRA reduced the shrinkage by approximately 50% compared to that of the reference. Both N-SRA and C-SRA showed the same degree of shrinkage reduction, despite the fact that N-SRA required only half the dosage of C-SRA. However, N-SRA was ineffective at reducing surface tension (Figure 3). Figure 8 shows the restrained stress of the reinforced steel in the concrete specimen as a function of the drying age. The results were obtained for two specimens in each case. The restrained stress of concrete also increases with that of the reinforced steel bar, and concrete cracking occurs when the restrained stress exceeds a specific fracture criterion such as the tensile strength of concrete. When compared to the others, concrete with N-SRA resists cracking for a longer period. Based on the results shown in Figure 8, the time required for the concrete to crack was determined, and the impact of SRAs on the occurrence of shrinkage cracking is shown in Figure 9. The cracking time was the average of two measurements. The determined cracking time was 13.8 days for reference, 32.0 days for Concrete-N, and 24.9 days for Concrete-C, indicating that N-SRA reduces the shrinkage cracking of restrained concrete.

The restrained shrinkage stress of concrete when cracking occurred was calculated from the restrained stress of reinforced steel as follows:

$$\sigma_c = (E_s \times \varepsilon_s \times A_s)/A_c \tag{2}$$

where σ_c is the restrained stress of concrete when cracking occurs (MPa), E_s is the elastic modulus of restrained steel (MPa), ε_s is the restrained stress of reinforced steel when cracking occurs, A_s is the area of restrained steel (mm^2), and A_c is the area of the concrete specimen (mm^2).

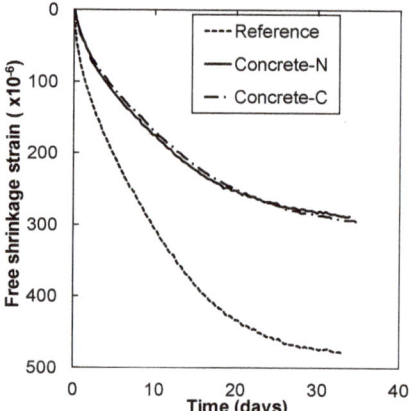

Figure 7. Free shrinkage strain of N-SRA and C-SRA.

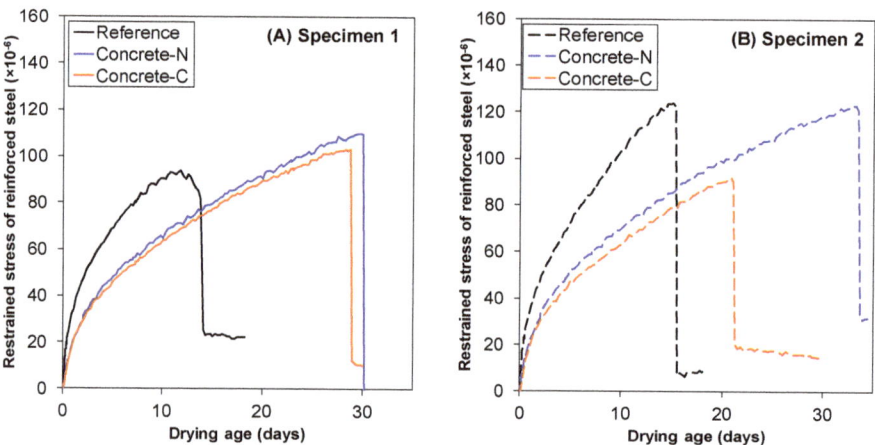

Figure 8. Relationships of restrained stress of reinforced steel and drying age. (**A**) Specimen 1 (**B**) Specimen 2.

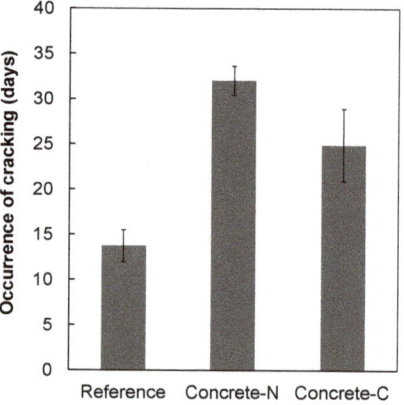

Figure 9. Resistance to restrained shrinkage cracking for concrete with N-SRA or C-SRA.

The calculated restrained stresses when cracking occurred were 2.0, 2.2, and 1.8 MPa for the reference concrete, Concrete-N, and Concrete-C, respectively. This indicates that the fracture criterion of cracking is almost the same despite the presence or absence of SRA.

3.4. Resistance to Freezing and Thawing Action

The results of the freezing and thawing tests are shown in Figure 10. The relative dynamic modulus of Concrete-N was maintained above 90%, which was similar to that of the reference concrete. However, as soon as testing started, the relative dynamics modulus of Concrete-C decreased rapidly. It should be emphasized that Concrete-N adequately meets the requirements for the specification of chemical admixture for concrete in accordance with Japanese Industrial Standard JIS R 6204: the relative dynamic modulus should be beyond 60% at 200 cycles of freezing and thawing. The freeze–thaw resistance is strongly influenced by the properties of air voids in concrete as well as its content [19]. Mindess et al. [24] reported that the spacing factor should not exceed 0.2 mm to ensure adequate freeze-thaw protection. In addition to the spacing factor, specific surface area is also an important factor for a protective air void system. The air void parameters measured by AVA are tabulated in Table 3. Superplasticizers may occasionally show good freezing and thawing resistance even when the air void parameter limits are exceeded; however, it can be clearly confirmed that N-SRA improves the air void system (Table 3).

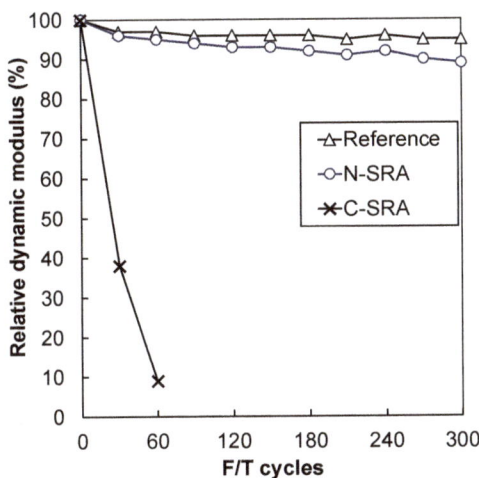

Figure 10. Freezing and thawing test results.

Table 3. Entrained air void system measured by air void analyser (AVA) for concrete with N-SRA, with C-SRA and without shrinkage-reducing agent (SRA).

Concrete Mix	SRA	Spacing Factor (μm)	Specific Surface Area (mm^2/mm^3)
Concrete-N	N-SRA	283	16.8
Concrete-C	C-SRA	439	13.9
Reference	-	316	15.6

4. Discussion

It is shown that a very small dosage of C-SRA or N-SRA (the dosage of N-SRA being half the amount compared to C-SRA) reduces shrinkage and shows a high resistance of shrinkage-induced cracking under restrained conditions. Various mechanisms have been proposed for shrinkage reduction by SRA, including lowering of the surface tension of the cement paste pore solution and reduction of capillary stress [2–9]. The water in the capillary pores acts as bulk water to create menisci. However, when the capillary pores are fully occupied with water or when they are fully dried, menisci are not created.

The shrinkage reduction in C-SRA can be explained by the reduction in surface tension. C-SRA is composed of an assembly of ethylene oxide (EO) chains, which show hydrophilicity, and hydrophobic part of methyl group or propylene oxide (PO). Therefore, C-SRA has surface-active properties that can decrease the surface tension in the capillary pores. The determined R_H value of C-SRA is 459 nm (higher than 10 nm) in the synthetic cement pore solution, which indicates that, C-SRA, when used as a surfactant, easily forms micelle in vain. It is suggested that the ineffectiveness in shrinkage reduction by creating micelles results in a higher dosage of C-SRA.

It is believed that the calcium silicate hydrate (C-S-H) microstructure affects the performance of N-SRA in concrete. Jennings highlighted that the colloidal behaviour of C-S-H influences the shrinkage strain [25,26]. C-S-H is composed of calcium silicate sheets separated by interlayer spaces filled with water and calcium ions. The space between the calcium silicate sheets is the interlayer porosity of C-S-H. In addition, the C-S-H contains gel pores (1–10 nm) and isolated capillary pores (larger than 10 nm). The interlayer spaces in the C-S-H gel contain water that is removed during drying and re-enters during rewetting, which causes shrinkage and swelling [25,26]. In the gel pores, the surfaces are too close to each other, which restricts menisci formation. The determined R_H value of N-SRA, which is a water-soluble compound, was 9 nm in the synthetic cement pore solution, and thus it can be easily stored in the gel pore (1–10 nm). It is well known that a hydrophilic compound can hold water molecules through interactions [27,28]. Therefore, it is suggested that N-SRA molecules are stored inside the gel pores to prevent the evaporation of capillary water, where N-SRA serves as a water-retainer. N-SRA does not create micelles in the capillary pores due to sufficient hydrophilicity, which is also one of the reasons why effective for shrinkage reduction is observed in Concrete-N even with a small dosage of N-SRA.

5. Conclusions

This study focused on the effect of a newly developed shrinkage-reducing agent (N-SRA) to mitigate the drying shrinkage, and the results were compared with those of a conventional shrinkage-reducing agent (C-SRA). N-SRA is a water-soluble compound, which helps N-SRA to act by a mechanism different from that of C-SRA. The measured and predicted surface tension of N-SRA is higher than that of C-SRA. Addition of SRA did not change seven-day compressive strength but reduced the early splitting tensile strength. Both SRAs induced strength development at a later age. The same degree of shrinkage reduction (approximately 50% reduction) was observed in both SRAs, where the dosage of N-SRA was half the amount of C-SRA. Furthermore, N-SRA significantly influences the occurrence of shrinkage cracking, that is, it withstands cracking for a longer period than the concrete with C-SRA or without SRA. With respect to durability, the addition of N-SRA significantly improves the freezing and thawing resistance due to the creation of a proper air void system.

Author Contributions: Conceptualization, M.M., T.H., and T.N.; Methodology, M.M., T.H., Y.M., and T.N.; Data curation, M.M. and Y.M.; Writing—original draft preparation, M.M., T.H., T.N. and Y.E.; Writing—review and editing, M.M., T.H., H.K., Y.M., T.N. and Y.E.; Visualization, M.M., T.H., and Y.E.; Supervision, T.H. and T.N.; Project administration, M.M., T.H., H.K. and T.N.; Funding acquisition, M.M., T.H., and H.K. All authors have read and agreed to the published version of the manuscript.

Funding: This research received no external funding.

Conflicts of Interest: The authors declare no conflict of interest.

References

1. Goto, T.; Sato, T.; Sakai, K.; Ii, M. Cement-Shrinkage-Reducing Agent and Cement Composition. U.S. Patent 4,547,223, 10 October 1985; United States Patent and Trademark Office.

2. Folliard, K.J.; Berke, N.S. Properties of high- performance concrete containing shrinkage-reducing admixture. *Cem. Concr. Res.* **1997**, *9*, 1357–1364. [CrossRef]
3. Rongbing, B.; Jian, S. Synthesis and evaluation of shrinkage-reducing admixture for cementitious materials. *Cem. Concr. Res.* **2005**, *35*, 445–448. [CrossRef]
4. Balogh, A. New admixture combats concrete shrinkage. *Concr. Constr.* **1996**, *41*, 546–551.
5. Rajabipour, F.; Sant, G.; Weiss, J. Interactions between shrinkage reducing admixtures (SRA) and cement paste's pore solution. *Cem. Concr. Res.* **2008**, *38*, 606–615. [CrossRef]
6. Roncero, J.; Gettu, R.; Martín, M.A. Evaluation of the Influence of a Shrinkage Reducing Admixture on the Microstructure and Long-Term Behavior of Concrete. In Proceedings of the 7th CANMET/ACI International Conference on Superplasticizers and Other Chemical Admixtures in Concrete (Supplementary Papers), Berlin, Germany, 20–23 October 2003; pp. 207–226.
7. Ribeiro, A.B.; Carrajola, A.; Gonçalves, A. Behavior of Mortars with Different Dosages of Shrinkage Reducing Admixtures. In Proceedings of the 8th CANMET/ ACI International Conference on Superplasticizers and Other Chemical Admixtures in Concrete (SP-239), Sorrento, Italy, 29 October–1 November 2006; American Concrete Institute: Farmington Hill, MI, USA; pp. 77–91.
8. Borsoi, A.; Collepardi, M.; Collepardi, S.; Troli, R. Influence of chemical admixtures on the drying shrinkage of concrete. In Proceedings of the 9th CANMET/ ACI International Conference on Superplasticizers and Other Chemical Admixtures in Concrete (Supplementary Papers), Seville, Spain, 21–24 October 2009; pp. 1–9.
9. Maruyama, I.; Beppu, K.; Kurihara, R.; Furuta, A. Action mechanisms of shrinkage reducing admixtures in hardened cement paste. *J. Adv. Concr. Technol.* **2016**, *14*, 311–323. [CrossRef]
10. Park, J.J.; Kim, S.; Shin, W.; Choi, H.J.; Park, G.J.; Yoo, D.Y. High performance photocatalytic cementitious materials containing synthetic fibers and shrinkage-reducing admixture. *Materials* **2020**, *13*, 1828. [CrossRef]
11. Ran, Q.; Gao, N.; Liu, J.; Tian, Q.; Zhang, J. Shrinkage action mechanism of shrinkage-reducing admixture based on the pore solution. *Mag. Concr. Res.* **2013**, *65*, 1092–1100. [CrossRef]
12. Yoo, D.Y.; Ryu, G.S.; Yuan, T.; Koh, K.T. Mitigating shrinkage cracking in posttensioning grout using shrinkage-reducing admixture. *Cem. Concr Compos.* **2017**, *81*, 97–108. [CrossRef]
13. Bílek, V., Jr.; Kalina, L.; Novotný, R.; Tkacz, J.; Pañze, L.K. Some issues of shrinkage-reducing admixtures application in alkali-activated slag system. *Materials* **2016**, *9*, 462.
14. Corinaldesi, V. Combined effect of expansive, shrinkage reducing and hydrophobic admixtures for durable self compacting concrete. *Constr. Build. Mater.* **2012**, *36*, 758–764. [CrossRef]
15. Bentz, D.P. Influence of shrinkage-reducing admixtures on early-age properties of cement paste. *J. Adv. Concr. Technol.* **2006**, *4*, 423–429. [CrossRef]
16. Qin, R.; Hao, H.; Rousakis, T.; Lau, D. Effect of shrinkage reducing admixture on new-to-old concrete interface. *Compos. Part. B* **2019**, *167*, 346–355. [CrossRef]
17. Mora-Ruacho, J.; Gettu, R.; Aguado, A. Influence of shrinkage-reducing admixtures on the reduction of plastic shrinkage cracking in concrete. *Cem. Concr. Res.* **2009**, *39*, 141–146. [CrossRef]
18. Brooks, J.J.; Johari, M.A.M.; Mazloom, M. Effect of admixtures on the setting times of high-strength concrete. *Cem. Concr. Compos.* **2000**, *22*, 293–301. [CrossRef]
19. Berke, N.S.; Li, L.; Hicks, M.C.; Bae, J. Improving Concrete Performance with Shrinkage-Reducing Admixtures. In Proceedings of the 7th CANMET/ ACI International Conference on Superplasticizers and Other Chemical Admixtures in Concrete (SP-217), Berlin,Germany, 20–23 October 2003; American Concrete Institute: Farmington Hill, MI, USA; pp. 37–50.
20. Zuo, W.Q.; Feng, P.; Zhong, P.; Tian, Q.; Gao, N.; Wang, Y.; Yu, C.; Miao, C. Effects of novel polymer-type shrinkage-reducing admixture on early age autogenous deformation of cement pastes. *Cem. Concr. Res.* **2017**, *100*, 413–422. [CrossRef]
21. Magua, D.D. Evaluation of the Air Void Analyzer. *Concr. Int.* **1996**, *18*, 55–59.
22. Yogarajah, E.; Natsumi, N.; Kazuki, M.; Yuka, M.; Takashi, C.; Hiroyoshi, K.; Toyoharu, N. Characteristics of ferrite-rich Portland cement: Comparison with ordinary Portland cement. *Front.Mater.* **2019**, *6*, 97.
23. Griffin, W.C. Classification of surface active agents by HLB. *J. Soc. Cosmet. Chem.* **1949**, *1*, 311–326.
24. Mindess, S.; Young, J.F.; Darwin, D. *Concrete*, 2nd ed.; Prentice Hall: Upper Saddle River, NJ, USA, 2003.
25. Jennings, H.M. Refinements to colloid model of C-S-H in cement:CM-II. *Cem. Concr. Res.* **2008**, *38*, 275–289. [CrossRef]

26. Jennings, H.M. Colloid model of C-S-H and implications to the problem of creep and shrinkage. *Mater. Struct.* **2004**, *37*, 59–70. [CrossRef]
27. Hauet, T.; Eugene, M.A. New approach in organ preservation: Potential role of new polymer. *Kidney Int.* **2008**, *74*, 998–1003. [CrossRef] [PubMed]
28. Shikata, T.; Okuzono, M.; Sugimoto, N. Temperature-dependent hydration/dehydration behavior of poly(ethylene oxide)s in aqueous solution. *Macromolecules* **2013**, *46*, 1956–1961. [CrossRef]

© 2020 by the authors. Licensee MDPI, Basel, Switzerland. This article is an open access article distributed under the terms and conditions of the Creative Commons Attribution (CC BY) license (http://creativecommons.org/licenses/by/4.0/).

Article

Strontium Retention of Calcium Zirconium Aluminate Cement Paste Studied by NMR, XRD and SEM-EDS

Dominika Madej

Faculty of Materials Science and Ceramics, AGH University of Science and Technology, al. A. Mickiewicza 30, 30-059 Krakow, Poland; dmadej@agh.edu.pl

Received: 18 April 2020; Accepted: 19 May 2020; Published: 21 May 2020

Abstract: This work concerns the hydration mechanism of calcium zirconium aluminate as a ternary compound appearing in the $CaO-Al_2O_3-ZrO_2$ diagram besides the calcium aluminates commonly used as the main constitutes of calcium aluminate cements (CACs). Moreover, a state-of-the-art approach towards significant changes in hydraulic properties was implemented for the first time in this work, where the effect of structural modification on the hydration behavior of calcium zirconium aluminate was proved by XRD, ^{27}Al MAS NMR and SEM-EDS. The substitution of Sr^{2+} for Ca^{2+} in the $Ca_7ZrAl_6O_{18}$ lattice decreases the reactivity of Sr-substituted $Ca_7ZrAl_6O_{18}$ in the presence of water. Since the original cement grains remain unhydrated up to 3 h ($Ca_7ZrAl_6O_{18}$) or 72 h ($Sr_{1.25}Ca_{5.75}ZrAl_6O_{18}$) of curing period in the hardened cement paste structures, strontium can be considered as an inhibition agent for cement hydration. The complete conversion from anhydrous $^{27}Al^{IV}$ to hydrated $^{27}Al^{VI}$ species was achieved during the first 24 h ($Ca_7ZrAl_6O_{18}$) or 7 d($Sr_{1.25}Ca_{5.75}ZrAl_6O_{18}$) of hydration. Simultaneously, the chemical shift in the range of octahedral aluminum from ca. 4 ppm to ca. 6 ppm was attributed to the transformation of the hexagonal calcium aluminate hydrates and Sr-rich $(Sr,C)_3AH_6$ hydrate into the cubic phase Ca-rich $(Sr,C)_3AH_6$ or pure C_3AH_6 in the hardened Sr-doped cement paste at the age of 7 d. The same ^{27}Al NMR chemical shift was detected at the age of 24 h for the reference hardened undoped $Ca_7ZrAl_6O_{18}$ cement paste.

Keywords: $Ca_7ZrAl_6O_{18}$; ^{27}Al MAS NMR; Sr-rich $(Sr,C)_3AH_6$; cement hydration; refractories; immobilization of radioactive Sr

1. Introduction

Calcium aluminate cements (CACs) [1–3] and other related cementitious systems [4–7] are believed to play an important role in a wide range of specialist are as from some construction areas and civil engineering, to refractory materials industry, due to their ability to gain strength rapidly in the initial days after casting and to withstand aggressive environments and high temperatures. The temperature–time weight ratio of water-to-cement (the w/c ratio) dependencies of the cement hydration processes have been widely investigated so far and are presented elsewhere in detail [8–10]. It must be made clear at this stage that both temperature and a weight ratio of water-to-cement (the w/c ratio) affect the performances of calcium aluminate cements, especially at an early age. The phase composition, microstructure and other properties of the cement pastes can easily be designed by choosing appropriate curing parameters. Nevertheless, upon curing of CAC-based paste at not adequately controlled, ambient conditions, these materials are very sensitive to humidity, CO_2, temperature and time. Due to the fact of not being able to predict their long-term behavior makes them practically not suitable for constructions but suitable for building chemistry. Nevertheless, the high-early-heat and high-early-strength gain makes CACs attractive, especially in the winter months and/or when rapid repairs are needed. The hydration product formation in CACs containing mainly monocalcium aluminate (CaAl$_2$O$_4$; CA) and monocalcium di aluminate (CaAl$_4$O$_7$; CA$_2$) is characterized by the dependence of C–A–H* (C = CaO, A = Al$_2$O$_3$, H = H$_2$O) phases on temperature. At a temperature below ca. 15 °C, calcium aluminates are hydrated to form metastable CAH$_{10}$. At room temperature, the hydration process proceeds through the formation of other metastable phases C$_2$AH$_8$ and/or C$_4$AH$_{13-19}$ (hexagonal hydrates). At higher temperatures (above 28 °C), the metastable hydrates will spontaneously convert to the cubic hydrogarnet C$_3$AH$_6$. Furthermore, according to insights into hydration of CACs, the hydration products during long hydration shall be presented as follows: C$_3$AH$_6$ + AH$_3$. The 10-year results, which have been added to existing results on the hydration of calcium aluminate compounds, are of interest as showing the CaO-Al$_2$O$_3$-ZrO$_2$ [4,6,11] as an interesting alternative to CACs-based materials. This can be attributed to the fact, that this category of special cement which contains both hydraulic phases (CA, CA$_2$, C$_{12}$A$_7$ and C$_7$A$_3$Z*), and Zr-bearing compounds (CZ and Z) [4] is synthesized by one environmentally friendly and effective technological process. This fact makes the attention-getter of a C-A-Z cements' users more important than those of other CACs users and worth considering from the point of view of economics. The principal interest here is concerned with the C$_7$A$_3$Z area including crystal structure, hydration behavior of C$_7$A$_3$Z and its solid solutions with SrO and BaO, hydration products and carbonation processes in cement paste structures [12,13]. Special interest was given to strontium. Because strontium is chemically closely related to calcium, it is easily introduced as a natural substitute for calcium in aluminate phases. Strontium has been shown to affect the hydration behavior of calcium zirconium aluminate cement [5,13]. Furthermore, strontium-containing cements attract the attention of materials scientists due to their higher refractory properties, increased resistance to thermal shock, increased resistance to chemical aggressive solutions and possible applications as special binders for shielding constructions in nuclear power plants [14]. Although this approach is unique, there are other approaches to affecting the hydration behavior of aluminates-bearing CACs and related materials. Results from other studies indicate that the metal chloride and nitrate salts are known in CAC science as agents affecting the CAC hydration behavior [15,16]. The exact nature of this behavior is different for alkali metal salts (NaCl, KCl, RbCl, CsCl, LiCl), alkaline earth metal salts (MgCl$_2$·6H$_2$O, CaCl$_2$, SrCl$_2$·6H$_2$O and BaCl$_2$·2H$_2$O) and transition metal salts (MnCl$_2$·H$_2$O, CoCl$_2$·6H$_2$O, CuCl$_2$·2H$_2$O and ZnCl$_2$). Generally, the addition of alkali metal salts accelerates the setting time of CAC, whereas the transition metal salts can be widely used as retarders. The effect of alkaline earth metal salts on setting the behavior of CAC is determined by the amount of addition [15]. Another approach in the acceleration effect is the addition of chlorides and nitrates of Li(I), Cr(III), Zn(II) and Cr(VI) (chromate), whereas Pb(II) and Cu(II) retarded the hydration [16].

Many various techniques including XRD, FT-IR, Raman spectroscopy, DSC-TG-EGA(MS), SEM-EDS, isothermal calorimetry and EIS were being accepted in this research. It is also well known that solid-state nuclear magnetic resonance (ssNMR) spectroscopy is an effective tool for the characterization of cement paste at the atomic scale in different stages of its aging. The application of NMR to monitor the progress of CACs hydration has a long history and can be found elsewhere [17–20]. In cement chemistry, four-fold coordinated (tetrahedral)^{IV}Al sites and six-fold coordinated (octahedral)^{VI}Al sites typically resonate within the regions 50–80 ppm and 0–20 ppm, respectively. The resonances assigned to five-coordinated ^{V}Al and highly distorted tetrahedral Al environments have been observed in the region 20–50 ppm [18,21,22]. For example, the ^{27}Al MAS NMR spectrum of CA shows a peak maximum at ca. 80 ppm and a shoulder at 76 ppm due to six crystallographically different aluminum tetrahedral sites [19,23–25]. For another example, the ^{27}Al MAS NMR spectrum of CA$_2$ exhibits the peaks between 50 and 75 ppm due to the two distinct ^{27}Al environments [18,25]. On the other hand, calcium aluminate hydrates (CAH$_{10}$, C$_2$AH$_8$, C$_4$AH$_{13-19}$, C$_3$AH$_6$), originating from the hydration process of CAC clinker phases, are octahedrally coordinated Al(OH)$_6$ and cause a ^{27}Al MAS NMR signal at about 0 ppm. Hence, ^{27}Al MAS NMR provides the relative changes of tetrahedral and octahedral Al sites of the hydrating cement paste at different stages of the curing. The resonance position from an octahedral ^{27}Al NMR signal varies slightly depending on the types of calcium aluminate hydrates present [19].

The resonance at ca. 12.36 ppm is due to tricalcium aluminate hexahydrate C_3AH_6 [18,20], at ca. 10.2 ppm due to CAH_{10} and C_4AH_{13} [18–20] and at ca. 10.3 ppm due to C_2AH_8 [17].

The aim of this work is to implement the NMR technique to monitor the progress of hydration of both undoped- and strontium-doped calcium zirconium aluminate cement. Hence, the influence of structural modification with SrO on the hydraulic activity of $Ca_7ZrAl_6O_{18}$ phase has been demonstrated. This work fills in a significant gap in the literature on the exploiting ssNMR spectroscopy to probe the early stages of hydration of new types of cements belonging to the $CaO-Al_2O_3-ZrO_2$ and $CaO-SrO-Al_2O_3-ZrO_2$ systems.

2. Experimental Procedure

2.1. Synthesis and Phase Identification

Low-cost synthesis of $Ca_7ZrAl_6O_{18}$, $Sr_{1.25}Ca_{5.75}Al_6O_{18}$ and $SrAl_2O_4$ cements via the solid-state reactive sintering technique was employed using the stoichiometric amount of the cationic ratios of Ca:Zr:Al = 7:1:6, Sr:Ca:Zr:Al = 1.25:5.75:1:6 and Sr:Al = 1:2, respectively. Strontium was incorporated into $Ca_7ZrAl_6O_{18}$ by replacing 1.25 atoms of calcium, since the maximum level of doping was experimentally established. Both $Ca_7ZrAl_6O_{18}$ and $SrAl_2O_4$ cements were synthesized as reference materials to obtain pure C-A-H and Sr-A-H phases. Starting raw materials were reagent-grade $CaCO_3$ (99.9%, POCH), $SrCO_3$ (98.0%, Merck), Al_2O_3 (99.0%, Acros Organics) and ZrO_2 (98.5%, Acros Organics). The 100 g mixtures of substrates were homogenized for 2 h, pressed into cylindrical pellets at pressures of 50 MPa and calcined at 1000 °C for 10 h (Sr-free sample) or at 1300 °C for 10 h (Sr-containing samples) in air. Phase pure cement clinker minerals were made from mixtures of prereacted powders via grinding, pressing and sintering at 1420 °C ($Ca_7ZrAl_6O_{18}$ and $Sr_{1.25}Ca_{5.75}Al_6O_{18}$) and 1550 °C ($SrAl_2O_4$) for 20 h in air.

Phase identification for the sintered samples was carried out using X-ray diffraction (XRD, PANalytical, Malvern PANalytical, Malvern, UK) on a ProPANalytical X'Pert X-ray diffractometer, with Cu Kα radiation (λ = 0.15418 nm), with 0.02° per step and 3s time per step (2theta range from 5° to 45°).

The NMR spectra were recordedat room temperature on Bruker Avance III 400WB (9.4T) spectrometer, (Bruker BioSpin, Rheinstetten, Germany) using 4 mm MAS (Magic Angle Spinning), dual-channel (1H/BB) probe-head, operating at a resonance frequency of 104.26 MHz for ^{27}Al. The sample was spun at a MAS frequency of 8 kHz in the rotors made of zirconium dioxide (4 mm). 32 K data points and 1024 scans FIDs were accumulated with a Single Pulse Excitation (SPE) pulse sequence using the observed 90° pulse (^{27}Al) set at 6.0 us with a relaxation delay of 200 ms. Note no proton decoupling was applied during the experiment. Prior to Fourier transformation, the data were zero-filled twice and 80 Hz apodization filter was applied. The ^{27}Al chemical shifts were referenced using a sample of $AlCl_3 \cdot 6H_2O$ in 1M solution as an external reference (0 ppm).

The microstructures of fracture surfaces of hydrated cement pastes were investigated using a scanning electron microscope (SEM, FEI Nova Nano SEM 200, Kyoto, Japan.). The chemical compositions of the samples were determined with electron-probe microanalysis using an energy-dispersive X-ray spectrometer (EDAX, Sapphire Si(Li) EDS detector, Mahwah, NJ, USA).

2.2. Preparation and Treatment of Cement Paste

Comparison study of hydration characteristics between $Ca_7ZrAl_6O_{18}$, $Sr_{1.25}Ca_{5.75}Al_6O_{18}$ and $SrAl_2O_4$ cements were determined for cement pastes prepared with water-to-cement (w/c) ratios of 1.0 or 0.5. The water-to-cement ratio of 0.5 was applied to achieve plastic properties without any undesirable sedimentation of neat $SrAl_2O_4$ paste. Whereas, both $Ca_7ZrAl_6O_{18}$ and $Sr_{1.25}Ca_{5.75}Al_6O_{18}$ cements require w/c = 1.0 to obtain the well-homogenized cement pastes without any undesirable phenomena of sedimentation. The cement powders which were obtained by grinding the sintered pellets and necessary mass of water were mixed together in a glass beaker to obtain three homogeneous neat cement pastes. Each neat cement paste was then placed in a polyethylene bag and sealed until 14 d at 50 °C. According to Litwinek and Madej [5], the optimal synthesis temperature for C_3AH_6 from different precursors through hydration is suggested to be 50 °C. Moreover, this period for curing cement pastes was accepted to attain the maximum degree of hydration, as concluded from the previous studies [13]. Moreover, as it was previously mentioned by Garcés et al. [26] and Zhang et al. [27], at temperatures as high as 60 °C, only the cubic phase and the gibbsite appear in the calcium aluminates-based cement pastes. At 24 h and 7 d, the microstructure of cement pastes was investigated by SEM. Acetone quenching was used to stop hydration at 15 min, 0.5 h, 1 h, 2 h, 3 h, 24 h, 48 h, 72 h, 7 d and 14 d (Table 1). The use of cold acetone, aiming to withdraw free water and inhibit further reactions within cement paste is known from the Ref. [28]. Cold acetone is related to acetone stored under laboratory conditions. Quenched pastes were characterized by XRD and ^{27}Al MAS NMR (Bruker BioSpin, Rheinstetten, Germany), according to the procedures presented in Section 2.1.

Table 1. Cement paste formulations, curing conditions and details of sample designation.

Sample Designation	Cement/Sample	Hydration Time	Water-to-Cement Ratio (w/c), Temperature
A	$Sr_{1.25}Ca_{5.75}ZrAl_6O_{18}$ (as a solid solution)	15 min, 0.5 h, 1 h, 2 h, 3 h, 24 h, 48 h, 72 h, 7 d and 14 d	w/c = 1.0, T = 50 °C
B	$SrAl_2O_4$ (as a reference undoped compound)	7 d	w/c = 0.5, T = 50 °C
C	$Ca_7ZrAl_6O_{18}$ (as a reference undoped compound)	15 min, 0.5 h, 1 h, 2 h, 3 h, 24 h, 48 h, 72 h, 7 d and 14 d	w/c = 1.0, T = 50 °C

3. Results and Discussion

3.1. X-ray Diffraction Analysis of Special Cements Hydration

According to X-ray diffraction analysis, the cement clinkers synthesized by the solid-state reactive sintering technique were all crystalline, single-phase aluminate phases $Ca_7ZrAl_6O_{18}$, Sr-doped $Ca_7ZrAl_6O_{18}$ and $SrAl_2O_4$. The positions of the characteristic diffraction peaks in the XRD patterns of the $Ca_7ZrAl_6O_{18}$ (Sample C) and $SrAl_2O_4$ (Sample B) cement clinkers have a strong agreement with the standards JCPDS No. 98-018-2622 and JCPDS No. 98-016-0296, respectively. This confirms that the powders are mainly composed of $Ca_7ZrAl_6O_{18}$ and $SrAl_2O_4$, respectively. As expected, the slight shift in diffraction peaks towards lower 2θ value in the XRD pattern of $Sr_{1.25}Ca_{5.75}ZrAl_6O_{18}$ cement clinker confirms an increase in lattice parameter of Sr-doped $Ca_7ZrAl_6O_{18}$ solid solution. Since no secondary phases containing Sr were detected, Sr was recognized as fully incorporated into the $Ca_7ZrAl_6O_{18}$ structure.

Compared with the XRD patterns of unhydrated phases, the decreasing intensity of peaks corresponding to $Sr_{1.25}Ca_{5.75}ZrAl_6O_{18}$ and $Ca_7ZrAl_6O_{18}$, and new peaks of low intensity corresponding to hydration products formation were recorded using XRD (Figures 1–3). A low degree of hydrates crystallinity was indicated by a poor XRD pattern, especially in the early stage of hydration. Moreover, the hydration processes result in a corresponding change in the XRD patterns of the initial cement clinker phases and formation of amorphous material, besides the crystalline hydrates, as it can be concluded from the severe intensity reduction and peaks broadening in the XRD pattern.

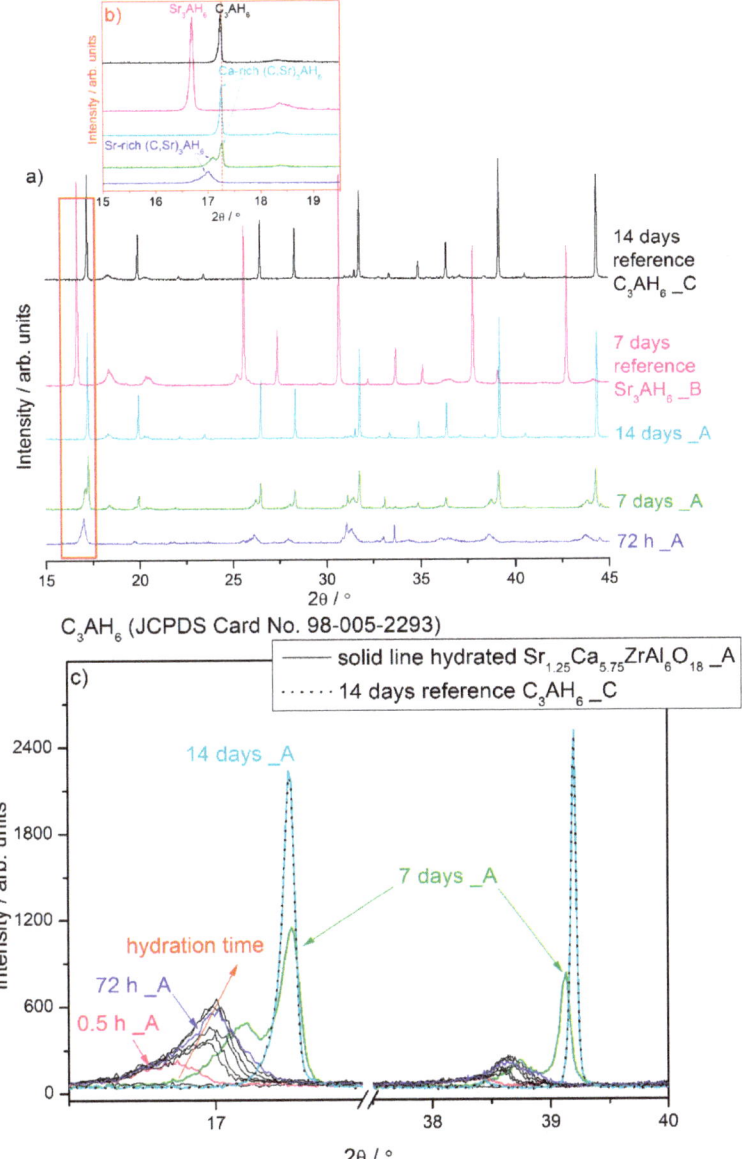

Figure 1. X-ray diffraction patterns of the $Sr_{1.25}Ca_{5.75}ZrAl_6O_{18}$ cement paste (Sample A) at different curing periods. (**a**,**b**) contains lines of two reference materials, i.e., Sr_3AH_6 synthesized through hydration from $SrAl_2O_4$ cement (Sample B) and C_3AH_6 synthesized through hydration from $Ca_7ZrAl_6O_{18}$ cement (Sample C). (**c**) contains a line (dot line) of the reference C_3AH_6 formed at 14 d (Sample C).

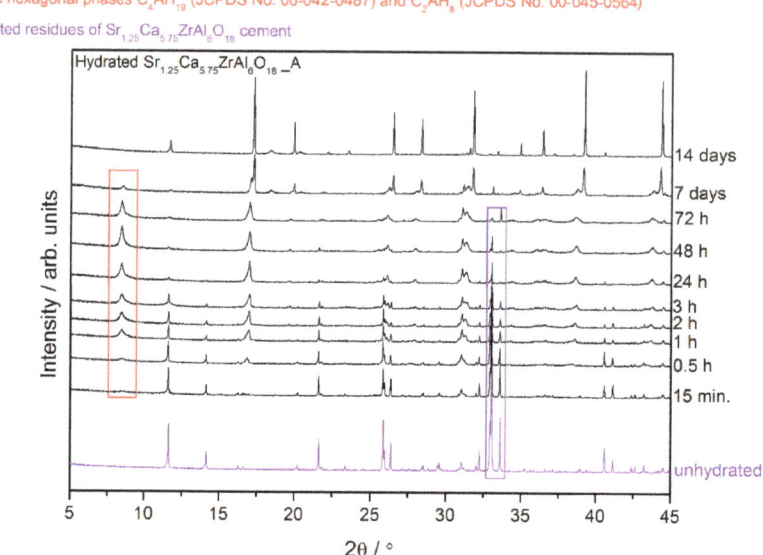

Figure 2. X-ray diffraction patterns of the $Sr_{1.25}Ca_{5.75}ZrAl_6O_{18}$ cement paste (Sample A) at different curing periods from 15 min to 14 d compared with the XRD pattern of the unhydrated compound.

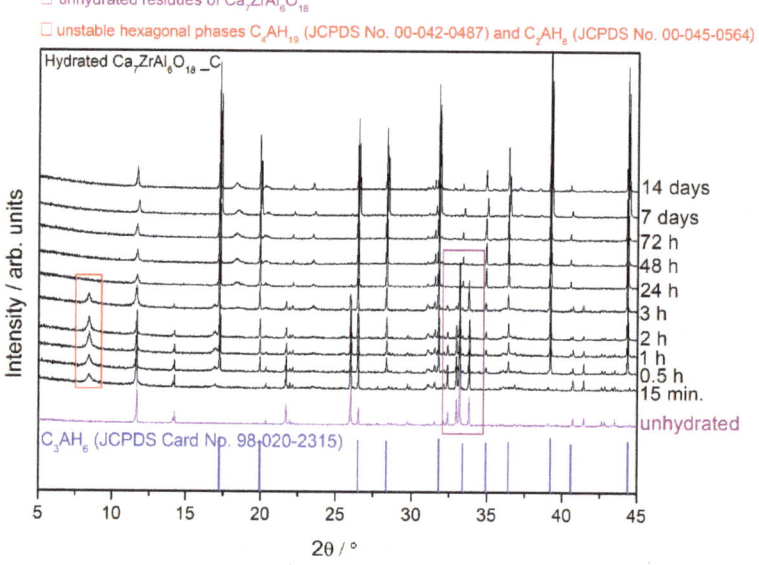

Figure 3. X-ray diffraction patterns of the $Ca_7ZrAl_6O_{18}$ cement paste (Sample C) at different curing periods from 15 min to 14 d compared with the XRD pattern of the unhydrated compound.

The results of powder X-ray diffraction patterns evaluation show the progress of $Sr_{1.25}Ca_{5.75}ZrAl_6O_{18}$ hydration in cement paste between 15 min and 14 d of curing which are more or less similar at an early stage of hydration (up to 72 h) Figure 1a–c. The XRD profiles of the cement pastes hydrated between 0.5 and 72 h demonstrate diffraction peak at ca. 17.00° 2θ which can now be interpreted as belonging to Sr-rich $(Sr,C)_3AH_6$ hydrate (Figure 1c). Hence, this XRD peak located at the position of 2θ = 17.00 which belongs to Sr-rich $(Sr,C)_3AH_6$ needs to be considered between the reference C_3AH_6 synthesized through $Ca_7ZrAl_6O_{18}$ hydration and other reference sample Sr_3AH_6 synthesized from $SrAl_2O_4$ precursor through hydration. As is evident from this figure, the formation of the intermediate Sr-rich $(Sr,C)_3AH_6$ hydrate precedes the formation of the stable Ca-rich $(Sr,C)_3AH_6$ hydrate at 7 d of curing. In this sample, two isostructural compounds with a hydrogarnet type crystal lattice were present. The position of the lower-intensity XRD line at ca. 17.00° 2θ is situated between lines belonging to pure phases Sr_3AH_6 (Sample B) and C_3AH_6 (Sample C). The second position of the higher intensity XRD line at ca. 17.26° 2θ is similar to that found for reference C_3AH_6 (Figure 1a–c). In addition, it is worth noting that the Sr-rich $(Sr,C)_3AH_6$ exists in the hardened cement paste between 0.5 h and 7 d of curing. This phase disappeared after longer curing times and became replaced by Ca-rich $(Sr,C)_3AH_6$ or C_3AH_6. This work has successfully shown the existence of the solid solution of strontium in the tricalcium hydrate C_3AH_6 lattice by direct verification using XRD. By reason of structural modification of C_3AH_6 through ionic substitution, the lattice parameter of the cubic phase was increased and the slight shift in XRD peaks belonging to $(Sr,C)_3AH_6$ solid solution towards lower 2θ value was observed (Figure 1b). This increase in the lattice parameter was due to the size of the ionic radius of Sr^{2+} (132 pm) which is bigger than the ionic radius of Ca^{2+} (114 pm).

A brief summary of XRD results is given as Figures 2 and 3. This overview XRD spectra recorded from the $Sr_{1.25}Ca_{5.75}ZrAl_6O_{18}$ cement paste (Sample A) at different curing periods from 15 min to 14 d showed a progressive reduction in the peaks associated with $Sr_{1.25}Ca_{5.75}ZrAl_6O_{18}$ due to its hydration process, which led to the formation of hydration products (Figure 2). The cement paste at the age of 15 min is a mixture of the unhydrated phase and amorphous or poorly crystalline hexagonal hydrates, whereas the cement paste at the age between 0.5 h and 72 h contained a mixture of the Sr-rich $(Sr,C)_3AH_6$ cubic phase, hexagonal hydrates and the still unhydrated residues of the $Sr_{1.25}Ca_{5.75}ZrAl_6O_{18}$ cement grains. It should be noted that the positions of the XRD peaks of C_4AH_{19} (JCPDS No. 00-042-0487; h k l = 0 0 6, d = 10.64350 Å, 2θ = 8.301°, I = 100%) are coincident well with those belonging to C_2AH_8 (JCPDS No. 00-045-0564, h k l = 0 0 6, d = 10.81270 Å, 2θ = 8.170°, I = 100%). Therefore, it is often difficult to clearly differentiate between C_4AH_{19} and C_2AH_8 in the XRD patterns, as is clearly demonstrated with a red rectangle (□) in Figure 2. However, at 7d second adjacent cubic phase, Ca-rich $(Sr,C)_3AH_6$ or pure C_3AH_6, exists together with the initially formed cubic phase Sr-rich $(Sr,C)_3AH_6$ and some residues of the hexagonal hydrates. As a general trend at the age of 14 d, XRD pattern of cement paste achieved profile similar to that of pure C_3AH_6 (Figures 1c and 2) without any metastable hydrates and unhydrated cement residues, i.e., unhydrated cement clinker mineral $Sr_{1.25}Ca_{5.75}ZrAl_6O_{18}$.

The overview XRD spectra recorded from the reference $Ca_7ZrAl_6O_{18}$ cement paste (Sample C) at different curing periods from 15 min to 14 d is shown in Figure 3. The hexagonal hydrates exist with the still unhydrated residues of the $Ca_7ZrAl_6O_{18}$ in cement paste between 15 min and 3 h of curing period, whereas C_3AH_6 hydrated phase is formed at the curing age of 0.5 h. At the age of 24 h, the XRD pattern of the cement pastes exhibits profile similar to that of pure C_3AH_6 without any traces of unhydrated cement particles $Ca_7ZrAl_6O_{18}$ and metastable hydrates.

From the X-ray diffraction results, it seems obvious that strontium doping affects the hydration behavior of the cement clinker mineral phase $Ca_7ZrAl_6O_{18}$, and leads to changes in the hydration products properties. There is a relationship between the proportion of residual unhydrated cement particles and the properties of the particular cement clinker mineral phases involved. After 24 h of curing at 50 °C, where the hardened cement paste (Sample C) consists primarily of C_3AH_6, the original $Ca_7ZrAl_6O_{18}$ cement particles are no longer evident. In the Sr-doping of $Ca_7ZrAl_6O_{18}$ case, there is inhibition of hydration, and the $Sr_{1.25}Ca_{5.75}ZrAl_6O_{18}$ cementitious particles exist in the hardened cement paste up to 72 h (Sample A). This material would need to cure over 72 h to reach complete hydration. Hence, XRD data for 0.5 h–7 d materials containing strontium indicates the appearance of additional peaks adjacent to each of the reference C_3AH_6 lines caused by the presence of an additional Sr-rich cubic phase. These results confirmed the strontium retention by calcium zirconium aluminate cement paste through the chemical bonding to C-A-H in the hydrated phase. The presence of strontium in the C-A-H matrix is also known to delay the transformation of hexagonal hydrates into the cubic phases.

3.2. Ex-Situ ^{27}Al NMR Study of the Hydration Reaction at 50 °C

Figure 4a,b presents the ^{27}Al NMR spectra of synthesized $Ca_7ZrAl_6O_{18}$ and Sr-doped $Ca_7ZrAl_6O_{18}$ cements together with their products of hydration. The ^{27}Al MASNMR spectra of the starting unhydrated samples are shown by the red lines. The intense and broad peak at ca. 50 ppm is due to $Ca_7ZrAl_6O_{18}$, which consists of orientationally disordered six AlO_4 tetrahedra linked together by sharing corners, to form $[Al_6O_{18}]$ rings [29]. The ^{27}Al MAS NMR spectra of the hydrated samples during the first 15 min all contain peaks near 4 ppm due to ^{VI}Al in the cement hydration reaction products (amorphous or poorly crystalline hexagonal hydrates C_4AH_{19} or C_2AH_8) (Figure 4a,b).

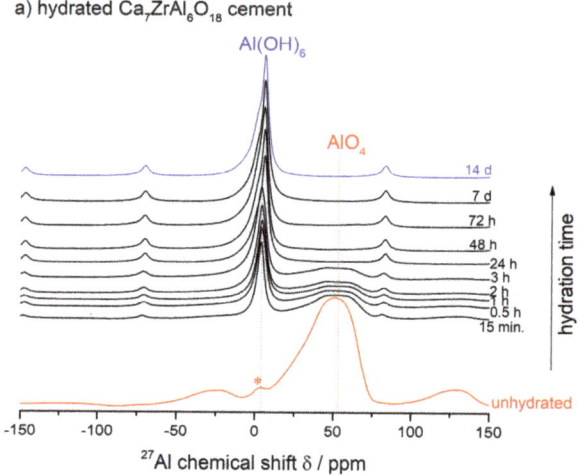

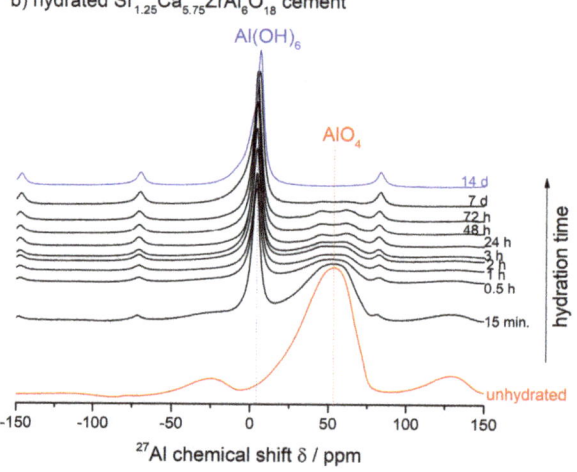

Figure 4. Cont.

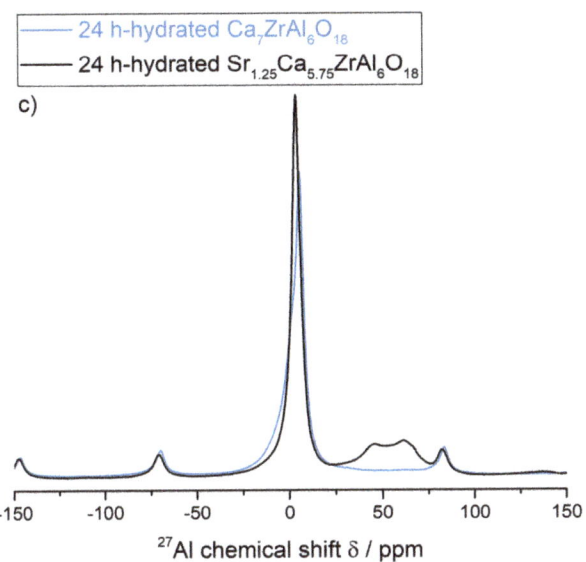

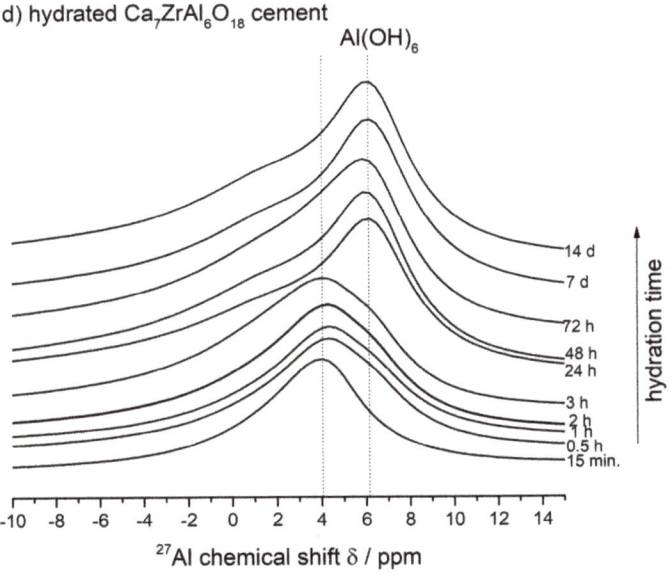

Figure 4. Cont.

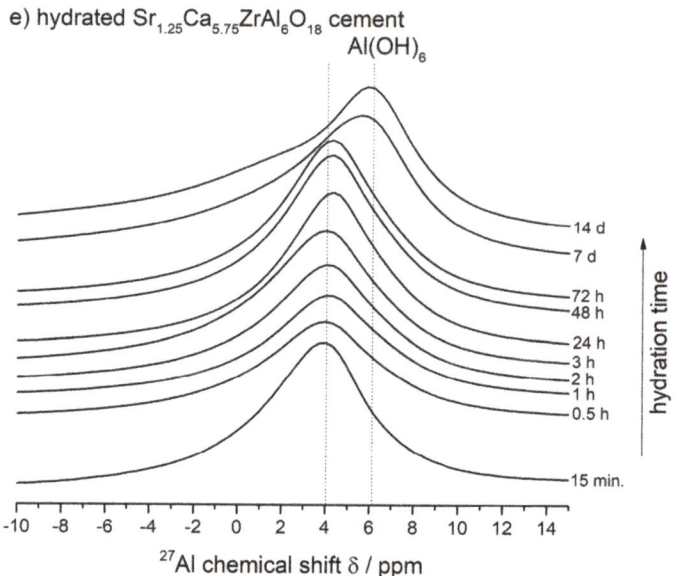

Figure 4. (a,b) The room-temperature ^{27}Al MASNMR spectra of unhydrated Ca$_7$ZrAl$_6$O$_{18}$ (Sample C) and Sr$_{1.25}$Ca$_{5.75}$ZrAl$_6$O$_{18}$ (Sample A) cements, and their products of hydration formed between 15 min and 14 d at 50 °C. The arrow means the changes according to the direction of increasing hydration time. (c) A comparison of NMR spectra of both cements at 24 h of hydration. (d,e) The lines at ca. 4–6 ppm for octahedrally coordinated lattice Al^{3+} in calcium aluminate hydrates.

For the partially reacted samples, the signals are in the ca. 4 ppm and ca. 46–61 ppm ranges due to both hydrates and unhydrated reactants, respectively (Figure 4c). For the totally hydrated (and converted) samples, all of the signalsarein the ca. 6 ppm range, consistent with total conversion of Al from tetrahedral coordination in the unhydrated Ca$_7$ZrAl$_6$O$_{18}$ and Sr$_{1.25}$Ca$_{5.75}$ZrAl$_6$O$_{18}$ cements to octahedral coordination [19] in the final hydrates formed at 24 h (undoped cement) or 7 d (Sr-doped cement), as expected from previous works on the calcium aluminate cement hydration processes investigated by solid-state^{27}Al MAS NMR studies [19,25]. The maximum for the ^{VI}Al peak alters slightly depending on the calcium aluminate hydrates present [18,19]. For the undoped Ca$_7$ZrAl$_6$O$_{18}$ cement hydrated between 15 min and 3 h, in which the detected crystalline hydrates are mainly hexagonal phases, the peak maximum is at ca. 4 ppm and shifts to 6 ppm for this cement hydrated between 24 h and 14 d (Figure 4d), which by XRD contain cubic C$_3$AH$_6$ as the predominant phase. This type of shift is delayed up to 7 d in the hydrated Sr-doped cement (Figure 4e), where Ca-rich (Sr,C)$_3$AH$_6$ or pure C$_3$AH$_6$ begin to form. Hence, it can be summarized that the chemical shift occurring at ca. 4 ppm was due to octahedrally coordinated framework aluminum atoms in Sr-rich (Sr,C)$_3$AH$_6$ (Sr$_{1.25}$Ca$_{5.75}$ZrAl$_6$O$_{18}$ cement paste), poorly crystalline C$_3$AH$_6$ (Ca$_7$ZrAl$_6$O$_{18}$ cement paste) and hexagonal hydrates (both cement pastes) formed at an early stage of hydration. The chemical shift occurring at ca. 6 ppm was due to octahedrally coordinated framework aluminum atoms in Ca-rich (Sr,C)$_3$AH$_6$ or pure C$_3$AH$_6$ formed in the totally reacted Sr-doped sample, as it can be referenced in for pure C$_3$AH$_6$ formed in the reference fully hydrated and converted undoped cement paste. It is worth discussing that the maximum for the Al(6) peak belonging to C-A-H phases varies slightly from data presented in Ref. [18,19]. In those works, and many others, the peak maximum belonging to C$_3$AH$_6$ was located at about 12 ppm, whereas the peak maxima belonging to hexagonal hydrates were located at ca. 10–11 ppm.

3.3. Microstructural Studies on the Hydrated $Sr_{1.25}Ca_{5.75}ZrAl_6O_{18}$ Cement Paste

The development of $Sr_{1.25}Ca_{5.75}ZrAl_6O_{18}$ cement paste microstructure in time can directly be linked to the evolution of phase composition presented in Sections 3.1 and 3.2. Figure 5 shows the typical microstructure of this cement paste fragment after 24 h hydration at 50 °C. The presence of Sr in $Sr_{1.25}Ca_{5.75}ZrAl_6O_{18}$ clinker affects formation of Sr-rich $(Sr,C)_3AH_6$ (Figure 5a—point 1) with a cubic/isometric crystal form [30,31]. The EDS spectrum presenting intensity vs. energy of the detected X-ray clearly identifies the peaks of Sr, Ca, Al and O (Figure 5b). The EDS spectrum from the hexagonal irregular flakes is shown in Figure 5c. The Ca, Al and O peaks are mainly due to C-A-H phases. EDS intensity ratio of calcium and aluminum peaks indicates the presence of C_4AH_{19} hydrate rather than C_2AH_8 hydrate.

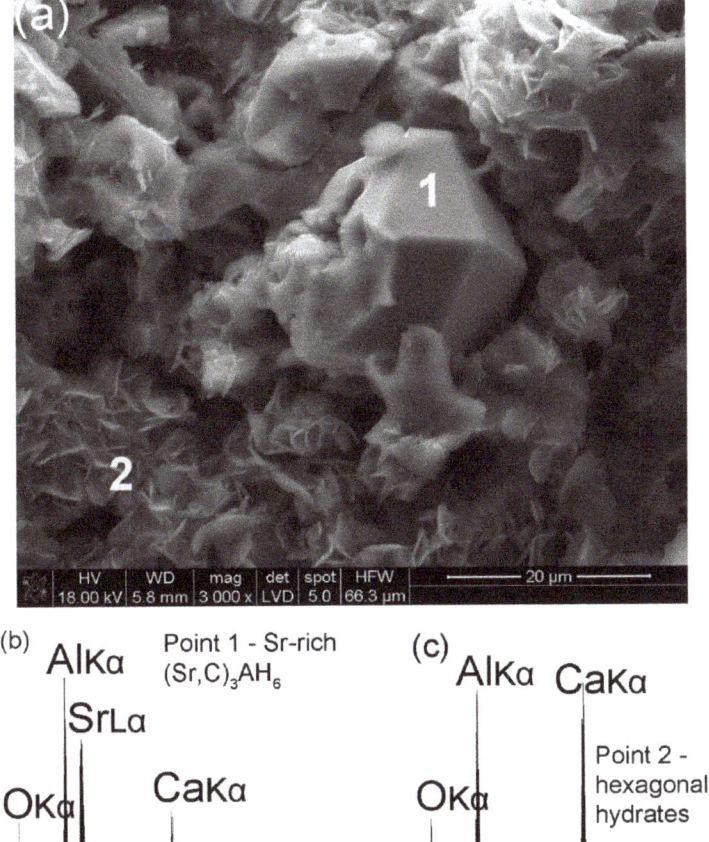

Figure 5. SEM image of the hydrated $Sr_{1.25}Ca_{5.75}ZrAl_6O_{18}$ cement (Sample A) at the age of 24 h (**a**). Spot 1–2 EDS analysis; (**b**,**c**) EDS spectra of the sample in the microarea 1 (Sr-rich $(Sr,C)_3AH_6$) and 2 (hexagonal calcium aluminate hydrates), respectively.

Most of the C_3AH_6 or Ca-rich $(Sr,C)_3AH_6$ crystals formed after a hydration time of 7 d attain the shape of cubes, pyritohedra or other more complex forms of the isometric system, which are reinforced with $Al(OH)_3$ crystals (Figure 6a,c). As observed before, those hydration products are strongly dependent on curing time and the Ca peak (Figure 6b) intensity increases relative to the Sr peak intensity (Figure 5b). Hence, as the curing time increases, the crystals belonging to a cubic or isometric system formed initially as a transient Sr-rich $(Sr,C)_3AH_6$ were replaced with Ca-rich $(Sr,C)_3AH_6$ or pure C_3AH_6. An unavoidable change of one form of calcium aluminate hydrate to another can be found elsewhere [32–34].

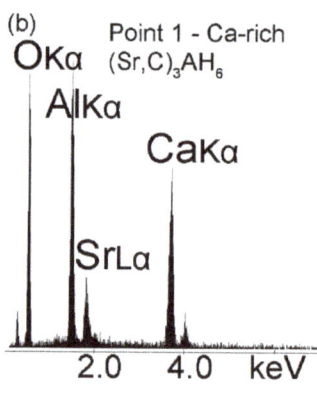

Figure 6. Cont.

Figure 6. SEM image of the hydrated $Sr_{1.25}Ca_{5.75}ZrAl_6O_{18}$ cement (Sample A) at the age of 7 d (**a**). Spot 1 EDS analysis; (**b**) EDS spectrum of the sample in the microarea 1—Ca-rich $(Sr,C)_3AH_6$; (**c**) overview of microstructure.

4. Conclusions

According to the current research, many conclusions can be drawn:

(1) The Sr-doped cement is developed through structural substitution for Ca ions by Sr ions in the $Ca_7ZrAl_6O_{18}$ clinker phase.
(2) Strontium was used as a retarding agent to block this cement clinker phase hydration at a curing temperature of 50 °C. Hence, the residual unhydrated cement particles in the hardened $Sr_{1.25}Ca_{5.75}ZrAl_6O_{18}$ cement paste were present for much longer than for the undoped $Ca_7ZrAl_6O_{18}$ clinker phase sample as it was observed by XRD.
(3) The hydration of both $Ca_7ZrAl_6O_{18}$ and $Sr_{1.25}Ca_{5.75}ZrAl_6O_{18}$ cements was also inspected using the ^{27}Al MAS NMR technique. This hydration is accompanied by a change of Al-coordination from tetrahedral to octahedral. This complete conversion from anhydrous $^{27}Al^{IV}$ to hydrated $^{27}Al^{VI}$ species was achieved during the first 24 h of hydration at 50 °C for $Ca_7ZrAl_6O_{18}$ and during 7 d of hydration at 50 °C for $Sr_{1.25}Ca_{5.75}ZrAl_6O_{18}$.
(4) The hexagonal phases were formed starting in the very first minutes of hydration of these cements. For each cement type tested, these unstable hydrates consist mainly of C_4AH_{19} and probably of C_2AH_8 as it was observed by XRD.
(5) The formation of a thermodynamically stable phase pure C_3AH_6 or Ca-rich $(Sr,C)_3AH_6$ in the hardened $Sr_{1.25}Ca_{5.75}ZrAl_6O_{18}$ cement paste was preceded by that of a number of less stable phases, i.e., Sr-rich $(Sr,C)_3AH_6$ hydrate and other hexagonal Ca-Al hydrates. The Sr-rich $(Sr,C)_3AH_6$ hydrate existing between 0.5 h and 7 d of curing was isostructural with the Ca-rich $(Sr,C)_3AH_6$ or pure C_3AH_6 formed at the age of 7 d.
(6) The transformation of the hexagonal calcium aluminate hydrates and Sr-rich $(Sr,C)_3AH_6$ hydrate into the cubic phase Ca-rich $(Sr,C)_3AH_6$ or pure C_3AH_6 was expressed in terms of chemical shift from ca. 4 ppm to ca. 6 ppm in the hardened $Sr_{1.25}Ca_{5.75}ZrAl_6O_{18}$ cement paste at the age of 7 d.

The same ^{27}Al NMR chemical shift was detected at the age of 24 h for the reference hardened $Ca_7ZrAl_6O_{18}$ cement paste.

Funding: This project was financed by the National Science Centre, Poland, project number 2017/26/D/ST8/00012 (Recipient: D. Madej). The sponsor had no role in the design, execution, interpretation, or writing of the study.

Conflicts of Interest: The author declares no conflicts of interest.

Nomenclature

C	CaO
A	Al_2O_3
Z	ZrO_2
Sr	SrO
H	H_2O

References

1. Nowacka, M.; Pacewska, B. Effect of structurally different aluminosilicates on early-age hydration of calcium aluminate cement depending on temperature. *Constr. Build. Mater.* **2020**, *235*, 117404. [CrossRef]
2. Cao, Y.-F.; Tao, Z.; Pan, Z.; Wuhrer, R. Effect of calcium aluminate cement on geopolymer concrete cured at ambient temperature. *Constr. Build. Mater.* **2018**, *191*, 242–252. [CrossRef]
3. Zhang, X.; He, Y.; Lu, C.; Huang, Z. Effects of sodium gluconate on early hydration and mortar performance of Portland cement-calcium aluminate cement-anhydrite binder. *Constr. Build. Mater.* **2017**, *157*, 1065–1073. [CrossRef]
4. Madej, D. Synthesis, formation mechanism and hydraulic activity of novel composite cements belonging to the system $CaO-Al_2O_3-ZrO_2$. *J. Therm. Anal. Calorim.* **2017**, *130*, 1913–1924. [CrossRef]
5. Litwinek, E.; Madej, D. Structure, microstructure and thermal stability characterizations of C_3AH_6 synthesized from different precursors through hydration. *J. Therm. Anal. Calorim.* **2019**. [CrossRef]
6. Madej, D.; Rajska, M.; Kruk, A. Synthesis and hydration behaviour of calcium zirconium aluminate powders by modifying co-precipitation method. *Ceram. Int.* **2020**, *46*, 2373–2383. [CrossRef]
7. Madej, D. A new implementation of electrochemical impedance spectroscopy (EIS) and other methods to monitor the progress of hydration of strontium monoaluminate ($SrAl_2O_4$) cement. *J. Therm. Anal.* **2019**, *139*, 17–28. [CrossRef]
8. Liu, K.; Chen, A.; Shang, X.; Chen, L.; Zheng, L.; Gao, S.; Zhou, Y.; Wang, Q.; Ye, G. The impact of mechanical grinding on calcium aluminate cement hydration at 30 °C. *Ceram. Int.* **2019**, *45*, 14121–14125. [CrossRef]
9. Kerienė, J.; Antonovič, V.; Stonys, R.; Boris, R. The influence of the ageing of calcium aluminate cement on the properties of mortar. *Constr. Build. Mater.* **2019**, *205*, 387–397. [CrossRef]
10. Garcia-Lodeiro, I.; Irisawa, K.; Jin, F.; Meguro, Y.; Kinoshita, H. Reduction of water content in calcium aluminate cement with/out phosphate modification for alternative cementation technique. *Cem. Concr. Res.* **2018**, *109*, 243–253. [CrossRef]
11. Madej, D.; Szczerba, J.; Nocuń-Wczelik, W.; Gajerski, R. Hydration of $Ca_7ZrAl_6O_{18}$ phase. *Ceram. Int.* **2012**, *38*, 3821–3827. [CrossRef]
12. Madej, D.; Boris, R. Synthesis, characterization and hydration analysis of Ba^{2+}-, Cu^{2+}- or Bi^{3+}-doped $CaO-Al_2O_3-ZrO_2$-based cements. *J. Therm. Anal. Calorim.* **2019**, *138*, 4331–4340. [CrossRef]
13. Madej, D. Hydration, carbonation and thermal stability of hydrates in $Ca_{7-x}Sr_xZrAl_6O_{18}$ cement. *J. Therm. Anal. Calorim.* **2018**, *131*, 2411–2420. [CrossRef]
14. Pöllmann, H.; Kaden, R. Mono- (strontium-, calcium-) aluminate based cements. In *Calcium Aluminates Proceedings of the International Conference*; Building Research Establishment: Watford, UK, 2014; pp. 99–108.
15. Ukrainczyk, N.; Vrbos, N.; Šipušić, J. Influence of metal chloride salts on calcium aluminate cement hydration. *Adv. Cem. Res.* **2012**, *24*, 249–262. [CrossRef]
16. Duran, A.; Sirera, R.; Nicolas, M.P.; Navarro-Blasco, I.; Fernández, J.M.; Alvarez, J.I. Study of the early hydration of calcium aluminates in the presence of different metallic salts. *Cem. Concr. Res.* **2016**, *81*, 1–15. [CrossRef]

17. Faucon, P.; Charpentier, T.; Bertrandie, D.; Nonat, A.; Virlet, J.; Petit, J.C. Characterization of calcium aluminate hydrates and related hydrates of cement pastes by ^{27}Al MQ-MAS NMR. *Inorg. Chem.* **1998**, *37*, 3726–3733. [CrossRef]
18. Skibsted, J.; Henderson, E.; Jakobsen, H.J. Characterization of calcium aluminate phases in cements by ^{27}Al MAS NMR spectroscopy. *Inorg. Chem.* **1993**, *32*, 1013–1027. [CrossRef]
19. Cong, X.; Kirkpatrick, R.J. Hydration of calcium aluminate cements: A solid-state ^{27}Al NMR study. *J. Am. Ceram. Soc.* **1991**, *76*, 409–416. [CrossRef]
20. Mercury, J.M.R.; Pena, P.; de Aza, A.H.; Turrillas, X.; Sobrados, I.; Sanz, J. Solid-state ^{27}Al and ^{29}Si NMR investigations on Si-substituted hydrogarnets. *Acta Mater.* **2007**, *55*, 1183–1191. [CrossRef]
21. Myers, R.J.; Bernal, S.A.; Gehman, J.D.; Van Deventer, J.S.J.; Provis, J. The role of Al in cross-linking of alkali-activated slag cements. *J. Am. Ceram. Soc.* **2014**, *98*, 996–1004. [CrossRef]
22. Walkley, B.; Provis, J.L. Solid-state nuclear magnetic resonance spectroscopy of cements. *Mater. Today Adv.* **2019**, *1*, 100007. [CrossRef]
23. Müller, D.; Rettel, A.; Gessner, W.; Scheler, G. An application of solid state magic-angle spinning ^{27}Al NMR to the study of cement hydration. *J. Magn. Reson.* **1984**, *57*, 152–156. [CrossRef]
24. Muller, D.; Gessner, W.; Samoson, A.; Lippmaa, E.; Scheler, G. Solid state ^{27}Al NMR studies of polycrystalline aluminates in the system CaO-Al$_2$O$_3$. *Polyhedron* **1986**, *5*, 779–785. [CrossRef]
25. Hughes, C.E.; Walkley, B.; Gardner, L.; Walling, S.A.; Bernal, S.A.; Iuga, D.; Provis, J.; Harris, K.D.M. Exploiting in-situ solid-state NMR spectroscopy to probe the early stages of hydration of calcium aluminate cement. *Solid State Nucl. Magn. Reson.* **2019**, *99*, 1–6. [CrossRef] [PubMed]
26. Garces, P.; Alcocel, E.; Chinchon, S.; Andreu, C.; Alcaide, J. Alcaide, Effect of curing temperature in some hydration characteristics of calcium aluminate cement compared with those of portland cement. *Cem. Concr. Res.* **1997**, *27*, 1343–1355. [CrossRef]
27. Zhang, Y.; Ye, G.; Gu, W.; Ding, D.; Chen, L.; Zhu, L. Conversion of calcium aluminate cement hydrates at 60 °C with and without water. *J. Am. Ceram. Soc.* **2018**, *101*, 2712–2717. [CrossRef]
28. Luz, A.P.; Pandolfelli, V.C. Halting the calcium aluminate cement hydration process. *Ceram. Int.* **2011**, *37*, 3789–3793. [CrossRef]
29. Fukuda, K.; Iwata, T.; Nishiyuki, K. Crystal structure, structural disorder, and hydration behavior of calcium zirconium aluminate, Ca$_7$ZrAl$_6$O$_{18}$. *Chem. Mater.* **2007**, *19*, 3726–3731. [CrossRef]
30. Das, S.K.; Kumar, S.K.; Das, P.K. Crystal morphology of calcium aluminates hydrated for 14 days. *J. Mater. Sci. Lett.* **1997**, *16*, 735–736. [CrossRef]
31. Kumar, S.; Das, S.K.; Daspoddar, P.K. Microstructure of calcium aluminates and their mixes after 3 days of hydration. *Trans. Indian Ceram. Soc.* **1999**, *58*, 115–117. [CrossRef]
32. Antonovič, V.; Keriené, J.; Boris, R.; Aleknevičius, M. The effect of temperature on the formation of the hydrated calcium aluminate cement structure. *Procedia Eng.* **2013**, *57*, 99–106. [CrossRef]
33. López, A.H.; Calvo, J.L.G.; Olmo, J.G.; Petit, S.; Alonso, M.C. Microstructural evolution of calcium aluminate cements hydration with silica fume and fly ash additions by scanning electron microscopy, and mid and near-infrared spectroscopy. *J. Am. Ceram. Soc.* **2008**, *91*, 1258–1265. [CrossRef]
34. Rashid, S.; Barnes, P.; Bensted, J.; Turrillas, X. Conversion of calcium aluminate cement hydrates re-examined with synchrotron energy-dispersive diffraction. *J. Mater. Sci. Lett.* **1994**, *13*, 1232–1234. [CrossRef]

© 2020 by the author. Licensee MDPI, Basel, Switzerland. This article is an open access article distributed under the terms and conditions of the Creative Commons Attribution (CC BY) license (http://creativecommons.org/licenses/by/4.0/).

Article

Thermal Performance of Mortars Based on Different Binders and Containing a Novel Sustainable Phase Change Material (PCM)

Antonella Sarcinella [1], José Luìs Barroso De Aguiar [2], Mariateresa Lettieri [3], Sandra Cunha [4] and Mariaenrica Frigione [1,*]

1. Innovation Engineering Department, University of Salento, Prov. le Lecce-Monteroni, 73100 Lecce, Italy; antonella.sarcinella@unisalento.it
2. Civil Engineering Department, University of Minho, Campus de Azurém, 4800-058 Guimarães, Portugal; aguiar@civil.uminho.pt
3. CNR—SPIN, via Giovanni Paolo II 132, 84084 Fisciano (Salerno), Italy; mariateresa.lettieri@cnr.it
4. Lusophone University of Humanities and Technologies, Campo Grande 376, 1749-024 Lisboa, Portugal; sandracunha86@gmail.com
* Correspondence: mariaenrica.frigione@unisalento.it; Tel.: +39-0832-297215

Received: 1 April 2020; Accepted: 25 April 2020; Published: 28 April 2020

Abstract: Increasing concerns about climate change and global warming bring about technical steps for the development of several energy-efficient technologies. Since the building sector is one of the largest energy users for cooling and heating necessities, the incorporation of a proper energy-efficient material into the building envelopes could be an interesting solution for saving energy. Phase change material (PCM)-based thermal energy storage (TES) seems suitable to provide efficient energy redistribution. This is possible because the PCM is able to store and release its latent heat during the phase change processes that occurs according to the environmental temperature. The purpose of this paper was the characterization of the thermal properties of a composite PCM (i.e., Lecce stone/poly-ethylene glycol, previously developed) incorporated into mortar compositions based on different binders (i.e., hydraulic lime and cement). The study was carried out using an experimental set up through which it was possible to simulate the different seasons of the years. It was observed that the addition of PCM in mortars leads to a decrease of the maximum temperatures and increase of the minimum temperatures. Furthermore, the results shown a reduction of the heating and cooling needs, thus confirming the capability of this material to save energy.

Keywords: phase change material (PCM); thermal energy storage (TES); sustainable materials for buildings; thermal properties; mortars; hydraulic lime; cement

---

1. Introduction

In the current world scenario, environmental issues as well as climate change represent a real problem that concerns all humanity. This awareness has led international policy to incentivize research pushing it in the development of new renewable energy solutions. On the other hand, the scientific community is focused on it and trying to make important steps forward to limit the increase in energy consumption. In the last few decades, it has been shown that the building sector contributes significantly to the increase the electric energy demands [1]. Major components of energy consumption in the building sector are in heating, cooling, air conditioning and ventilation systems for comfort demands [2]. For this reason, buildings should be designed to ensure the thermal comfort of the occupants, with minimum auxiliary energy for heating and cooling equipment [3].

Another problem associated with energy consumption is due to old and ancient constructions characterized by a lack of building envelope insulation; in the case of historical buildings, moreover, the installation of modern devices for heating and cooling necessities is limited. For this reason, it becomes essential to find solutions able to improve the energy efficiency of buildings preserving, at the same time, the environment [4].

A potential tool for energy conservation, able to store the excess energy and to release it, filling the gap between the energy supply and demand, is a system called thermal energy storage (TES).

This latter system, using the principle of latent heat thermal energy storage (LHTES), combined with a proper phase change material (PCM) is a promising technology to save energy, improving building efficiency [5]. As the temperature increases, the PCM has the capability to change its phase from solid to liquid. The reaction is endothermic and the PCM absorbs heat; when the temperature decreases, the material changes its phase from liquid to solid. At this point, the process is exothermic and the PCM desorbs heat. Thus, the addition of a PCM in building service equipment is a way to enhance energy storage capacity. Consequently, building energy performance could be optimized and also indoor thermal comfort can be improved [6]. The incorporation of a PCM in construction materials (passive building system) has proven to be the most interesting approach [7].

Thereby, wallboards, floors, roof, mortar or concrete and other parts are integrated with PCMs in order to improve the thermal performance of the building [8–12]. The most common solution for implementing PCMs in buildings is the installation of PCM into the interior side of the building envelope. Thus, the use of suitable PCMs in the interiors of the construction allows absorbing and releasing heat in any room during a large part of the day. Several experimental investigations have shown how this strategy positively affects indoor climate and energy use [6,7,13,14]. On the other hand, among all these possible applications, the incorporation of phase-change materials in mortars employed in the interiors of buildings appears the most attractive solution in an attempt to minimize the massive energetic consumption related to building conditioning. Such an approach allows the regulation of the temperature inside buildings through latent heat energy storage, using only solar energy as a resource, thus, reducing the need for heating/cooling equipment [15,16]. Incorporating PCMs in mortar and concrete can be an efficient method due to the large heat exchange area surfaces; in addition, the final functional material can be adapted in a wide variety of shapes and sizes. Being mortar and concrete widely used as construction materials, such PCM composites can be employed in any practical application. Moreover, quality control can be easily achieved in the materials produced [17].

Among the available methods to incorporate an effective PCM into a building material, such as mortar or concrete, the "form stable" is among the simpler and more efficient technologies [3,17–19]. In this method, a porous matrix of inorganic (such as silica-based material, perlite, diatomite, clay, etc.) can act as the inert support that will contain the true PCM [20,21]. A form-stable PCM composite can then be obtained by immersing the matrix in the liquid PCM, employing a vacuum system to force the impregnation. After the impregnation process, the porous matrix is able to retain the optimal percentage of PCM, with no leakage [22,23]. The production of form-stable PCMs involves very cheap and simple equipment. Several studies have been recently published on the use of construction materials incorporating a form-stable PCM composite [21,24–26].

In this work, this route (i.e., the form-stable method) was followed, employing as support matrix small pieces of Lecce stone (LS) supplied by a local quarry as a waste product and selecting low toxic and low flammable PEG (poly(ethylene glycol)) as real PCM. The use of waste stone as support for a PCM able to improve the thermal efficiency of the buildings, and to reduce the consumption of petroleum-derived energy, would represent a double advantage for the environment. PEG was chosen because of its favorable properties, such as suitable phase change temperatures and large phase change enthalpy, elevated long-term thermal/chemical stability, low toxicity and resistance to corrosion, and limited volume change during solid–liquid phase change [27,28]. Furthermore, the range of melting/crystallization temperatures of the selected grade of PEG (i.e., 1000) were considered particularly suitable for a PCM included in mortars to be used in Mediterranean warm countries.

The produced form-stable PCM system was then used as an aggregate for the production of some mortar formulations based on different binders. In our previous works [29,30], the production and the optimization process to achieve a suitable form-stable composite PCM have been described, as well as its chemical, physical and thermal characterization. The influence of this PCM composite system included in different mortar compositions on their mechanical properties was also investigated [30,31].

Starting from the previous research, taking into account the results obtained, the main aim of the study described in this paper was the analysis of the thermal performance of some of the produced mortars exposed to temperatures currently recorded in the South of Italy. To this purpose, an experimental setup was implemented: a small-scale test cell, on which the produced mortars (with and without PCM) were poured, was submitted to a preset temperature program able to simulate the different seasons of the year. The obtained thermal performance for each binder mix, unmodified or containing the innovative PCM, was recorded and analyzed.

2. Materials and Methods

2.1. Production of the Form-Stable Phase Change Material (PCM)

Different mortar compositions to be used indoors were developed in previous studies [30,31].

Lecce stone (LS), a biocalcarenite typical of the Salento area (South Italy) with a high open porosity [32], was selected as support matrix to contain the PCM. LS was chosen also because it is widely available as a waste product from extraction and production of stone components. Lecce stone was reduced in small pieces and sieved up to a granulometry ranging between 1.6 and 2.0 mm, to be used as aggregate in the mortar formulations. The PCM was based on poly (ethylene glycol), supplied in solid form (Sigma–Aldrich company, Germany) with the trade name PEG 1000.

According to the data sheet, the density of PEG 1000 at 20 °C is 1.2 g/cm^3. PEG 1000 displays a mix of favorable properties (cheapness, low toxicity, low flammability) along with suitable melting and crystallization range of temperatures. In order to be used in building applications, in fact, a PCM must display an appropriate range of phase transition temperatures. Since PEG shows melting and crystallization temperatures of 43 °C and 23 °C, respectively, this polymer has been considered favorable to be used as a Phase Change Material included in mortars to be employed in buildings located in warm regions, for instance in the Mediterranean area [27,28].

To produce the PCM composite system, namely LS/PEG, a simple vacuum impregnation process was employed, according to the form-stable principle. As reported in our previous work [30], a proper quantity of LS granules, that constituted the support for the PCM, was positioned in a flask linked to a vacuum pump (at a vacuum pressure of 0.1 MPa). This flask was placed on a magnetic stirrer, kept at 60 °C, containing also a magnetic stir bar. The PEG (originally in solid form), constituting the true phase change component, was heated at 80 °C and then added to the flask where the LS granules were placed. At that point, air was allowed to enter in the flask, forcing the penetration process of the liquid PEG into the support, i.e., in the pores of LS. This process was continued for 60 min. The results achieved with this method, reported in [30], have shown that the maximum percentage of PEG absorbed in LS is 23% by weight. It has been demonstrated that in this form stable PCM displays appropriate LHTES properties to be used in indoor mortars employed in buildings located in warm (for instance, Mediterranean) regions [27].

2.2. Compositions of Mortar Formulations

In our previous papers, different mortar formulations based on aerial lime, hydraulic lime, gypsum and cement, containing the composite LS/PEG have been produced and investigated [30,31], with the aim of identifying, for each binder, the most convenient composition able to produce mortars with adequate mechanical properties, in terms of both flexural and compressive modes. The mix designs based on hydraulic lime and cement mortars were then identified as those best performing. These latter were, therefore, employed to assess the true capability of LS/PEG to improve the thermal regulation of a building, decreasing the indoor maximum temperature and increasing the indoor minimum temperature. The selected binders, moreover, can be used in different applications: for historical constructions the hydraulic lime, the cement-based mortars in modern buildings and for any other common application [30,31].

A natural hydraulic lime (NHL), with density of 2700 kg/m^3 supplied by CIMPOR (Lisbon, PT) was selected. A CEM I 42.5 R cement, with a density of 3030 kg/m^3, supplied by SECIL (Lisbon, PT), was employed in the present work. The inert support for the PEG-based PCM is the stone (LS). Additional information of these mortars and its raw materials can be found in [31].

A superplasticizer (SP) was added to both mortar compositions, in order to reduce the amount of water required for the mixing. The SP was a polyacrylate (MasterGlenium SKY 627, supplied by the BASF company), with density of 1050 kg/m^3. In Table 1, the composition of all the mortars realized and analyzed, produced according to the European Standard EN 998-1 [33], are reported.

Table 1. Mortar compositions (reported as kg/m^3 of produced mortar).

System	Binder/Content	Aggregates		SP	Water Saturation	Water	Water/Binder
		Lecce Stone (LS) Content	Poly Ethylene Glycol (PEG) Content				
HL$_{800}$LS	Hydraulic Lime/800	1092	0	15	275	320	0.40
HL$_{800}$LS/PEG	Hydraulic Lime/800	1729	398	15	0	375	0.47
C$_{800}$LS	Cement/800	1070	0	15	269	296	0.37
C$_{800}$LS/PEG	Cement/800	1347	310	15	0	360	0.45

Four compositions were developed: two of them were produced by adding the composite LS/PEG to the binders (i.e., HL$_{800}$_LS/PEG and C$_{800}$_LS/PEG), in order to evaluate the effect of the presence the PCM on the thermal properties of different mortars. For comparison purposes, two control formulations were prepared by introducing only LS as aggregate (i.e., HL$_{800}$_LS and C$_{800}$_LS). As for the mortar compositions containing the composite PCM (i.e., LS/PEG), the amount of this latter is different in the two produced mortars (based on hydraulic lime and cement, respectively) since the amount of each component in the mortars was determined according to Equation (1), where each component is related to its density:

$$\frac{Binder\ (kg)}{\rho_b} + \frac{Aggregates\ (kg)}{\rho_a} + \frac{Superplasticizer\ (kg)}{\rho_s} + \frac{Water\ (kg)}{\rho_w} = 1\ \text{m}^3 \qquad (1)$$

where ρ_b, ρ_a, ρ_s and ρ_w are the densities of the binder, the aggregates, the superplasticizer and the water, respectively.

The indication "water saturation" in Table 1 accounts for the water used to saturate the LS aggregates, possessing a high porosity, to prevent them from absorbing the water necessary for the mortar's manufacture. This additional water was not required when LS/PEG composite was added to mortars, since PEG was able to saturate the pores of Lecce stone [30].

2.3. Methods

The morphology and microstructure of the developed mortars were analyzed using a scanning electron microscope (SEM, Carl Zeiss Auriga40 Crossbeam instrument, Jena, Germany). The investigations were performed under vacuum on samples without metallization, using a beam accelerating voltage of 20 kV and the secondary electron detector. The SEM images were further analyzed using the Scanning Probe Image Processor (SPIP) software package (Image Metrology A/S, Kgs. Lyngby, Denmark), version 6.2.6. Three images at magnification of 2500× were analyzed for each material and the results were averaged. The "Particle and Pore analysis" function was used to detect the pores and to measure their size. The threshold segmentation method was applied to quantify the total pores percentage; the watershed method was used to evaluate the distribution of the pores on the basis of their dimension.

The thermal behavior of the produced mortars was tested in a climatic chamber, setting a temperature program representative of each season of the year, in order to study the thermal behavior of the developed mortars in summer, spring, autumn and winter. The temperature program was fixed based on climatic data collected at the weather station installed in the Salento region, South Italy.

For each composition, a small-scale test cell (Figure 1a) was constructed with an insulating material (polystyrene) 3-cm thick, coated in the internal side with a mortar layer of 1 cm (Figure 1b). Each cell was a cube, with a side equal to 200 mm. A thermocouple was placed in the center of each small-scale cell, as illustrated in Figure 1c. Each small-scale test cell was placed inside a climatic chamber, equipped with thermocouples for the temperature control (Figure 1d). Each thermocouple used during the tests was connected to a data acquisition system of high sensibility (AGILENT 34970A), measuring every minute the temperature of the climatic chamber and that inside the small-scale test cells; the measured temperatures were then recorded by software (BenchLinkDataLogger3). Type K thermocouples were used. The described set-up is able to supply important and reliable information on the thermal behavior of the mortars, as previously demonstrated [34,35].

Figure 1. Set up to investigate the thermal performance of the different mortars: (**a**) six faces to build a small-scale test cell; (**b**) mortar layer of 1 cm; (**c**) small-scale test cell with a thermocouple in the center; (**d**) small-scale test cells placed inside a climatic chamber with temperature control.

3. Results and Discussion

3.1. Microstructure

The scanning electron microscope observations were performed to evaluate possible incompatibilities between the different phases present in the mortars. Figure 2 shows the microstructure of the hydraulic lime and cement based mortars with incorporation of the composite system, i.e., PEG-based PCM, compared with the reference mortars without PCM.

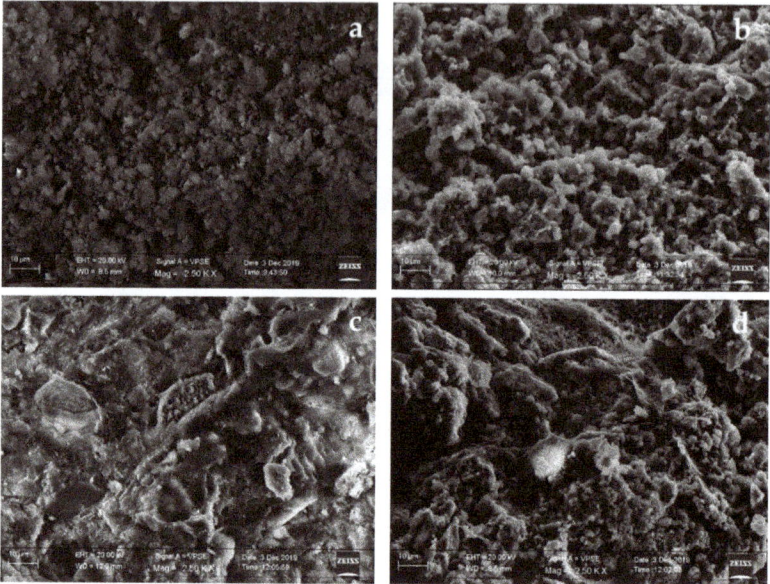

Figure 2. Scanning electron microscope (SEM) images of (**a**) hydraulic lime-based mortar without phase change material (PCM) (HL$_{800}$_LS (Lecce stone)); (**b**) hydraulic lime-based mortar with PCM (HL$_{800}$_LS/PEG (poly(ethylene glycol))); (**c**) cement-based mortar without PCM (C$_{800}$_LS) and (**d**) cement-based mortar with PCM (C$_{800}$_LS/PEG).

The absence of voids and cracks in the microstructure of the developed mortars, irrespective to their composition, suggests a good connection between the different components (LS or LS/PEG and each binder). Differences in the pore size and pore distribution were found when comparing the mortars with and without PCM, as can be observed from the results in Figure 3. A slight increase in the total pores was measured when the PCM was added to each mortar (Table 2). Taking into account that the total pores account for the porosity of the material, the mortars based on hydraulic lime were generally more porous than the cement-based ones, as expected. The most frequent pore size was similar for all the analyzed materials; it ranged between 0.761 and 0.894 µm (Figure 3a and Table 2). In addition, the mortars containing the PCM exhibited a higher percentage of larger pores, irrespective of the kind of binder. The fraction of pores larger than 2 µm was approximately 15% in HL$_{800}$_LS/PEG; it decreased by half in C$_{800}$_LS/PEG, while it was negligible in the most compact mortar (i.e., C$_{800}$_LS).

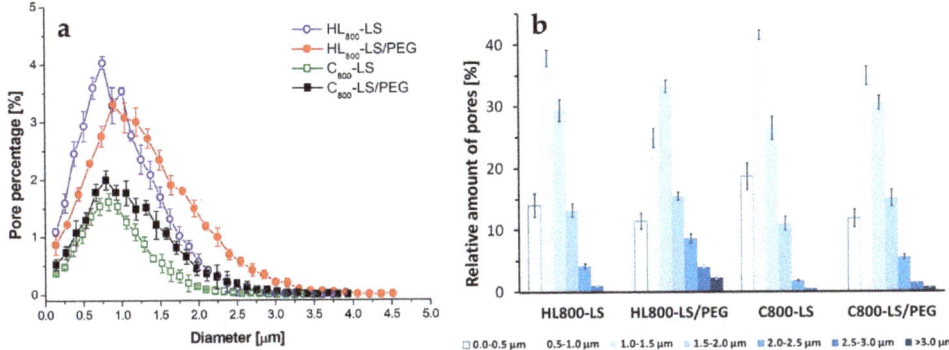

Figure 3. Porosimetric features by image analysis of the SEM micrographs in Figure 2: (**a**) percentage of pores as a function of their diameter; (**b**) number of pores per classes of pore-size.

Table 2. Total pores as a percentage and the most frequent pore size obtained by image analysis of SEM micrographs taken on the investigated mortars.

System	Total Pores (%)	Most Frequent Pore Size (μm)
HL_{800}_LS	33.57 ± 3.53	0.761 ± 0.062
HL_{800}_LS/PEG	36.57 ± 3.18	0.894 ± 0.075
C_{800}_LS	14.57 ± 1.35	0.832 ± 0.057
C_{800}_LS/PEG	19.66 ± 3.24	0.794 ± 0.076

In general, the presence of both higher porosity and larger pore size can be explained in terms of a higher water content of the mortar's formulation. As reported in Table 1, the mortar compositions containing LS/PEG were prepared using a higher amount of water. Thus, both hydraulic lime and cement-based mortars with PCM exhibited increased porosity and pores larger in dimension than the corresponding controls due to the higher water content employed. The observed porosimetric features can also account for the changes in mechanical properties; in fact, the addition of the PCM brought about a certain decrease in mechanical strength if compared to the mortars not containing the LS/PEG composite.

3.2. Thermal Behavior

Since the presence of a PCM is expected to positively influence the internal temperature of a room where it is applied [3,11,34], thermal tests were conducted with the aim of evaluating the thermal behavior of mortars with incorporation of PCM. During the tests, all seasons of the year were evaluated, taking as reference the climate recorded in Salento region (South Italy). Poly-ethylene glycol 1000 was, in fact, selected as PCM since its range of melting/crystallization temperatures is suitable for the intended purpose. The summer conditions were simulated employing a temperature range from 22 °C to 32 °C. For the spring climate, the analyzed temperature ranged between 12 °C and 24 °C. In the autumn, the temperatures ranged between 16 °C and 24 °C. To simulate the winter, a temperature interval between 8 °C and 15 °C was selected. Figure 4 shows the characteristic temperatures used to simulate a typical summer, spring, autumn and winter season in Salento region. Each season was simulated with three cycles, each one with a duration of 24 h.

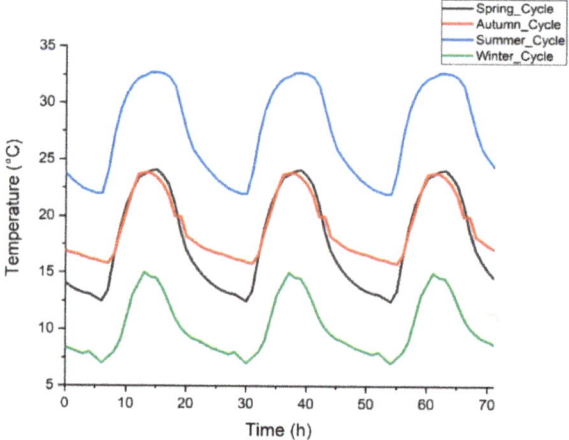

Figure 4. Temperature programs used to simulate the different season of the year.

During these tests, the PEG-based PCM reached the phase transition (between 27 and 30 °C during the heating process and between 10 and 13 °C during the cooling process) storing and releasing energy from the environment, respectively, regardless of the kind of binder. These temperatures are reported to be, in theory, favorable to obtain a PCM-based mortar to be employed as thermal energy storage system included in the exterior and/or in indoor walls of buildings located in warm regions [30,31]. In the case of winter season, it was not possible to evaluate the PCM effect since its melting point was hardly achieved [31], therefore, these data were not reported.

In Figures 5–7 report the behavior of the developed mortars, i.e., lime-based and cement-based mortars, with and without the novel PCM, in spring, summer and autumn climatic conditions. In each graph, the fluctuations of the typical temperatures for each season, set in the climate chamber where the test cells were located, are also reported. Figure 5 shows the thermal behavior of the developed mortars during summer. It can be observed that temperatures above 25 °C are registered but not below 20 °C. Thus, it was concluded that during the summer there are cooling needs. Figure 6 shows the thermal behavior of the mortars during the spring. Cooling is not necessary since, for the mortars containing the PCM, the maximum temperature is lower than 25 °C. Finally, Figure 7 shows the thermal behavior during the autumn: even in the presence of the PCM-based mortar, slight heating is needed, since the minimum temperatures were lower than 20 °C. In all the performed tests it was observed that when the temperature inside the test scale-cell achieves the range between 20 °C and 25 °C, a slight PCM phase change occurs: the thermal behavior of the PCM-based mortars evolves in a different way compared to the temperature program used for the simulation of the season. It was verified that the temperatures inside the test cells did not reach such extreme temperatures if compared to the temperature program. Furthermore, the temperature fluctuations of the mortar compositions containing the PCM are always narrow if compared to those measured on the mortar compositions without PCM, conforming the efficient behavior of the LS/PEG composite as phase change material for different mortars.

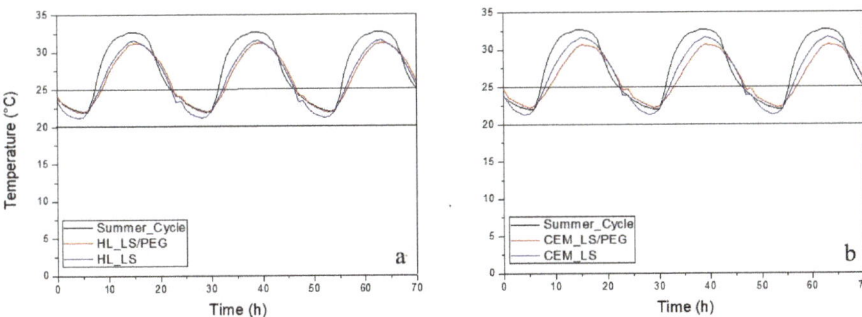

Figure 5. Thermal behavior in summer of the developed mortars: (**a**) hydraulic lime-based mortar and (**b**) cement-based mortar. The thermal comfort zone is between 20–25 °C.

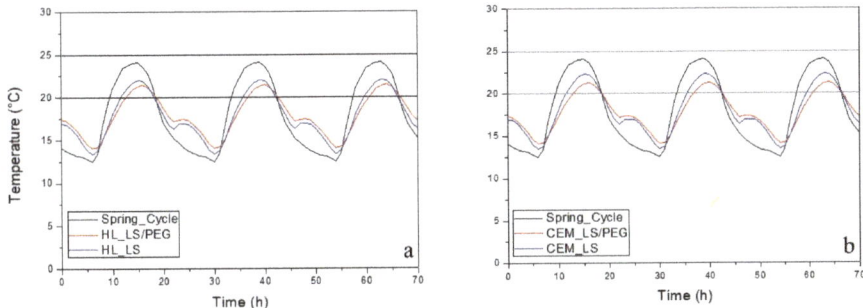

Figure 6. Thermal behavior in spring of developed mortars: (**a**) hydraulic lime-based mortar and (**b**) cement-based mortar. The thermal comfort zone is between 20–25 °C.

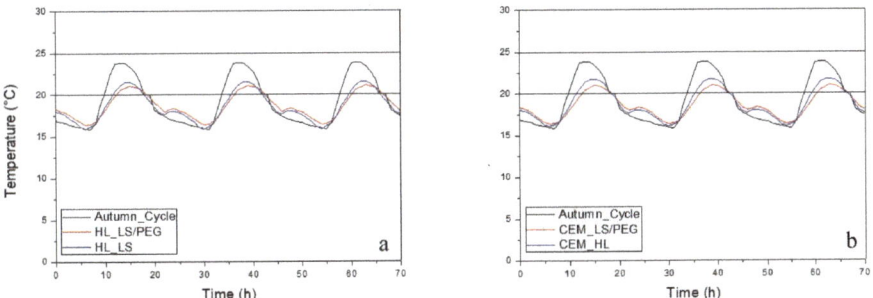

Figure 7. Thermal behavior in autumn of developed mortars: (**a**) hydraulic lime-based mortar and (**b**) cement-based mortar. The thermal comfort zone is between 20–25 °C.

As for the summer climatic condition (Figure 5), it was observed that in the heating step, i.e., when the temperature exceeds 25 °C, the PCM-based mortars showed a slower heating and a lower maximum temperature, this behavior being more marked in the case of the cement-based mortar. When the temperature lies near the indoor thermal comfort zone (20–25 °C), the cells exhibit similar temperature values. The effect of heat storage/release is only detected when the temperature diverges from the thermal comfort zone. As for the cement-based mortar, it was observed that the higher temperature of $C_{800\_}$LS/PEG is lower than 2 °C compared to the temperature program, while the temperature of $C_{800\_}$LS is lower than 1 °C if compared to the temperature program. Thus, starting from a decrease of the maximum temperature greater than 3% in the cooling stage achieved for the cement mortar, a decrease of the maximum temperature greater than 6% is achieved when the PCM is added to the same mortar formulation. Passing to analyzing the hydraulic lime-based mortar, on the other hand, the differences between the mortar formulation with and without PCM were negligible. During the same tests, it was also observed a lag time of the minimum temperatures of 60 min during the heating stage, while no lag time was recorded during cooling stage (Table 3). During the summer the greater part of residential buildings electricity consumption is used for cooling needs, thus a certain shift to off-peak periods of this consumption can represents an economical advantage.

Table 3. Lag time between the maximum and minimum temperatures in summer climate.

Summer	Lag Time (min)	
	Cooling Stage	Heating Stage
Cement	0	60
Hydraulic Lime	0	60

According to Figure 6, in the spring climate, the maximum temperature did not exceed 25 °C, meaning that the indoor environment does not require any cooling. On the other hand, the minimum temperatures are lower than 20 °C and, consequently, there are heating necessities. As can be seen in the graphs shown in Figure 6, the lowest temperature, recorded inside the climatic chamber and representative of the spring cycle, diverges from the temperature recorded inside the test cells. In general, there are not significant differences between the different binders (i.e., hydraulic lime and cement), since for both of them the incorporation of PCM leads to an increase in the minimum temperature of 11%. In particular, it was observed that the lower temperature of C_{800}_LS/PEG is greater than 2 °C when compared to the temperature program, while the temperature of C_{800}_LS is greater than 1 °C with respect to the temperature program. The same can be said for the hydraulic lime-based mortar. However, in this tested season, the effect of the energy storage/release of the PCM is found also in the thermal comfort zone (20–25 °C). The incorporation of PCM into mortars leads to a decrease of 12% in the maximum temperature. For the cement-based mortar as well as for the hydraulic lime-based one containing the PCM, the higher temperature was found to be lower than 3 °C compared to the temperature program, while the temperatures of the same mortar formulations without the PCM are lower than 2 °C if compared to the temperature program.

These results demonstrated that the PCM produced positively influences to a similar extent both high- and low-temperature external conditions. For the spring season, since there are not substantial differences between the different mortar compositions in terms of thermal regulation, the only difference can be highlighted by the thermal gradient, as can be seen in Figure 8, where the better thermal behavior of the cement-based mortar appears remarkable.

Table 4 shows the lag time of the maximum and minimum temperatures verified in the different mortars with and without PCM incorporation. It was observed that there is a lag time of the maximum temperature of 60 min in the cooling stage, no lag time of the minimum temperature is observed in the heating stage.

Table 4. Lag time between the maximum and minimum temperatures in spring climate.

Spring	Lag Time (min)	
	Cooling Stage	Heating Stage
Cement	60	0
Hydraulic Lime	60	0

Figure 7 shows the thermal performance of the different mortars in an autumn climate. It can be observed that the lowest temperatures are outside the comfort temperature zone since they are lower than 20 °C. This observation suggests that it could be necessary to use heating equipment, leading to higher energetic consumption in buildings. Starting from the mortar compositions containing the PCM (i.e., C_{800}_LS/PEG and HL_{800}_LS/PEG), it was possible to obtain an increase in the lowest temperature of 4%. This means that the difference between the temperature program, representative of the autumn season, and the test cells containing the mortars with the PCM is about 1 °C. On the other hand, the increase in temperature for the mortar compositions without the PCM is irrelevant if compared to the temperature program. However, a decrease in the maximum temperature of 12% was also measured for both mortar formulations containing the PCM. For the cement-based mortar as well as for the hydraulic lime-based one containing the PCM, the higher temperature was found to be lower than 3 °C when compared to the temperature program, while the temperature of the same formulations without the PCM is lower than 2 °C with respect to the temperature program. Thus, it was concluded that the PCM is able to more greatly influence a cooling stage if compared to a heating one.

According to the results reported in Table 5, the lag time of the maximum temperature in the cooling stage is 60 min; no lag time of the minimum temperature was recorded in the heating stage.

Table 5. Lag time between the maximum and minimum temperatures in autumn climate.

Autumn	Lag Time (min)	
	Cooling Stage	Heating Stage
Cement	60	0
Hydraulic Lime	60	0

The analysis of the experimental temperature curves described does not provide enough information about the energy performance of the different mortars subjected to heating and cooling cycles. Therefore, it is necessary to measure the temperature differences within each cell relative to the PCM-based mortars with respect to the reference ones (0% PCM). The thermal gradient was determined for each hour of the thermal tests and calculated by Equation (2).

$$\Delta T = T_{ref} - T_{PCM} \qquad (2)$$

where T_{ref} is the temperature of reference mortars (0% PCM) (°C); T_{PCM} is the temperature inside the cell with PCM-based mortars (°C).

Figure 8 shows the variation of the thermal gradient for each mortar composition in each season. As the temperature cycle runs, the thermal gradient increases as a result of the cyclic heat storage process. Hence, the gradient decreases until it reaches the point where the cells have the same temperature ($\Delta T = 0$). The cement-based mortars exhibit better thermal regulation with the greater difference in the temperatures observed in all the season tested. On the other hand, hydraulic lime-based mortar exhibit a smaller thermal gradient and, as a consequence, lower thermal performance.

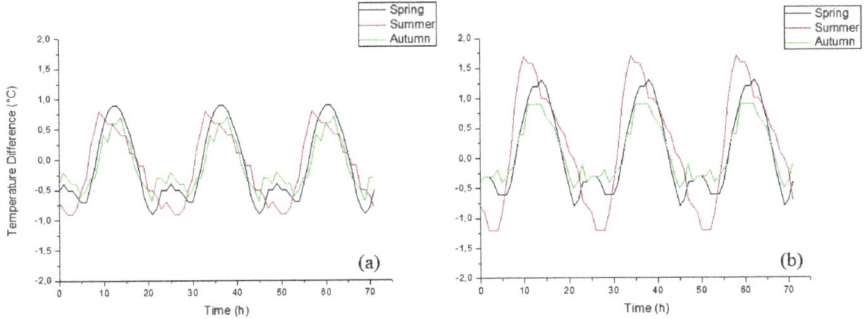

Figure 8. Thermal gradient between unmodified and PCM-based mortars: (**a**) hydraulic lime-based mortars and (**b**) cement-based mortars.

Based on the temperature curves (Figures 5–7), the reduction of energy consumption during the cooling and heating cycles was also quantified, with the aim of evaluating the energy saving actually achieved with the use of the novel PCM. The amount of energy required to maintain the temperature of the cell inside the temperature comfort range during one day was then calculated for each season. In this way, it was possible to have an estimation of the cooling and heating needs to maintain the interior temperature of a building within the comfort temperature range, i.e., between 20 °C and 25 °C. Table 6 presents the cooling and heating needs for the different mortars in the different seasons.

Table 6. Cooling and heating needs during one day.

Formulation	Cooling Needs (J/m^3)			Heating Needs (J/m^3)		
	Summer	Spring	Autumn	Summer	Spring	Autumn
HL$_{800}$LS	265,855	0	0	0	254,044	255,321
HL$_{800}$LS/PEG	265,702	0	0	0	253,572	254,947
C$_{800}$LS	266,003	0	0	0	254,017	255,266
C$_{800}$LS/PEG	265,230	0	0	0	253,636	255,051

It was observed that the incorporation of the novel PCM, through the aggregates LS/PEG, into mortars caused a slight decrease for the cooling needs in summer season. In this season, cement-based mortar shows a better thermal behavior, since lower cooling needs are required using this mortar composition.

As for the spring climate, a decrease was calculated in the heating needs for both the formulations, hydraulic lime, and cement-based mortar, containing the PCM. The differences between the two different binders appear to be insignificant. In the autumn conditions, a very small decrease was calculated in the heating needs due to the incorporation in the mortars of the LS/PEG composite. Once again, the differences between HL$_{800}$_LS/PEG and C$_{800}$_LS/PEG are not significant. From the performed thermal tests was, then, possible to confirm that the PCM incorporation in both mortars formulations reduces, even if to a low extent, the cooling and heating needs of a building located in Salento region depending on the simulated season.

In Table 7 the values of energy saving that could be achieved using both mortar formulations containing the PCM composite are reported. This latter was determined according to the difference of the energy required for cooling and heating needs, starting from the results reported in Table 6 and calculated by Equation (3):

$$\Delta NE = NE_{0PCM} - NE_{PCM} \tag{3}$$

where NE is the reduction of energy needs (J/m^3); NE$_{0PCM}$ is the energy required for mortar without PCM (J/m^3); NE$_{PCM}$ is the energy required for mortar with PCM (J/m^3).

Table 7. Energy savings per day for mortars containing PCM composite.

Formulation	Cooling Needs (J/m^3)	Heating Needs (J/m^3)	
	Summer	Spring	Autumn
Hydraulic Lime	152.7	472.0	374.8
Cement	772.9	381.1	214.8

In summer, when only refreshment is required, the cement-based mortar displays a better thermal behavior, confirmed by the much greater amount of the energy saving. In spring, on the other hand, the energy saving for the hydraulic lime-based mortar is almost 20% greater than that calculated for the cement-based mortar. The advantage of hydraulic lime-based mortar is even more appreciable during the autumn season, with a difference of 40% in energy saving compared to cement-based mortar. This behavior could be ascribed to the greater PEG content and porosity of the hydraulic lime mortar formulation (Figures 1 and 2). Thus, it can be concluded that the presence of a greater amount of macro pores improves the temperature regulation effect.

4. Conclusions

In this work, the influence of a composite phase change material system included into two mortar compositions, based on cement and on hydraulic lime, on thermal properties of the same mortars was evaluated. The PCM represents a novel sustainable composite system composed of eco-friendly PEG included, through the "form-stable method", in waste natural stone flasks, i.e., Lecce stone. This system, prepared with a low-cost equipment and low-cost raw materials, is characterized by other favorable properties, such as low flammability and low toxicity that constitute advantageous features for applications in the building sector.

The microscopic analysis of the microstructure of each mortar revealed a good connection between the different components of the mortars, with no voids or cracks in a homogeneous structure. From the same analysis it was also possible to observe that the incorporation of LS/PEG led to an increase in the porosity in the tested binders in comparison with the reference unmodified mortars. This observation was explained by the higher amount of water required by the mortars containing the PCM, with the hydraulic lime-based mortar containing LS/PEG showing the greatest porosity.

The results obtained in the thermal tests, performed with a small-scale test cells, simulating the temperatures typically experienced in the Salento region (South Italy) in the four seasons, have proven that the incorporation of the PCM under analysis in the mortars leads to changes in their thermal properties. In particular, during the summer season, a decrease in the cooling needs was measured, while during spring and autumn, a decrease in the heating needs can be achieved upon the addition of the PCM in both mortars. The best thermal performance was achieved in the summer season by the cement-based mortar containing the LS/PEG, testified by the highest thermal gradient, the maximum temperature reduction, the minimum temperature increase and time delay. In spring and autumn, the best thermal performance was displayed by the hydraulic lime-based mortar containing the LS/PEG.

Author Contributions: Conceptualization, M.F. and J.L.B.D.A.; experimental work, A.S. and M.L.; formal analysis, A.S., S.C. and M.L.; data curation, A.S. and S.C.; writing—original draft preparation, A.S.; writing—review and editing, M.F. and J.L.B.D.A.; supervision, M.F. and J.L.B.D.A.; project administration, M.F. All authors have read and agreed to the published version of the manuscript.

Funding: This research received no external funding.

Conflicts of Interest: The authors declare no conflict of interest.

References

1. Dean, B.; Dulac, J.; Petrichenko, K.; Graham, P. *Towards Zero-Emission Efficient and Resilient Buildings*; UN Environment Programme: Nairobi, Kenya, 2016.
2. Bhamare, D.K.; Rathod, M.K.; Banerjee, J. Passive cooling techniques for building and their applicability in different climatic zones—The state of art. *Energy Build.* **2019**, *198*, 467–490. [CrossRef]
3. Rao, V.V.; Parameshwaran, R.; Ram, V.V. PCM-mortar based construction materials for energy efficient buildings: A review on research trends. *Energy Build.* **2018**, *158*, 95–122. [CrossRef]
4. Lu, S.; Li, Y.; Kong, X.; Pang, B.; Chen, Y.; Zheng, S.; Sun, L. A review of PCM energy storage technology used in buildings for the global warming solution. In *Energy Solutions to Combat Global Warming*; Zhang, X., Dincer, I., Eds.; Springer International Publishing: Cham, Switzerland, 2017; Volume 33, pp. 611–644, ISBN 978-3-319-26948-1.
5. Du, K.; Calautit, J.; Wang, Z.; Wu, Y.; Liu, H. A review of the applications of phase change materials in cooling, heating and power generation in different temperature ranges. *Appl. Energy* **2018**, *220*, 242–273. [CrossRef]
6. Song, M.; Niu, F.; Mao, N.; Hu, Y.; Deng, S. Review on building energy performance improvement using phase change materials. *Energy Build.* **2018**, *158*, 776–793. [CrossRef]

7. Soares, N.; Costa, J.J.; Gaspar, A.R.; Santos, P. Review of passive PCM latent heat thermal energy storage systems towards buildings' energy efficiency. *Energy Build.* **2013**, *59*, 82–103. [CrossRef]
8. Kuznik, F.; Virgone, J.; Johannes, K. In-situ study of thermal comfort enhancement in a renovated building equipped with phase change material wallboard. *Renew. Energy* **2011**, *36*, 1458–1462. [CrossRef]
9. Kośny, J.; Biswas, K.; Miller, W.; Kriner, S. Field thermal performance of naturally ventilated solar roof with PCM heat sink. *Sol. Energy* **2012**, *86*, 2504–2514. [CrossRef]
10. Lu, S.; Xu, B.; Tang, X. Experimental study on double pipe PCM floor heating system under different operation strategies. *Renew. Energy* **2020**, *145*, 1280–1291. [CrossRef]
11. Kheradmand, M.; Azenha, M.; de Aguiar, J.L.B.; Krakowiak, K.J. Thermal behavior of cement based plastering mortar containing hybrid microencapsulated phase change materials. *Energy Build.* **2014**, *84*, 526–536. [CrossRef]
12. Cunha, S.; Aguiar, J.; Ferreira, V.; Tadeu, A. Mortars based in different binders with incorporation of phase-change materials: Physical and mechanical properties. *Eur. J. Environ. Civ. Eng.* **2015**, *19*, 1216–1233. [CrossRef]
13. Kalnæs, S.E.; Jelle, B.P. Phase change materials and products for building applications: A state-of-the-art review and future research opportunities. *Energy Build.* **2015**, *94*, 150–176. [CrossRef]
14. Huang, X.; Alva, G.; Jia, Y.; Fang, G. Morphological characterization and applications of phase change materials in thermal energy storage: A review. *Renew. Sustain. Energy Rev.* **2017**, *72*, 128–145. [CrossRef]
15. Lecompte, T.; Le Bideau, P.; Glouannec, P.; Nortershauser, D.; Le Masson, S. Mechanical and thermo-physical behaviour of concretes and mortars containing phase change material. *Energy Build.* **2015**, *94*, 52–60. [CrossRef]
16. Franquet, E.; Gibout, S.; Tittelein, P.; Zalewski, L.; Dumas, J.-P. Experimental and theoretical analysis of a cement mortar containing microencapsulated PCM. *Appl. Therm. Eng.* **2014**, *73*, 32–40. [CrossRef]
17. Frigione, M.; Lettieri, M.; Sarcinella, A. Phase change materials for energy efficiency in buildings and their use in mortars. *Materials* **2019**, *12*, 1260. [CrossRef]
18. da Cunha, S.R.L.; de Aguiar, J.L.B. Phase change materials and energy efficiency of buildings: A review of knowledge. *J. Energy Storage* **2020**, *27*, 101083. [CrossRef]
19. Akeiber, H.; Nejat, P.; Majid, M.Z.; Wahid, M.A.; Jomehzadeh, F.; Zeynali Famileh, I.; Calautit, J.K.; Hughes, B.R.; Zaki, S.A. A review on phase change material (PCM) for sustainable passive cooling in building envelopes. *Renew. Sustain. Energy Rev.* **2016**, *60*, 1470–1497. [CrossRef]
20. Zhang, P.; Xiao, X.; Ma, Z.W. A review of the composite phase change materials: Fabrication, characterization, mathematical modeling and application to performance enhancement. *Appl. Energy* **2016**, *165*, 472–510. [CrossRef]
21. Li, M.; Shi, J. Review on micropore grade inorganic porous medium based form stable composite phase change materials: Preparation, performance improvement and effects on the properties of cement mortar. *Constr. Build. Mater.* **2019**, *194*, 287–310. [CrossRef]
22. Fallahi, A.; Guldentops, G.; Tao, M.; Granados-Focil, S.; Van Dessel, S. Review on solid-solid phase change materials for thermal energy storage: Molecular structure and thermal properties. *Appl. Therm. Eng.* **2017**, *127*, 1427–1441. [CrossRef]
23. Lv, P.; Liu, C.; Rao, Z. Review on clay mineral-based form-stable phase change materials: Preparation, characterization and applications. *Renew. Sustain. Energy Rev.* **2017**, *68*, 707–726. [CrossRef]
24. Liu, Y.; Xu, E.; Xie, M.; Gao, X.; Yang, Y.; Deng, H. Use of calcium silicate-coated paraffin/expanded perlite materials to improve the thermal performance of cement mortar. *Constr. Build. Mater.* **2018**, *189*, 218–226. [CrossRef]
25. Wan, X.; Chen, C.; Tian, S.; Guo, B. Thermal characterization of net-like and form-stable ML/SiO$_2$ composite as novel PCM for cold energy storage. *J. Energy Storage* **2020**, *28*, 101276. [CrossRef]
26. Costa, J.A.C.; Martinelli, A.E.; do Nascimento, R.M.; Mendes, A.M. Microstructural design and thermal characterization of composite diatomite-vermiculite paraffin-based form-stable PCM for cementitious mortars. *Constr. Build. Mater.* **2020**, *232*, 117167. [CrossRef]
27. Karaman, S.; Karaipekli, A.; Sarı, A.; Biçer, A. Polyethylene glycol (PEG)/diatomite composite as a novel form-stable phase change material for thermal energy storage. *Sol. Energy Mater. Sol. Cells* **2011**, *95*, 1647–1653. [CrossRef]

28. Kou, Y.; Wang, S.; Luo, J.; Sun, K.; Zhang, J.; Tan, Z.; Shi, Q. Thermal analysis and heat capacity study of polyethylene glycol (PEG) phase change materials for thermal energy storage applications. *J. Chem. Thermodyn.* **2019**, *128*, 259–274. [CrossRef]
29. Frigione, M.; Lettieri, M.; Sarcinella, A.; de Aguiar, J.B. Mortars with Phase Change Materials (PCM) and stone waste to improve energy efficiency in buildings. In *International Congress on Polymers in Concrete (ICPIC 2018)*; Taha, M.M.R., Ed.; Springer International Publishing: Cham, Switzerland, 2018; pp. 195–201, ISBN 978-3-319-78174-7.
30. Frigione, M.; Lettieri, M.; Sarcinella, A.; de Aguiar, J.B. Sustainable polymer-based phase change materials for energy efficiency in buildings and their application in aerial lime mortars. *Constr. Build. Mater.* **2020**, *231*, 117149. [CrossRef]
31. Frigione, M.; Lettieri, M.; Sarcinella, A.; de Aguiar, J.L.B. Applications of sustainable polymer-based phase change materials in mortars composed by different binders. *Materials* **2019**, *12*, 3502. [CrossRef]
32. Andriani, G.F.; Walsh, N. Petrophysical and mechanical properties of soft and porous building rocks used in Apulian monuments (south Italy). *Geol. Soc. Lond. Spec. Publ.* **2010**, *333*, 129. [CrossRef]
33. CEN. *Specification for Mortar for Masonry—Part. 1: Rendering and Plastering Mortar*; EN 998–1; CEN: Brussels, Belgium, 2010.
34. Cunha, S.; Aguiar, J.B.; Tadeu, A. Thermal performance and cost analysis of mortars made with PCM and different binders. *Constr. Build. Mater.* **2016**, *122*, 637–648. [CrossRef]
35. Kheradmand, M.; Aguiar, J.; Azenha, M. *Assessment of the Thermal Performance of Plastering Mortars within Controlled Test. Cells*; University of Minho: Guimaraes, Portugal, 2014.

© 2020 by the authors. Licensee MDPI, Basel, Switzerland. This article is an open access article distributed under the terms and conditions of the Creative Commons Attribution (CC BY) license (http://creativecommons.org/licenses/by/4.0/).

Article

Analytic Hierarchy Process-Based Construction Material Selection for Performance Improvement of Building Construction: The Case of a Concrete System Form

Dongmin Lee [1], Dongyoun Lee [1], Myungdo Lee [2], Minju Kim [1] and Taehoon Kim [1,*]

[1] School of Civil, Environmental and Architectural Engineering, Korea University, 145, Anam-ro, Seongbuk-gu, Seoul 02841, Korea; ldm1230@korea.ac.kr (D.L.); dy_lee@korea.ac.kr (D.L.); Minju830@korea.ac.kr (M.K.)
[2] Research and Development Center, Yunwoo Technology Co. Ltd., 128, Beobwon-ro, Songpa-gu, Seoul 058054, Korea; md.lee@yunwoo.co.kr
* Correspondence: kimth0930@korea.ac.kr; Tel.: +82-2-921-5920; Fax: +82-2-923-4229

Received: 11 March 2020; Accepted: 7 April 2020; Published: 8 April 2020

Abstract: Selecting the best materials that ensure maximum performance is crucial in the construction engineering design of any construction project. However, this is challenging and usually not properly considered because of the lack of systematic and scientific evaluation methods for the performance of materials. This paper proposes a new approach of selecting material to satisfy the performance goal of material designers in building constructions based on the analytic hierarchy process method. To validate the suggested model, a case study was conducted for a concrete system form, the performance of which is susceptible to its materials and has a strong effect on overall project productivity. The newly developed form comprising polymers and alloys showed that the proposed material selection model provided a better combination of materials, and the solution was technically more advanced and ensured better performance. This paper contributes to the body of knowledge by expanding the understanding of how construction material properties affect project performance and provides a guideline for material engineers to select the best-performing building materials while considering a performance goal.

Keywords: material selection; project performance; material property; analytic hierarchy process (AHP); building construction; concrete system form

1. Introduction

Material selection is one of the most important yet complex tasks encountered by construction engineers, because it is directly related to overall project performance (e.g., time, cost, and quality) [1]. Construction engineers must select the best-performing materials based on the mechanical (e.g., specific strength and elasticity modulus), functional (e.g., noise reduction, corrosion resistance, and nonadhesiveness), and physical (e.g., density, color, and thermal conductivity [TC]) properties of the materials in the selection process in association with cost [2,3]. These evaluation criteria are often in conflict with each other, because an optimal selection for one criterion could sacrifice other criteria [2]. Therefore, construction material selection should be conducted through a systematic decision-making process, investigating how each criterion has an impact on project performance.

Despite the importance of material selection in construction projects, in practice, it mainly has been conducted with a heuristic approach based on personal experience because of the lack of a systematic evaluation model and an ambiguous measuring criterion for considering the potential performance of materials. Previously, some research has provided building material selection models using various methodologies, such as the ranking and scoring method [4], analytical network process

(ANP) and technique for order of preference by similarity to ideal solution (TOPSIS) [5], quality function deployment (QFD) [6], and fuzzy-extended analytic hierarchy process (AHP) [7]. Those models provide useful guidance for the multicriteria decision-making (MCDM) of construction materials. However, previous research on material selection in the construction industry has mainly focused on the life cycle cost and sustainability of materials, especially for completed buildings. In addition, the characteristics of construction material and how they affect the potential performance of products have not been studied thoroughly.

To fill the gap in existing research, this paper aims to develop a general material selection model to help researchers and practitioners select optimal materials for construction products or facilities in terms of the performance goal. The performance goal can be defined by a material designer considering the user's requirements. A systematic procedure through translating, screening, and rating processes provides a guideline to approach the performance goal in a systematic and scientific manner. The AHP method is a key technique during a rating process that quantifies all the qualitative properties of the material in terms of performance goal in building construction projects.

The aim of this study is to establish a generalized material selection model for construction engineers and practitioners for easy application. We tested the model in a case study based on selecting formwork materials such as panels and frames. Formwork plays a key role in concrete building construction because formwork cost accounts for as much as 15% of the total construction cost and approximately 25% of duration in a reinforced concrete structure [8,9]. In addition, there are many formwork materials commercially available in the market, enabling the manufacture of a real prototype and mock-up product based on the result of the case study. The remainder of the paper is arranged as follows. Section 2 explains a material selection process for performance improvement. Section 3 shows a case study on a concrete form, and we have reviewed it in Section 4. Section 5 describes the results, and Section 6 discusses the results. Finally, Section 7 concludes the study entirely.

2. Material Selection Process for Construction Materials

2.1. Material Selection Process

Material selection is one of the most significant and confusing tasks encountered by construction practitioners. In the construction materials market, there are 85,000 different commercially available polymers to choose from, at least 14 types of general-purpose engineering plastics, and at least 20 kinds of applicable alloys that are already commercialized. Available materials also may differ according to the country, market, and time, so the material engineer should consider allowable material candidates first before making a selection. The process is conducted in the order of translating, screening, and rating (Figure 1).

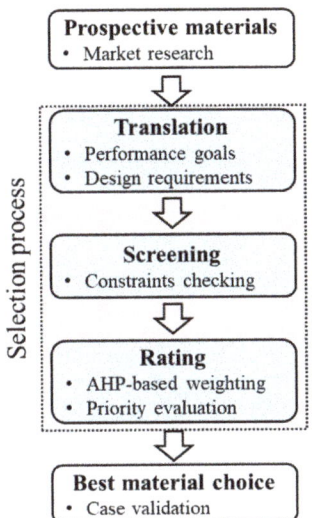

Figure 1. Material selection process for construction materials.

The first step for material selection is a translation. The material designer should define a technical performance goal (e.g., improved labor productivity and high maintainability). The performance goal should be testable and measurable like an index which is usually an accumulation of scores. The second step is a screening. Constraints (e.g., waterproof and high tensile strength) should be defined along with performance goals, and the constraints are used to eliminate ineligible candidates so we can save time for the rating process. The third step is a rating. Through this step, we evaluate what properties of the materials affect the performance goal. In this step, AHP is a key method that calculates the relative weight of material properties for maximizing the potential performance goal. AHP is a powerful yet simple method for making decisions even when important elements of the decision are difficult to quantify or compare.

2.2. Analytic Hierarchy Process (AHP)

The AHP method, conceptualized by Saaty [10], is one of the most popular MCDM methods [11]. The purpose of MCDM is to select the best alternative from a set of competitive alternatives and evaluate it with a set of criteria [12]. The AHP method can be successfully applied to analyze qualitative data quantitatively. It transforms a complex and multicriteria problem into a structured hierarchy [13]. The AHP requires minimal mathematical calculations and is the only methodology that can consider consistency in decision-making [14]. In addition, it has been applied in construction industry to select suppliers [15], construction method [16], and equipment [17]. The general AHP procedure is described in Figure 2 [10].

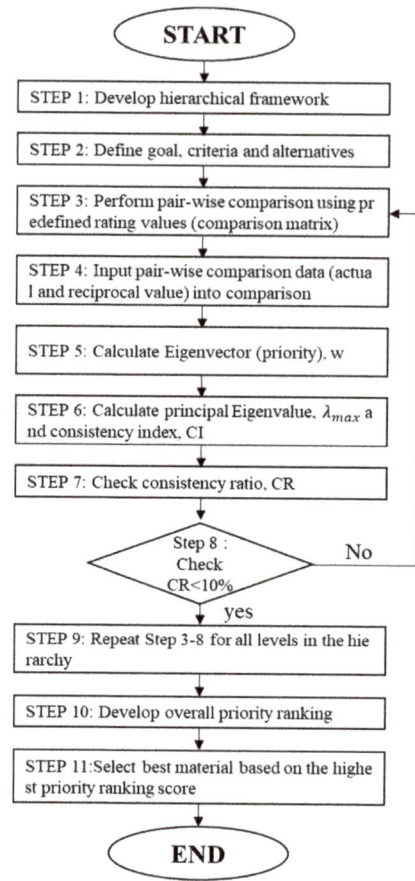

Figure 2. Analytic hierarchy process (AHP) methodology for material selection.

The hierarchy levels are structured in such a way that there is a set of alternatives at the minimal level and a general goal is placed at the top level. In between the minimal and top levels, the general criteria and sub-criteria are placed [18]. After this, logical hierarchy levels are constructed, and the decision-maker can systematically assess the alternatives based on pair-wise comparison judgments. The pair-wise comparisons are conducted using Saaty's [10] predefined scale (generally a nine-point scale [Table 1]).

Table 1. Nine-point scale suggested by Saaty [10].

Definition	Intensity of Importance
Equally important	1
Moderately more important	3
Strongly more important	5
Very strongly more important	7
Extremely important	9
Intermediate values	2, 4, 6, 8

If there are n criteria, the pair-wise comparison of criteria i and j yields an $(n \times n)$ dimension matrix A, where a_{ij} denotes the comparative importance of criterion i with respect to criterion j.

$$A = \begin{bmatrix} a_{11} & \cdots & a_{1n} \\ \vdots & \ddots & \vdots \\ a_{n1} & \cdots & a_{nn} \end{bmatrix}, \ a_{ij} = 1, \ when \ i = h \ and \ a_{ji} = \frac{1}{a_{ij}} \qquad (1)$$

After the pair-wise comparison, relative weights of criteria can be computed. The computation also includes the calculation of a normalized principal eigenvector from the given matrix A. The relative weights are derived by the eigenvector (w) corresponding to the largest eigenvalue (λ_{max}) such as:

$$Aw = \lambda_{max} \times w. \qquad (2)$$

The evaluation requires a certain level of matrix consistency and can be checked by the consistency index (CI) as follows.

$$CI = \frac{\lambda_{max} - n}{n - 1} \qquad (3)$$

If the matrix is perfectly consistent, then $CI = 0$. Consistency is an important factor in AHP, so it must be checked for each pair-wise comparison matrix at each stage.

Finally, the consistency ratio (CR) can be calculated such as:

$$CR = \frac{CI}{RI}. \qquad (4)$$

The RI is a random index that can be derived from the number of criteria n. Usually, a CR of 0.1 (10%) or less is considered acceptable (enough to be trustworthy).

3. Case Study: Material Selection for System Form

Formwork has a direct effect on the quality of the concrete surface, concrete framework cost, noise level, environmental issues, labor productivity, and even worker safety, because it is labor-intensive and physically demanding work [19]. From among the different types of forms, in this paper, a system form was selected for case verification because it is the most widely used form in mid- or high-rise building construction around the world. The term system form denotes a standard prefabricated form unit that normally consists of a panel and inner and outer frames. The system form can secure a high-quality concrete surface and high productivity with more recycle times than conventional concrete form (e.g., hand-set wood form), so the material selection of the system form can be more critical for the improved performance of the total construction project. The scope of this case is to select materials for the panel, and inner and outer frames.

3.1. Prospective Materials

In the formwork materials market in South Korea, steel, aluminum, wood, magnesium, and titanium alloys are commercialized, and 13 engineering plastics (polymers) are available. Thus, the prospective materials in this case study were limited to these 18 materials. Each material has different characteristics in terms of mechanical, functional, and physical properties, and each of the properties has three different criteria (Table 2). These criteria may not be limited to formwork material selection, and they can broadly be applied to any other construction material selection. Allowable formwork material properties according to the performance criteria is provided in Table 2.

Table 2. Criteria for material selection of concrete form.

Criterion	Properties	Explanation
Flexural strength (FS)	Mechanical	Because the formwork must withstand the loads of concrete, the minimum FS required by the design must be secured. Generally, when the concrete, live, and dead loads are applied, the minimum FS value is set so that it is below the allowable deflection.
Flexural modulus (FM)	Mechanical	The higher the FM, the better the quality of the concrete surface, as the deflection of the form decreases as the concrete is poured.
Impact resistance (IR)	Mechanical	Strength to withstand breakage of the form when dismantling and dropping of formwork. It should secure sufficient strength to prevent breakage.
Weather resistance (WR)	Functional	It should not be deformed or corroded by weather such as ultraviolet rays, snow, or rain.
Alkali resistance (AR)	Functional	Because the concrete exhibits strong alkaline properties, the material in contact with the concrete must have AR.
Noise generation (NG)	Functional	Noise created during the installation, disassembly, and dropping of the formwork causes psychological damage to the operator and the site, so it is necessary to use a material with low noise.
Density (DE)	Physical	During installation and disassembly, forms are carried by the workers, so the lowest density material should be used to reduce the weight as much as possible. Reducing the weight of the form not only increases the productivity and constructability of formwork, but also reduces the incidence of work accidents.
Water absorption (WA)	Physical	Forms should be made of a material that absorbs as little moisture as possible because they are continuously exposed to a wet environment and affected by rain.
Thermal conductivity (TC)	Physical	To achieve uniform quality in the curing process during hot and cold weather, materials that come into contact with concrete should have low TC. In addition, materials with low TC are particularly important when not using a release agent because they have an advantage for making relatively smooth surfaces.

3.2. Translation

The starting point of material selection for a system form is to identify the performance goal. There are many different performance criteria for formwork, but in this study, the authors set the goal with respect to maximizing user requirements while satisfying the technical requirements of formwork.

3.2.1. User Requirements (Performance Goal)

There is insufficient research on how to improve overall formwork performance [19]. This paper tried to test a form after constructing it in such a way that it satisfied the requirements of the users (i.e., workers and engineers) who use it in practice on a construction site. To derive the user requirements, interviews were first conducted with two supervisors of two high-rise building construction projects, four heads of formwork companies, and six experts in system formwork on-site. A detailed interview process is shown in the author's previous work [19,20]. The results of these interviews are summarized in Table 3. In Table 3, the importance index is a value obtained by dividing the importance value by the current performance value by examining the users' requirements with importance and performance. Therefore, as the number increases, the more important or urgent the issue to be improved is for increasing formwork performance.

Table 3. User requirements for form.

No.	Category	User Requirements	Importance Index	Rank
1	Constructability	Easy assembly and disassembly (can be fit and fastened together with reasonable ease)	91.3	1
2		Low noise during dismantlement or assembly and disassembly	87.4	2
3		Easy separation from concrete	76.2	8
4		Efficient lifting and carrying	79.6	6
5	Safety	Not distorted or deflected during concrete casting	56.3	15
6		Reduced work accidents (struck by object)	57.7	13
7	Durability	High repeat use with constant module size	83.2	4
8		Recyclable material usage	61.4	12
9		Durable against falling and external impacts	64.2	10
10		Easy maintenance and cleaning	69.9	9
11	Reliability	Low TC (low temperature sensitivity)	59.1	14
12		High concrete surface quality	78.8	7
13	Conformance	Compatible (size, height, fixing method) with existing formwork units (e.g., Euro form, aluminum form, and Skydeck)	86.7	3
14		Hybrid (concurrent usage) usage for vertical (wall and column) and horizontal (slab) forms	63.1	11
15		Provides various module sizes to minimize on-site work (filler and conventional formwork)	81.9	5

3.2.2. Function and Components

Apart from the performance goal presented, a form must be designed to meet the required technical performance level as a product. A system form can be defined as a temporary structure that helps to hold the fluid concrete in place until it hardens and acquires a particular shape [21]. The system form is dismantled after the finishing formwork, but its technical performance strongly affects the subsequent tasks with respect to cost and duration. Therefore, the system form should be well designed not only to create the rigid, strong conditions required during concrete casting to avoid loss of concrete or collapse, but also to secure high performance (e.g., high constructability and a smooth concrete surface).

The general system form is divided into an inner frame (Figure 3c), an outer frame (Figure 3a), and a panel (Figure 3b). The panel is in contact with the concrete and transfers the load (e.g., live load and wind load) to the inner frame. The inner frame receives the load from the panel, transfers it to the outer frame, and is installed at regular intervals to prevent deformation of the panel. The outer frame receives the load from the inner frame and delivers it to the shore or beam. Two or more different materials can be applied as composite (heterogeneous) materials in one concrete form because the technical requirements differ for each part in a system form (Table 4). For example, aluminum inner frames and steel outer frames with a wooden panel can be combined in a system form.

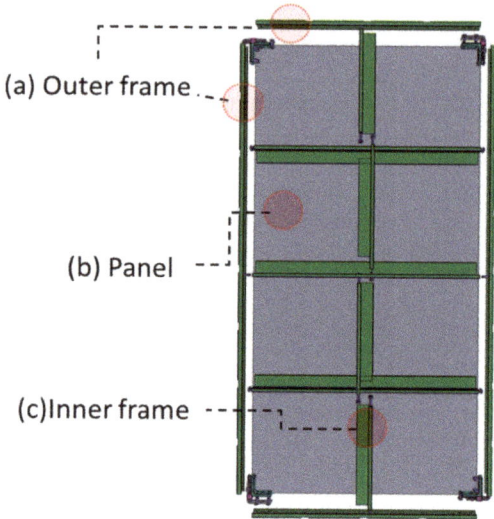

Figure 3. Basic concrete system form module used in South Korea. An outer frame (**a**), a panel (**b**), an inner frame (**c**).

Table 4. Technical requirements for form materials.

Factors for Consideration	Inner Frame	Outer Frame	Panel
Mechanical consideration	High FS [a] and FM [b]	High FS and FM, high IR [c]	High FM, high IR
Functional consideration	Reduced noise, WR [d]	Reduced noise, continuous use temperature, AR [e], WR	Reduced noise, low TC [f], easy to nail, easy to change, easy to strip off concrete, AR
Physical consideration	Lightweight, low WA [g]	Lightweight, low WA	Lightweight, low WA

[a] Flexural Strength, [b] Flexural Modulus, [c] Impact Resistance, [d] Weather Resistance, [e] Alkali Resistance, [f] Thermal Conductivity, [g] Water Absorbtion.

3.3. Screening

Constraints

A concrete form should endure concrete pressure, live load, and dead load; thus, during concrete casting, there are constraints in mechanical properties to avoid accidents caused by deflection. In addition, from previous research by the author [19,20], there are functional and physical requirements as well for a concrete form. If a material does not satisfy the requirements for each component (i.e., outer frame, inner frame, and panel), it is excluded from the list of alternative materials. Table 5 shows the constrains and possible materials for each form part. In this table, the moment of inertia (I) is assumed to be 6.0357 for the inner frame and 27.77 for the outer frame, considering the size of the normal Euro form, which is one of the most widely used system forms (600 mm × 1200 mm) in Europe and Asia. However, the required specifications differ between the manufacturers, countries, and times, so the material engineer should make constraints considering the potential users, purpose of the product, and performance goal.

Table 5. Possible materials for each form part.

Properties	Inner Frame	Outer Frame	Panel
Functional requirements (strong constraints)	NG [a] ≥ B WR [b] ≥ B	NG ≥ B WR ≥ A AR [c] ≥ A	NG ≥ A Easy to nail
Mechanical requirements (strong constraints)	FS [d] > 75 Mpa for the wall, 45 Mpa for the slab FM [e] > 167,751 kgf/cm^2 for wall, 95,059 kgf/cm^2 for slab If I is 6.0357	FS > 64 Mpa for the wall, 42 Mpa for the slab IR [f] ≥ A FM > 218,761 kgf/cm^2 for wall, 123,965 kgf/cm^2 for slab If I is 27.77	FS > 64 Mpa for the wall, 42 Mpa for the slab IR ≥ A FM > 73,242 kgf/cm^2 for wall, 41,504 kgf/cm^2 for slab
Physical requirements (weak constraints)	WA [g]	DE, WA	DE, WA, low TC [h]
Possible materials (weak constraints)	Aluminum, steel, plastic, alloys	Plastic, alloys	Plywood, plastic

[a] Noise Generation, [b] Water Resistance, [c] Alkali Resistance, [d] Flexural Strength, [e] Flexural Modulus, [f] Impact Resistance, [g] Water Absorption, [h] Thermal Conductivity.

3.4. Rating

In the rating procedure, the priority ranking is quantitatively calculated to improve user requirements according to the goal, criteria, and alternatives.

3.4.1. Technical Performance Judgment for Each Alternative Material

There are four qualitative features in the evaluation criteria: impact resistance (IR), noise generation (NG), weather resistance (WR), and alkali resistance (AR). These features can only be evaluated relatively according to the performance goals. For example, in urban city projects, NG is very important because there are many residents near the construction site. In this situation, the designer may evaluate NG as a critical factor for performance. For this reason, these features should be evaluated by material designers or users according to their needs. In contrast, there are five quantitative features: flexural strength (FS), FM, density (DE), water absorption (WA), and TC. These features can be evaluated quantitatively according to the technical requirements of the project. In this study, 12 formwork experts simply evaluated them in the following manner: 7, 5, 3, and 1 point are given for excellent (S), good (A), fair (B), and poor (C) performance for qualitative features. In the case of quantitative features, the values of the properties were divided into four sections in the order of the highest values, and 7, 5, 3, and 1 are allocated, respectively. Based on this evaluation process, candidate materials in Table 6 were evaluated. The total summation of the scores from these four features is defined as the performance score (PS). The PS for nine quantitative and qualitative features are provided in Table 7.

Table 6. Allowable formwork material properties according to the performance criteria.

Materials	Flexural Strength (FS) (MPa)	Flexural Modulus (FM) (GPa)	Impact Resistance (IR)	Weather Resistance (WR)	Alkali Resistance (AR)	Noise Generation (NG)	Density (DE) (kg/m²)	Water Absorption (WA) (%)	Thermal Conductivity (TC) (W/(m·K))	Price (€/kg)
Steel	400	210	S	B	A	B	7850	-	45	0.45
Aluminum	386	70	S	A	C	C	2712	-	205	1.95
Wood (oak)	60	11	C	C	C	S	650	>8	0.16	0.35–0.9
Magnesium alloys	150	45	S	B	A	B	1738	-	165	5–15
Titanium alloys	800	110	S	B	A	B	4500	-	15	15–20
CFRP[1]	900	89	S	A	A	A	1550	-	0.5–3.0	13–22
ABS[2]	75–128	2.5–8	S	C	A	A	1070	0.3	0.1	2.05
Acetal (POM)[3]	85	2.5–11	A	C	A	A	1410	0.25	0.22	0.7
PVC[4]	35	3.1–8	C	A	A	A	1470	0.06	0.19	0.95
Nylon 6 (PA6)[5]	85–405	2.4–20	A	A	A	A	1130	1.2	0.25	1.8
PA66[5]	103–420	3.1–18	A	B	A	A	1183	1.2	0.26	1.92
Polyimide	175	5–32	A	S	A	A	1420	0.2	0.11	3.5
Polycarbonate	90–138	2.3–4.4	S	A	C	A	1200	0.15	0.20	2.8
Polyethylene	40	0.7–6	S	B	A	A	970	0.01	0.11	0.8
PET[6]	80	1	A	A	A	A	1380	0.1	0.15	1.4
PBT[7]	79–270	2.6–13	A	A	A	A	1310	0.08	0.29	2.95
Polypropylene	40–190	1.5–8	A	A	S	A	946	Slight	0.12	0.7
Polystyrene	70	2.5–13	A	B	A	A	1040	-	0.11	0.76

S: Excellent; A: Good; B: Fair; C: Poor. CFRP[1]: carbon fiber reinforced plastic; ABS[2]: acrylonitrile butadiene styrene; POM[3]: polyoxymethylene; PVC[4]: polyvinyl chloride; PA6, PA66[5]: polyamides 6, 66; PET[6]: polyethylene terephthalate; PBT[7]: polybutylene terephthalate.

Table 7. Scoring for priority index (PI) using material properties and relative weights of performances.

Materials	FS	FM	IR	WR	AR	NG	DE	WA	TC	PI for the Outer Frame	PI for the Inner Frame	PI for Panel
Steel	7	7	7	3	5	3	1	7	1	970.9	840.4	1072.4
Aluminum	7	5	7	5	5	1	1	7	1	209.1	154.6	241.2
Wood (oak)	1	5	5	1	1	7	1	1	5	563.9	553.9	552.3
Magnesium alloys	5	7	7	3	5	3	1	7	1	42.5	37.0	45.0
Titanium alloys	7	7	7	3	5	3	1	7	1	25.0	21.6	27.6
CFRP[1]	7	7	7	5	5	5	3	7	3	30.5	29.1	32.2
ABS[2]	3	3	7	1	5	5	7	3	7	243.4	196.1	206.3
Acetal (POM)[3]	3	3	5	1	5	5	3	3	5	515.0	504.6	511.1
PVC[4]	1	3	1	5	5	5	3	5	5	354.6	461.5	390.9
Nylon 6 (PA6)[5]	7	5	5	5	5	5	7	1	3	280.8	284.3	239.6
PA66[5]	7	5	5	3	5	5	5	3	3	239.3	231.6	230.4
Polyimide	5	7	7	7	5	5	3	3	7	137.3	155.4	159.8
Polycarbonate	3	5	7	5	1	5	5	3	5	149.3	143.8	113.0
PE[6]	1	1	7	3	5	5	7	5	7	640.9	558.0	494.8
PET[7]	3	1	5	5	5	5	5	5	7	330.5	343.9	305.7
PBT[8]	5	1	5	5	5	5	5	5	3	166.2	168.2	154.8
PP[9]	5	5	5	5	7	5	7	1	7	795.9	784.9	722.3
Polystyrene	1	3	5	3	5	5	7	1	7	591.2	570.3	479.7

CFRP[1]: carbon fiber reinforced plastic; ABS[2]: acrylonitrile butadiene styrene; POM[3]: polyoxymethylene; PVC[4]: polyvinyl chloride; PA6, PA66[5]: polyamides 6, 66; PE[6]: polyethylene; PET[7]: polyethylene terephthalate; PBT[8]: polybutylene terephthalate; PP[9]: polypropylene;

- Disqualified for outer frame
- Disqualified for inner frame
- Disqualified for panel
- Highest priority

The required characteristics may differ according to the form parts (i.e., inner frame, outer frame, and panel). For example, the noise generated during installation and dismantlement is crucial for the outer frame, but not for the inner frame, because it is not hit when dropping. AR is also more important for the outer frame and panel than for the inner frame because only a small amount of concrete sticks to the inner frame compared with the outer frame and panel. WR is crucial for the inner and outer frames because they are permanent-use parts exposed to outdoor conditions with wind and rain, but not for the panel because the latter is a disposable item that is replaced periodically.

3.4.2. Material Selection Methodology Using AHP

Generally, AHP consists of three main principles, including the hierarchy framework, priority analysis, and consistency verification [10]. The first stage in applying the AHP method for material selection is to develop an AHP hierarchical framework that shows a systematic overview of goals, criteria, sub-criteria, and alternatives. The hierarchical framework looks like a tree from level-1, which represents the goal of selection, to level-2, which represents the criteria or factors that affect the goal, and level-3, which consists of the components of each criterion at level-2; their weights are quantitatively calculated to select optimal materials at level-4 (alternatives) (Figure 4). We referred the criteria and sub-criteria from Ashby's [22] model that is one of the most widely used material design models.

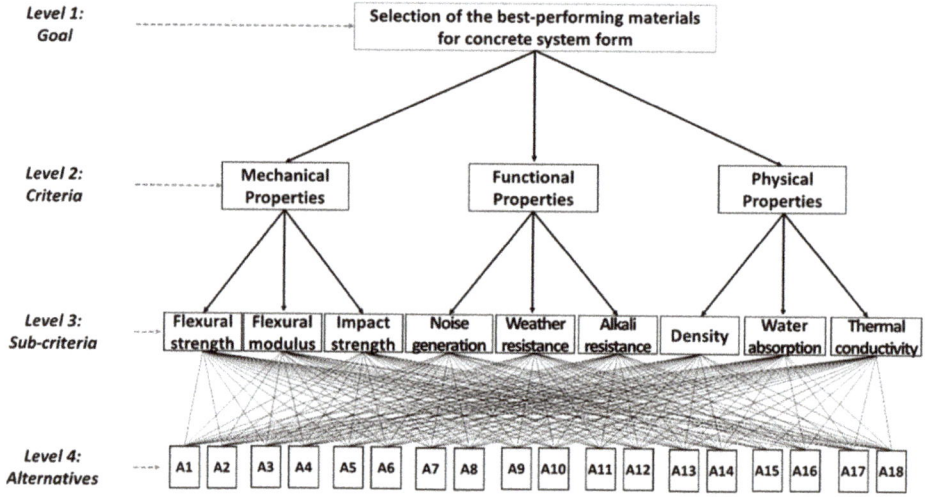

Figure 4. Hierarchical diagram for material selection criteria for concrete form.

At AHP hierarchy level-1, the goal of the project (the case study) is defined, and it is to select the most suitable materials hybridized with different materials for each component in the concrete form, outer frame, inner frame, and panel. Afterward, the structure is expanded to level-2, where the main criteria (mechanical, functional, and physical properties of the technical requirements) are represented. These are divided into several sub-criteria (FS, FM, IR, WR, AR, NG, DE, WA, and TC). Once a hierarchy framework has been constructed, users (i.e., workers and engineers) are requested to participate in a survey for a pair-wise comparison matrix at each hierarchy. In the priority analysis stage, each comparison matrix is calculated by an eigenvector to determine the weight of each criterion and the performance of alternatives [23]. The final stage is to calculate a CR to measure the consistency of the judgments in the survey. This is a comparison between the CI and random consistency index (RI). AHP allows assessment inconsistencies but they should not exceed 10%.

The AHP questionnaire was a pair-wise comparison of the questionnaires for each component from a total of 10 formwork experts (five engineers, five users, with 18 years of experience on average). For the outer frame, the λ value was 9.491 and the CR value was 4.2%. For the inner frame, the λ value was 10.084 and the CR value was 9.4%. For the panel, the λ value was 10.263 and the CR value was 9.1%. Figure 5 shows the overall relative weights for each part of the concrete form for nine sub-criteria. This relative weight, which is the priority value (w) of each material, can be calculated to select the optimum material for each constituent member of concrete form in terms of the performance goal.

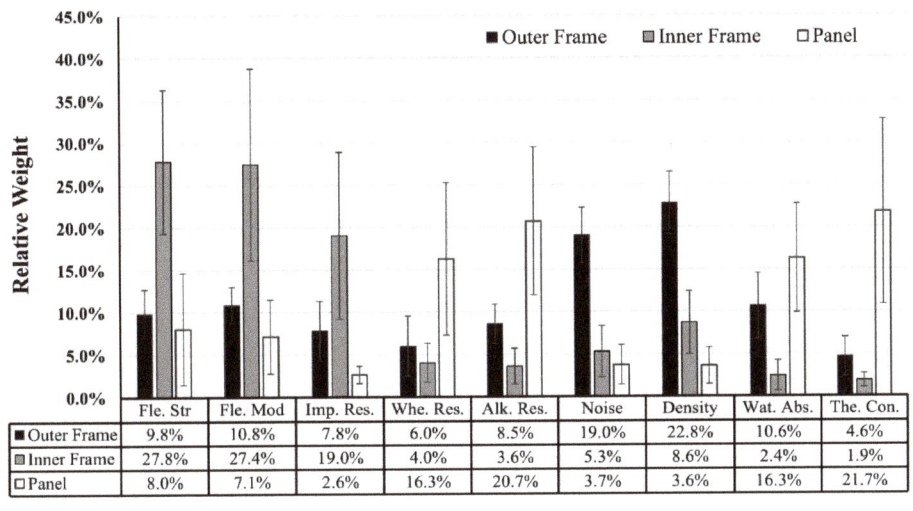

Figure 5. Overall relative weights of concrete form materials.

3.4.3. Best Materials Choice

The priority index (PI), which is defined as the priority value (w) × PS divided by cost, is newly suggested for quantitative comparison among materials. The PI quantifies the effect of material properties on a user's performance goal. In other words, a material that has a high PI should be selected to optimize the performance of materials. The PI can be designed and defined differently depending on the design goals. Table 7 shows the calculated PI for the outer frame, inner frame, and panel. In addition, the disqualified and highest score alternatives are selected.

4. Development of Composite System Form

After the material selection process, the detailed configuration of a new composite system form (CSF) was drawn (Figure 6). The highest PI materials for the outer frame, inner frame, and panel were polyamide 6 (PA6), steel, and polypropylene (PP), respectively. The connecting bracket between the members was made of PA6, and the rivets and bolts were made of steel. In the case of the panel, a PP sandwich panel was used, but for the low FM value of PP and the ease of nailing, a second prior material, plywood coated with PP or high-pressure laminate (HPL) film was also tested. The design of the inner frame, outer frame, and panel is described in more detail in the following section.

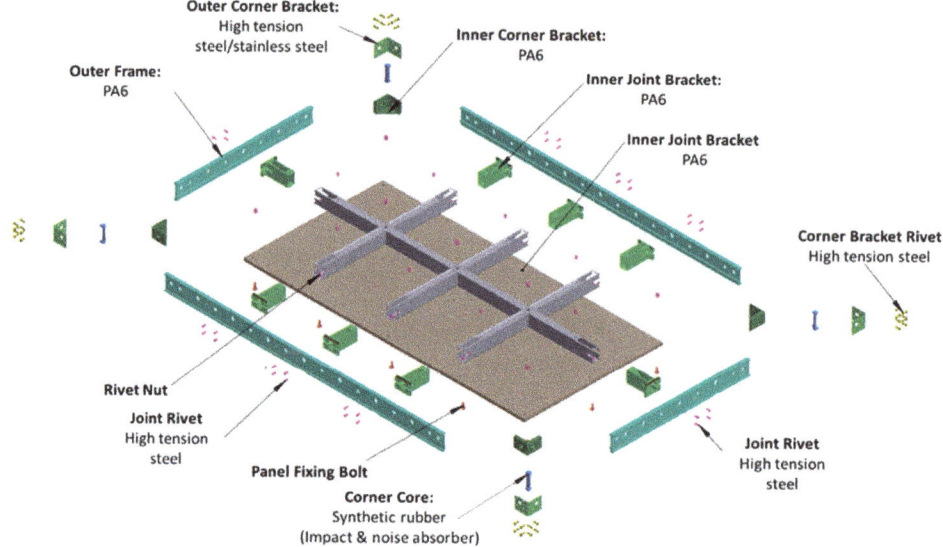

Figure 6. Detailed configuration of the composite system form (CSF).

4.1. Material Selection for Each Part of the CSF

4.1.1. Outer Frame

The outer frame is in direct contact with the concrete and is exposed to the weather and external impacts. Therefore, it should be alkali and weather resistant with low adhesion and high IR to withstand impact during installation and dismantlement. There are two alternative outer frame materials, namely, plastics and alloys. Given that there are various kinds of plastics, among the materials satisfying the mechanical requirements in Tables 6 and 7, PA6 has the highest PI value. In this study, PA6 was selected as the outer frame material considering the PI value, but it may be changed at any point based on the material cost. PA6 does not shrink, is resistant to ultraviolet (UV) radiation, and is insensitive to knocks and scrapes. In addition, even if some concrete sticks to the frame, it is easy to clean because it does not react with concrete.

4.1.2. Inner Frame

As shown in Table 5, the possible materials for the inner frame are aluminum, steel, alloys, and plastic. Compared with an external frame, there are many alternatives for the inner frame because it has less contact with concrete or weather, and the possibility of external impact is low. Considering the PI, the best material is steel. Because the strong modulus of elasticity of steel prevents deflection, it is the most suitable material for the inner frame, which is most affected by deflection, at a comparatively low price.

4.1.3. Panel

The panel material should be plywood or plastic, as shown in Table 5, as it should be lightweight and capable of being released from concrete even when no release agent is applied. Considering the PI, PP ranked the highest; however, because of the weak FM of PP and the difficulty of nailing, we tested both sandwich panels of PP and HPL-coated wood panel, which are on the second tier on PI for panels. These special panels ensure chemical resistance, moisture resistance, and UV resistance.

4.1.4. Corners and Brackets

PA6 was used for the joint brackets, and synthetic rubber was applied for the core corners because it can reduce noise as well as impact damage during installation and dismantling, which were strong constraints for all assembly parts. In addition, the CSF, which is assembled from several separable structures, has advantages as a temporary resource because it is easily replaced part-by-part when damaged.

4.2. Verification of the CSF

After designing the product with the selected materials and applying the design to the actual site, we compared and analyzed the performance of the CSF versus existing forms. There are many performance criteria, but the authors validated only the recycle time, NG, and work efficiency because these are the three most important performance criteria for concrete formwork.

4.2.1. Three-Dimensional (3D) Modeling of the CSF

After checking the structural analysis, a detailed 3D model of the CSF was drawn for the purpose of fabricating real prototypes (Figure 7). Through computer analysis, the frame structure was optimized to minimize the amount of material input and deflection.

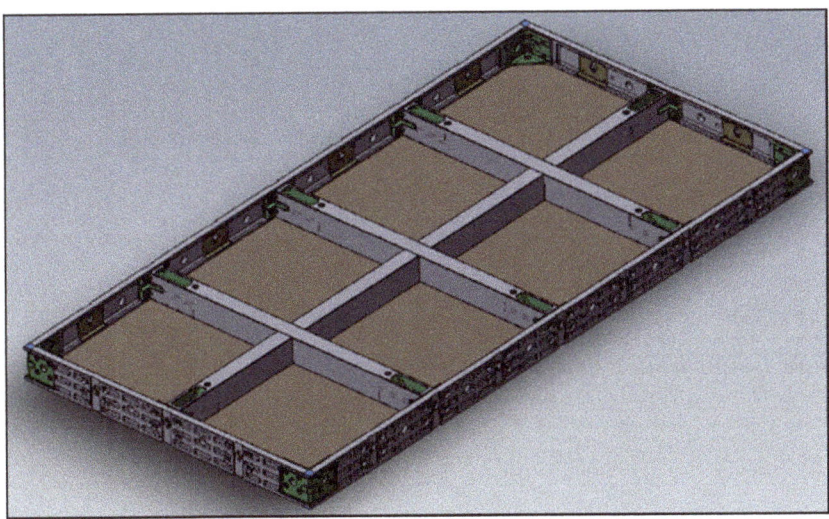

Figure 7. Three-dimensional model of the CSF.

4.2.2. CSF Prototype

After the 3D design of the CSF, a prototype was developed including the panel, corner panel, and accessories to build a mock-up model house (Figure 8). Installation and assembly tests were then conducted to check the applicability of the prototype (Figure 9).

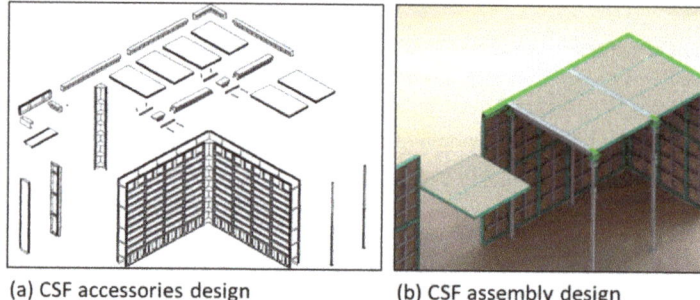

(a) CSF accessories design   (b) CSF assembly design

Figure 8. The CSF prototype design ((**a**) accessories components design, (**b**) assembly method design).

(a) CSF mock-up construction   (b) CSF connection test

Figure 9. A mock-up assembly house made up of CSF ((**a**) constructability test, (**b**) connectivity test).

4.2.3. Structural Analysis (Standard Loading Procedure)

A standard loading test was performed on the fabricated CSF prototype to calculate the deflection of the form during concrete casting. Because there is no official performance standard for composite concrete forms in Korea, the loading was performed based on the Korea Standard (KS) criterion F 8006 (Figure 10) with a maximum load (P) of 14,400 N. KS F 8006 is a very strong standard for steel form, and if the maximum deflection is less than 1.4 mm, it ensures that the form is strong enough to endure concrete pouring in any position. The CSF deflection did not exceed 1.4 mm during 14,400 N loading, verifying the safety of the CSF during concrete casting.

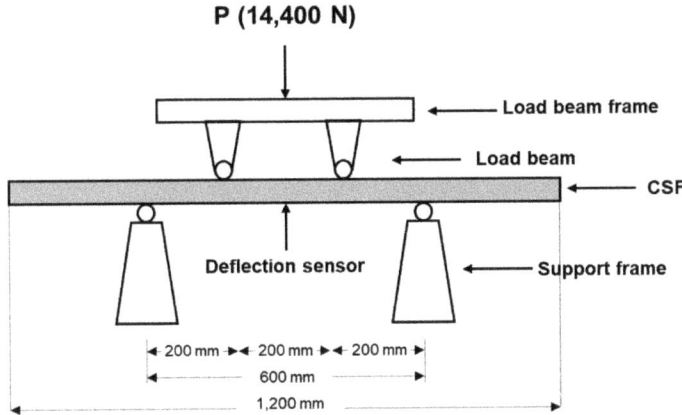

Figure 10. KS F 8006 (standard loading method for concrete form in Korea).

4.2.4. Recycle Time Test

The CSF consists of an assembly structure (i.e., panel + frames), and the panel can be changed if it is worn out or damaged. Current plywood, the most widely used formwork panel, can be reused (recycled) 7 to 10 times without replacement. The panel replacement is a cost- and time-consuming task, so a higher recycle time ensures better formwork performance. In this study, recycle time tests were performed for the PP sandwich panel and HPL-coated wood panels. Because there is no official certification test for the number of times a form can be reused, the evaluation was performed subjectively by comparing the change to the surface of the panel with the surface of the concrete after removing the concrete. After running a concrete casting test 50 times without changing the panel, the surfaces of the CSF and the concrete remained clean, even without cleaning, because the concrete did not stick to the PP or thin film-covered panel. In addition, the new panels are neither worn out or damaged by concrete because they have enough IR and AR. This result means that the CSF is suitable for high-rise building construction requiring the repeated use of panels.

4.2.5. Noise Test

Two kinds of tests were conducted to measure the noise produced when using the CSF, generated when dropping the CSF from a certain height to the floor, and generated while installing and dislodging pins with a hammer. The average value was measured after four tests using a noise meter. For comparison with existing forms, the same experiment was performed on the Euro and aluminum forms, and the measurement results are shown in Figure 11. The CSF dampens noise through shock absorption and the separate frame structures, resulting in less noise than other forms.

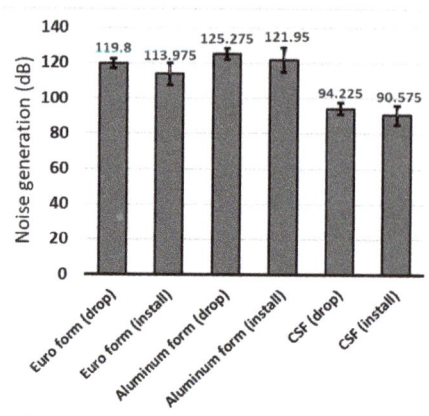

Figure 11. Test result of noise generation (NG) during dropping and installation.

Table 8 shows a comparison of the characteristics of the proposed CSF with existing aluminum and Euro forms.

Table 8. Comparison of characteristics of the CSF and existing forms.

Item	CSF	Aluminum Form	Euro Form
Material	Composite (PA6 GF60) + Steel + HPL-coated plywood panel	100% Aluminum	Steel + coated plywood
Image			
Specification	600 mm × 1200 mm	600 mm × 1200 mm	600 mm × 1200 mm
Weight	10 kg	15 kg	19 kg
Number of recycling cycles	Above 50 times	Above 50 times	Below 10 times
Use of form oil	None	Use	Use
Noise creation	Below 95 dB	Above 120 dB	Above 110 dB
Applicability	Wall + Slab	Wall + Slab	Wall
Systemization	Table form, gang form	-	-
Compatibility	100% compatible with both Aluminum and Euro forms	Not compatible with Euro form	Not compatible with Aluminum form

4.2.6. Field Application

The developed CSF was applied to an actual building construction site to validate the performance and quality of the CSF. The average formwork time between the aluminum form and CSF were compared and analyzed during a task of wall, slab, and stair formwork (Table 9). Each of the formwork tasks consists of four work tasks: stripping, lifting, plaster form oil, and installation. The average task (i.e., installation and dismantlement) time of 20 units of CSF and Al-form were measured. In addition, to validate the concrete quality, CSF was applied alongside the existing Euro form (plywood panel), and the concrete surface quality was compared and analyzed after stripping forms (Figure 12).

Table 9. Performance comparison between CSF and Al-form.

Task Distribution		Avg. Task Time (Al-form) (s)	Avg. Task Time (CSF) (s)
Wall formwork	Stripping	30	26
	Lifting	21	16
	Plaster form oil	10	-
	Installation	48	34
Stair formwork	Stripping	32	27
	Lifting	16	12
	Plaster form oil	10	-
	Installation	54	39
Slab formwork	Stripping	18	17
	Lifting	14	14
	Plaster form oil	10	-
	installation	24	22
Total		287	208

Measurement time range

Stripping: (start) When the tools for dismantling begin to reach a form; (end) when the dismantled form is placed on the slab.

Lifting: (start) When workers started to hold the form to lift; (end) when putting down the lifted form in the work area.

Plaster form oil: (start) When workers start to grip the roller and take form oil; (end) when the roller is put down after plastering.

Installation: (start) When workers start to hold the form by hand; (end) when the tool is put in after installation.

Figure 12. Concrete surface quality comparison between the CSF and the Euro form.

5. Results

Figure 5 shows the overall relative weights of the material performances. The authors calculated the PI by multiplying the material property data in Table 6 by the relative weights of the performances. Table 7 shows the results of PI for each material and allowable candidates, considering the technical requirements. The outer frame has higher importance for NG and DE, the inner frame has higher importance for FS and FM, and the panel has a higher importance for TC and AR. When considering the PI, PA6, steel, and PP have the highest values for the outer frame, inner frame, and panel, respectively, but PP has several limitations on deflection and is difficult to nail. The HPL-coated plywood panel is the second-highest alternative for the panel, so we tested it with PP panel during the field test.

A new composite system form (CSF) was developed based on the result of the case study, and the authors verified its structural safety and then tested applicability in terms of the panel's recycle time, NG, and work efficiency, and these are the most important factors for formwork performance.

In the developed CSF, PP- and HPL-thin film coated on the plywood panel adds impermeability and does not stick to concrete, and they can be reused more than 50 times without applying form oil. In addition, the noise level of the CSF was lower than 95 dB during the installation and dismantling work, a remarkable reduction compared with the aluminum (121.95–125.275 dB) and Euro (113.975–119.8 dB) forms. Most importantly, CSF was 33% lighter than the conventional aluminum form (15 kg) and 47% lighter than the Euro form (19 kg). The weight reduction, compared with the aluminum form, provided a 27.5% increase in work productivity during wall, stair, and slab formwork. In addition, CSF shows 100% compatibility with Euro form and Al-form, and it can be systemized as Table form or Gang form by assembly.

6. Discussion

This paper aims to develop an MCDM technique for the construction material selection model to help researchers and practitioners select optimal materials in terms of their performance goals. In this regard, this study proposed an AHP-based MCDM procedure for construction materials and provided a case as a guideline for further applications. A newly developed CSF is structurally safe during and after concrete casting because it passed the KS standard. This does not simply describe safety as a small-scale form; the same structure of the form module can be assembled into a large size and various shapes of concrete forms in building construction such as gang form and slip and truss tables.

The result of the recycle times shows that the CSF panel and the PP- and HPL-thin film coated on the plywood can be reused more than 50 times without applying form oil. Replacing the formwork panel is very costly and time-consuming work especially in mid- and high-rise building construction. In particular, when constructing irregularly shaped buildings, the installation location of the form is fixed, and it is very inefficient work to replace the formwork panel. In this respect, the suggested CSF panel can be an innovative and state-of-the-art technology to reduce formwork productivity (e.g., duration and cost).

In contrast, CSF reduced NG during the formwork process by up to 30 dB compared with aluminum form. Consequently, the amount of noise generated is reduced by 1000 times. The NG of the formwork is the main cause of complaints around the construction site, along with the hearing damage of workers. This often lowers the productivity of the construction projects and creates a bad social perception, in turn inhibiting the influx of skilled workers. Therefore, reducing noise can play a very important role in increasing the productivity and sustainability of construction in the construction industry.

The CSF was 33% lighter than the conventional aluminum form (15 kg) and 47% lighter than the Euro form (19 kg). It helped reduce formwork time by 27.5% by increasing the formwork performance. Because the formwork is transported, installed, and dismantled by a worker, its weight has a significant impact on performance. Weight reduction can also play a major role in reducing the risk of safety accidents caused by formwork dropping and the operator's physical fatigue. In addition, because more forms can be loaded on a truck at once, transportation costs can be reduced. Thus, using the CSF not only saves on cost, but also reduces construction time.

Finally, CSF is 100% compatible with aluminum form and Euro form because it is designed with high compatibility, so that workers may not be confused when they work with this new system form. This is significant in construction sites where very large amounts of forms are required. The formwork company can use the part where new material needs to be applied optionally without having to purchase all CSFs covering the total formwork area.

These verifications show that the optimal choice of materials in the field of building construction could have an effect on the construction methods and could even change the technical paradigm of the construction as a whole.

7. Conclusions

This study proposed a material selection model for construction materials and applied the model to a concrete form, a key temporary resource in building construction. The CSF consists of several separately designed parts and each component has different technological requirements with respect to properties such as IR, WR, and AR. As each index is related to a specific performance parameter (e.g., productivity, concrete surface quality, and corrosion), careful selections should be made according to user requirements. Using a systematic and scientific design through AHP methodology, a highly advanced concrete form was fabricated that satisfied both the user requirements and the technical requirements of the system formwork for performance improvement.

The results of this study suggest that the appropriate selection of construction materials is very effective as a method for increasing construction performance. Moreover, problems that involve productivity decrease, and safety accidents and environmental damage can be addressed; such issues have been identified in the construction field as being in need of improvement with the supplementing of materials that have been made in an empirical and intuitive manner.

Several limitations of this study should be addressed in future work. First, the cost parameter considered was only that of raw materials without the processing and recycling costs. Second, there were not enough people to be surveyed on AHP, so more practitioners should have participated for accurate and general implications. In addition, it was difficult to apply the new concrete form because several construction companies declined to apply the CSF, which was not verified earlier. Third, the shape optimization of the CSF frame was insufficient. In future research, a topological optimization method for designing an optimal CSF shape that satisfies the demand load condition should be explored. In a further study, more generalized functions and validation should be provided.

Author Contributions: D.L. (Dongmin Lee) and T.K. designed the research process using AHP selection model and wrote the draft of the manuscript. D.L. (Dongyoun Lee) and M.L. collected the CRs data and analyzed the data. M.K. checked and revised final manuscript. All authors contributed to the analysis of the data and read the final paper. All authors have read and agreed to the published version of the manuscript.

Funding: This research was supported by a grant (19AUDP-B106327-06) from the Architecture & Urban Development Research Program funded by Ministry of Land, Infrastructure and Transport of the Korean Government, and (2016R1A2B3015348) from the National Research Foundation of Korea grant funded by the Korean government.

Acknowledgments: This research was supported by a grant (19AUDP-B106327-06) from the Architecture & Urban Development Research Program funded by Ministry of Land, Infrastructure and Transport of the Korean Government, and (2016R1A2B3015348) from the National Research Foundation of Korea grant funded by the Korean government.

Conflicts of Interest: The authors declare no conflict of interest.

Availability of Data and Material: All data, models, and code generated or used during the study appear in the submitted article.

References

1. Mehmood, Z.; Haneef, I.; Udrea, F. Material selection for Micro-Electro-Mechanical-Systems (MEMS) using Ashby's approach. *Mater. Des.* **2018**, *157*, 412–430. [CrossRef]
2. Ashby, M.F. Engineering materials and their properties. In *Materials Selection in Mechanical Design*, 4th ed.; Ashby, M.F., Ed.; Butterworth-Heinemann: Oxford, UK, 2011; pp. 31–56.
3. Farag, M.M. Quantitative methods of materials selection. In *Handbook of Materials Selection*; Kutz, M., Ed.; Wiley Online Books: New York, NY, USA, 2002; pp. 3–24.

4. Maskell, D.; Thomson, A.; Walker, P. Multi-criteria selection of building materials. *Proc. Inst. Civ. Eng. Constr. Mater.* **2018**, *171*, 49–58. [CrossRef]
5. Govindan, K.; Madan Shankar, K.; Kannan, D. Sustainable material selection for construction industry—A hybrid multi criteria decision making approach. *Renew. Sustain. Energy Rev.* **2016**, *55*, 1274–1288. [CrossRef]
6. Chen, Z.S.; Martínez, L.; Chang, J.P.; Wang, X.J.; Xionge, S.H.; Chin, K.S. Sustainable building material selection: A QFD- and ELECTRE III-embedded hybrid MCGDM approach with consensus building. *Eng. Appl. Artif. Intell.* **2019**, *85*, 783–807. [CrossRef]
7. Akadiri, P.O.; Olomolaiye, P.O.; Chinyio, E.A. Multi-criteria evaluation model for the selection of sustainable materials for building projects. *Autom. Constr.* **2013**, *30*, 113–125. [CrossRef]
8. Ko, C.-H.; Wang, W.-C.; Kuo, J.-D. Improving formwork engineering using the Toyota Way. *EPPM-J.* **2011**, *1*, 13. [CrossRef]
9. Ling, Y.Y.; Leo, K.C. Reusing timber formwork: Importance of workmen's efficiency and attitude. *Build. Environ.* **2000**, *35*, 135–143. [CrossRef]
10. Saaty, T.L. Analytic hierarchy process. In *Wiley StatsRef: Statistics Reference Online*; Balakrishnan, N., Colton, T., Everitt, B., Piegorsch, W., Ruggeri, F., Teugels, J.L., Eds.; John Wiley & Sons, Ltd.: Hoboken, NJ, USA, 2014. [CrossRef]
11. Lee, G.K.L.; Chan, E.H.W. The analytic hierarchy process (AHP) approach for assessment of urban renewal proposals. *Soc. Indic. Res.* **2008**, *89*, 155–168. [CrossRef]
12. Fu, Y.K. An integrated approach to catering supplier selection using AHP-ARAS-MCGP methodology. *J. Air Transp. Manag.* **2019**, *75*, 164–169. [CrossRef]
13. Triantaphyllou, E.; Chi-Tun, L. Development and evaluation of five fuzzy multiattribute decision-making methods. *Int. J. Approx. Reason.* **1996**, *14*, 281–310. [CrossRef]
14. Durán, O.; Aguilo, J. Computer-aided machine-tool selection based on a Fuzzy-AHP approach. *Expert Syst. Appl.* **2008**, *34*, 1787–1794. [CrossRef]
15. Wang, T.K.; Zhang, Q.; Chong, H.Y.; Wang, X. Integrated Supplier Selection Framework in a Resilient Construction Supply Chain: An Approach via Analytic Hierarchy Process (AHP) and Grey Relational Analysis (GRA). *Sustainability* **2017**, *9*, 289. [CrossRef]
16. Pan, N.F. Fuzzy AHP approach for selecting the suitable bridge construction method. *Automat. Constr.* **2008**, *17*, 958–965. [CrossRef]
17. Shapira, A.; Goldenberg, M. AHP-Based Equipment Selection Model for Construction Projects. *J. Constr. Eng. Manag.* **2005**, *131*, 1263–1273. [CrossRef]
18. Rajak, M.; Shaw, K. Evaluation and selection of mobile health (mHealth) applications using AHP and fuzzy TOPSIS. *Technol. Soc.* **2019**, *59*, 101186. [CrossRef]
19. Lee, D. Hybrid System Formwork and AI Based Construction Planning Model for High-Rise Building Construction. Ph.D. Dissertation, Korea University, Seoul, Korea, 2019.
20. Lee, D.; Kim, T.; Lee, D.; Lim, H.; Cho, H.; Kang, K.I. Development of an Advanced Composite System Form for Constructability Improvement through a Design for Six Sigma Process. *J. Civ. Eng. Manag.* **2020**, *26*. [CrossRef]
21. Arjun, N. Concrete forms—Types and Selection of Concrete Forms. Available online: https://theconstructor.org/building/concrete-forms-types-selection/26232/ (accessed on 12 February 2020).
22. Ashby, M.F. The design process. In *Materials Selection in Mechanical Design*, 4th ed.; Ashby, M.F., Ed.; Butterworth-Heinemann: Oxford, UK, 2011; pp. 15–29.
23. Cheng, S.-C.; Chen, M.-Y.; Chang, H.-Y.; Chou, T.-C. Semantic-based facial expression recognition using analytical hierarchy process. *Expert Syst. Appl.* **2007**, *33*, 86–95. [CrossRef]

© 2020 by the authors. Licensee MDPI, Basel, Switzerland. This article is an open access article distributed under the terms and conditions of the Creative Commons Attribution (CC BY) license (http://creativecommons.org/licenses/by/4.0/).

Article

Flexural Behavior of a Precast Concrete Deck Connected with Headed GFRP Rebars and UHPC

Won Jong Chin [1,2], Young Hwan Park [2], Jeong-Rae Cho [2], Jin-Young Lee [3,*] and Young-Soo Yoon [1,*]

1. School of Civil, Environmental, and Architectural Engineering, Korea University, Seoul 02841, Korea; wjchin@kict.re.kr
2. Department of Infrastructure Safety Research, Korea Institute of Civil Engineering and Building Technology, Goyang 10223, Korea; yhpark@kict.re.kr (Y.H.P.); chojr@kict.re.kr (J.-R.C.)
3. School of Agricultural Civil & Bio-industrial Engineering, Kyungpook National University, Daegu 41566, Korea
* Correspondence: jinyounglee@knu.ac.kr (J.-Y.L.); ysyoon@korea.ac.kr (Y.-S.Y.); Tel.: +82-53-950-5733 (J.-Y.L.); +82-2-3290-3320 (Y.-S.Y.)

Received: 2 October 2019; Accepted: 23 January 2020; Published: 29 January 2020

Abstract: Steel bent reinforcing bars (rebars) are widely used to provide adequate anchorage. Bent fiber-reinforced polymer (FRP) rebars are rarely used because of the difficulty faced during the bending process of the FRP rebars at the construction site. Additionally, the bending process may cause a significant decrease in the structural performance of the FRP rebars. Therefore, to overcome these drawbacks, a headed glass fiber-reinforced polymer (GFRP) rebar was developed in this study. The pull-out tests of the headed GFRP rebars with diameters of 16 and 19 mm were conducted to evaluate their bond properties in various cementitious materials. Moreover, structural flexural tests were conducted on seven precast concrete decks connected with the headed GFRP rebars and various cementitious fillers to estimate the flexural behavior of the connected decks. The results demonstrate that the concrete decks connected with the headed GFRP rebar and ultra-high-performance concrete (UHPC) exhibited improved flexural performance.

Keywords: glass fiber-reinforced polymer (GFRP) rebar; ultra-high-performance concrete (UHPC); concrete headed GFRP rebar; bond strength; development length; flexural strength; precast concrete deck

1. Introduction

1.1. General

The corrosion of reinforcing steel in structures decreases its life expectancy and causes extensive maintenance costs. In this context, fiber-reinforced polymer (FRP) reinforcing bars (rebars) have been extensively used as an alternative to steel reinforcement owing to their non-corrosive characteristics. Moreover, FRP has additional advantages as a construction material, such as high specific strength, relatively lightweight, and non-conductivity. However, the substitution of a steel rebar with an FRP rebar has several difficulties owing to the differences between them. While steel bent rebars are used to provide sufficient anchorage, bent FRP rebars are rarely used because the bending process of FRP rebars at construction sites is difficult. Additionally, the bending process can cause a significant decrease in the structural performance of FRP rebars. These problems have resulted in the development of FRP bars with a headed end to satisfy the required development length. Therefore, the Korea Institute of Civil engineering and Building Technology (KICT) developed new headed FRP rebars to improve their practical applications.

Although several researchers have studied the bond strength of FRP rebars, a comprehensive relationship between the strength of concrete and FRP rebars has not been established because various companies manufacture FRP rebars using their own distinct methods. In other words, FRP rebars have not been standardized yet. Therefore, if a new type of FRP rebar is developed, the mechanical properties and bond characteristics need be evaluated as the bond characteristics of FRP rebars are important factors that govern the design of FRP reinforced structural members.

Thus, in this study, newly developed glass fiber-reinforced polymer (GFRP) rebars and headed GFRP rebars are introduced. A total of 44 specimens were fabricated and tested to estimate the pull-out behavior of GFRP rebars considering various aspects, such as the diameter of the rebars, the strength of concrete, and the types of headed end. Moreover, flexural tests were conducted on seven precast concrete decks connected with headed GFRP rebars and various cementitious fillers to determine their flexural behavior.

1.2. Manufacturing of a GFRP Rebar

The easiest way to increase the tensile strength of a GFRP rebar is to use high-performance constituent materials and reduce the amount of materials to overcome the problem of a price increase. To achieve superiority over ordinary rebars, the structural performance of the GFRP rebar was enhanced along with price reduction through efficient manufacturing. In this study, e-glass fiber and vinylester resin were used to manufacture an optimized GFRP rebar while maintaining a balance between cost and strength properties. A modified braidtrusion process, which is a combination of braiding and pultrusion, was applied to develop the GFRP rebar, as illustrated in Figure 1. The modified braidtrusion process was used to impart special features by pre-tensioning the glass fibers. This technique further includes a strand for forming ribs on the surface of the rebar using braiding fibers to obtain better bonding through a single process. The tightening of fibers to form the ribs results in the reduction of voids in the cross-section and increases the tensile strength of the GFRP rebar up to 900 MPa through the pre-tensioning of the core fiber bundle, as depicted in Figure 2. Figure 3 depicts the shape of the efficiently manufactured GFRP rebar in this study [1].

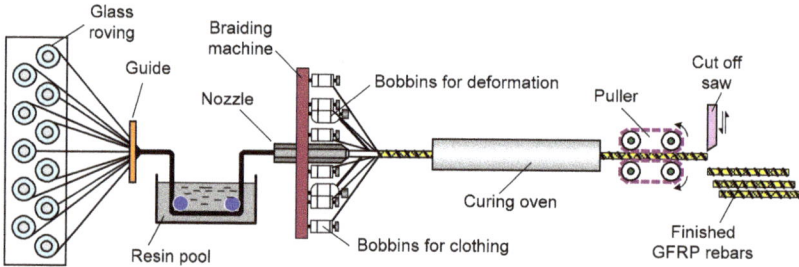

Figure 1. Modified braidtrusion process.

Figure 2. Schematic depicting the manufacturing of glass fiber-reinforced polymer (GFRP) reinforcing bars (rebars). (**a**) general braiding. (**b**) modified braiding.

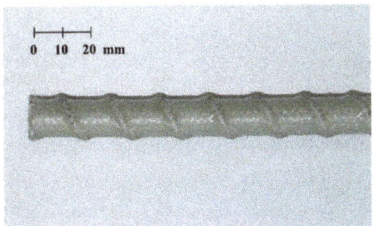

Figure 3. Shape of the Korea Institute of Civil engineering and Building Technology-(KICT) developed GFRP rebar.

1.3. Concrete Headed GFRP Rebar

In this study, a new concrete-headed GFRP rebar is proposed, in which the head is formed by cutting the end of the GFRP rebar in the longitudinal direction and casting a concrete-like filler (mortar, ultra-high-performance concrete (UHPC), etc.) (Figure 4). Our aim is to increase the interfacial area and reduce the adhesive length. In comparison with the plastic headed GFRP rebar [2], it has a lower production cost and can be integrated with concrete. Our objective is to achieve a pullout strength capacity greater than 60% of the maximum strength of the GFRP rebar with a minimal head length. The proposed concrete-headed GFRP rebar causes bond failure when the concrete head length is short and causes FRP tensile failure when the head length is long. However, the tensile strength is lower than that of ordinary GFRP bars owing to the cross-sectional loss caused by longitudinal cutting (Figure 5) [3–5].

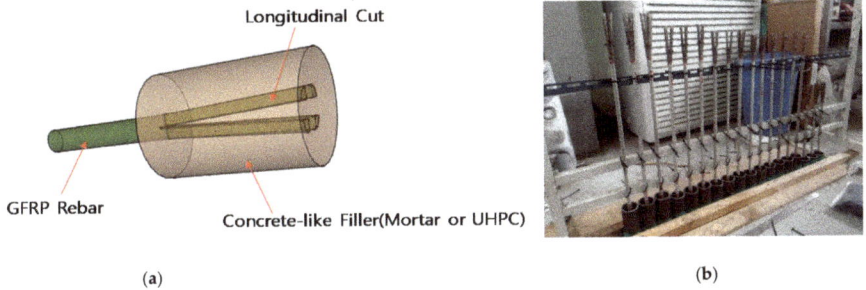

Figure 4. Newly developed concrete headed GFRP rebar. (**a**) Conceptual diagram. (**b**) Photograph of concrete headed GFRP rebar.

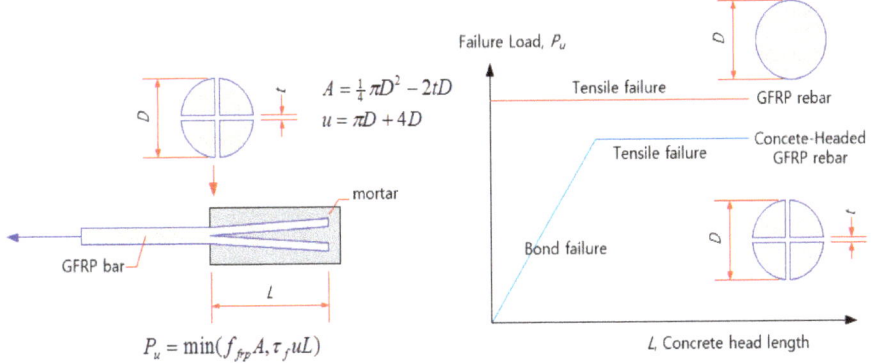

Figure 5. Schematic diagram of the failure load and failure mode.

2. Experimental Program

2.1. Pull-out Test of GFRP Rebars

2.1.1. Material Properties and Test Specimens

To evaluate the pull-out behavior of the normal GFRP and headed GFRP rebars, 44 specimens were fabricated and tested in this experimental study. As presented in Table 1, the parameters considered in this study are the compressive strength of concrete: the normal strength concrete (27, 35 MPa), high-strength mortar (83 MPa), and ultra-high-performance concrete (165 MPa), and the type of head: longitudinal cut, concrete head, and bar diameters: 16 mm, 19 mm.

Table 1. Mechanical properties of test specimens.

Notation	d (mm)	C	f_{ck} (MPa)	Type of Head	No. of Specimens
D16-NC35-N	16	NC	35.0	W/O head	10
D16-NC27-LC	16	NC	27.0	LC	4
D16-NC35-LC		NC	35.0		4
D19-NC27-LC	19	NC	27.0		4
D19-NC35-LC		NC	35.0		4
D16-NC35-CH	16	NC	35.0	CH	3
D16-HM83-CH		HM	83.2		3
D16-UH165-CH		UH	165.6		3
D19-NC35-CH	19	NC	35.0		3
D19-HM83-CH		HM	83.2		3
D19-UH165-CH		UH	165.6		3

D: diameter of FRP bar; C: Type of cementitious filler; f_{ck}: Compressive strength of filler; NC: normal strength concrete; HM: high strength mortar; UH: ultra-high performance concrete; LC: longitudinal cut without concrete head; CH: longitudinal cut with concrete head.

The tensile strength of the GFRP rebars was estimated by the direct tensile test according to ACI 440.1R standard [6]. The mechanical properties of the GFRP rebars used in this study are shown in Figure 6 and Table 2. The tensile strength of the rebars used was 1050 MPa, approximately.

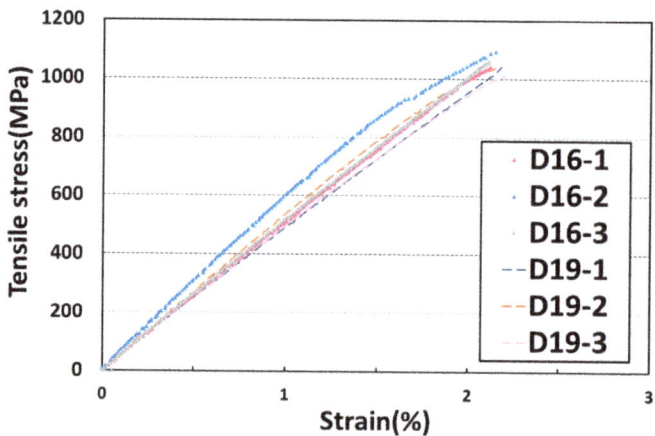

Figure 6. Tensile strength–strain curves of GFRP rebars.

Table 2. Tensile properties of GFRP rebars.

Tensile Properties	D16	D19
Maximum tensile strength, $f_{u,ave}$, MPa	1066.39	1030.38
Nominal tensile strength, $f^*_{fu} = f_{u,ave} - 3\sigma$, MPa	987.55	981.38
Design tensile strength, $f_{fu} = C_E f^*_{fu} (C_E = 0.8)$, MPa	691.28	686.97
Elastic modulus, E_{frp}, GPa	47.84	46.71

σ: standard deviation; C_E: environmental reduction factor.

The concrete blocks were cast with normal strength concrete, high strength mortar, and ultra-high-performance concrete (UHPC). The compressive strengths at 28 days were measured by three concrete cylinders for each mix in accordance with ASTM C39 [7], and the results are presented in Table 1. The mix proportions of these concretes and mortars are presented in Table 3.

Table 3. Mix designs of specimens.

Variables	W/B (%)	Unit Weight (kg/m³)						SP (g)
		W	C	FA	CA	AD	Steel Fiber	
NC	40.0	226	562	598	786	0	0	-
HM	42.5	170	200	1640 *	0	200 (Fly ash)	0	5.25
UHPC	20.0	183	799.5	880	0	84 (Silica fume)	78 (19.5 mm) 39 (16.3 mm)	18.4

W/B: water to binder ratio; W: water; C: cement; FA: fine aggregate; CA: coarse aggregate; AD: admixture, SP; superplasticizer; * proportion of fine aggregates: silica sand: crushed sand = 800:840.

In accordance with ACI 440.1R [6], the GFRP rebar was embedded in a 200 mm cubic block. Two types of heads were considered in this study. The first one was a longitudinal cut head (LC), which was cut and embedded directly in the cube. Another one was a concrete head (CH), which had a concrete head, as shown in Figure 4a. The concrete heads were made of high strength mortar. To avoid rupturing at the gripping part of the GFRP rebar, the steel jackets covered a length of 700 and 1000 mm on the 16 and 19 mm diameter bar specimens, respectively, because the traditional wedge-shaped frictional grip can cause damage at the gripping part and the FRP rebar is vulnerable to shear force. The details of the specimens are depicted in Figure 7. Ten straight GFRP rebars (D16) without any type of head were prepared and tested to evaluate the bond strength of the GFRP rebar as a preliminary test. These specimens have an embedment length of 80 mm (5d) according to the ACI 440.1R standard. Additionally, a total of 34 headed GFRP rebars were fabricated and tested. Sixteen specimens were made of GFRP rebars with the longitudinal cut head and normal concrete (27, 35 MPa). Eighteen specimens were made of GFRP rebars with the concrete head and three types of concrete blocks (NC, HM, UH).

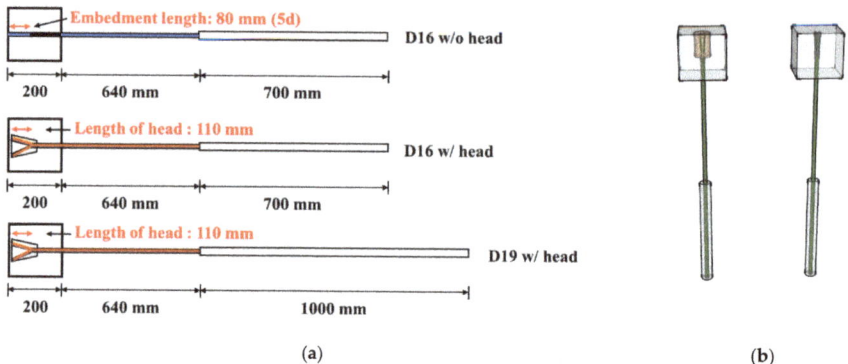

Figure 7. Details of specimens for the pull-out test: (**a**) specimen profiles and (**b**) a schematic view of the longitudinal cut head (LC) and concrete head (CH) specimens.

2.1.2. Testing Method and Setup

The pull-out test setup is illustrated in Figure 8. The tests were performed using a universal testing machine with a capacity of 1000 kN. LVDTs were installed to measure the pull-out displacement: A, as depicted in Figure 8. The load was applied to the GFRP rebar with a rate of 0.02 mm/s.

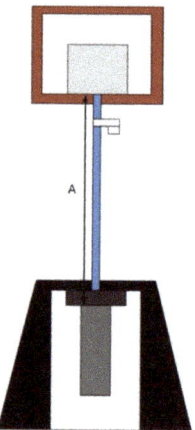

Figure 8. Test setup.

2.2. Flexural Test of a Connected Precast Concrete Deck

2.2.1. Material Properties and Test Specimens

Many researchers have focused on improving the precast concrete deck connections [2,8–11]. To decrease the connected length and improve flexural performance, headed GFRP rebars were developed. The flexural test on the connected precast concrete decks using two types of headed GFRP bars were conducted to verify the performance and practical effectiveness of the developed headed GFRP rebars. As depicted in Figure 9, using the headed GFRP rebar can decrease the development length in comparison with the ordinary splice connection. The length of the ordinary splice connection and that of the connection with the headed GFRP rebar were calculated according to Equations (1) and (2) [12].

The tensile strength f_{fuh} of the headed FRP bar can be expressed as the product of the effective strength factor γ_h of the headed bar and the tensile strength of the FRP bar, as shown in Equation (3) [12].

$$l_d = \frac{\alpha \frac{f_{fr}}{0.083\sqrt{f_{ck}}} - 340}{13.6 + \frac{C}{d_b}} d_b \quad (1)$$

$$l_d = \frac{\alpha \frac{f_{fr} - f_{fuh}}{0.083\sqrt{f_{ck}}} - 340}{13.6 + \frac{C}{d_b}} d_b \quad (2)$$

$$f_{fuh} = \gamma_h f_{fu} \quad (3)$$

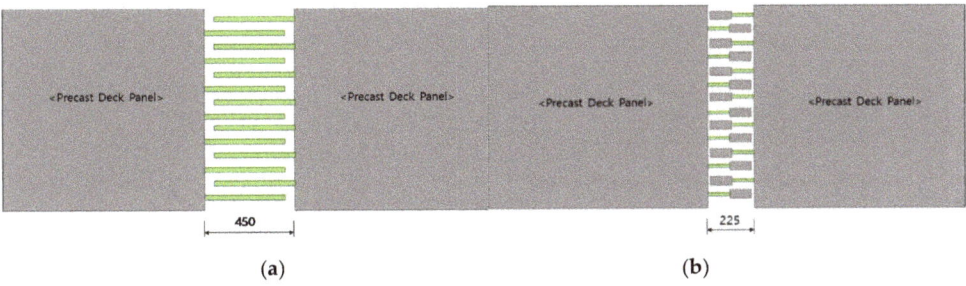

(a)   (b)

Figure 9. Schematic diagram of (a) the connection joint of the splice of GFRP rebars and (b) the connection joint of headed GFRP rebars.

Here, f_{fr} = the required strength of FRP reinforcement; f_{fuh} = the tensile strength of the headed FRP bar; α = the rebar-position modification factor; C = the smallest value of cover thickness from the centroid of rebar and 1/2 of the central spacing of rebar; d_b = the nominal diameter of rebar.

A total of seven specimens of size 240 × 450 × 2500 mm were fabricated in this study. Table 4 lists the test variables. One non-connected specimen was fabricated using straight GFRP rebars and normal strength concrete as a reference specimen. Two head types, LC and CH, filled in the connected zone, and normal strength concrete, high strength mortar, and UHPC were considered as test variables. Figure 10 depicts the details of specimens, and Figure 11 depicts the photograph of the specimens.

Table 4. Test variables for flexural test on the connected precast concrete deck.

Specimen	Connection Joint	Connection Joint Length	Filler
GSR-R	None	Reference	Reference
GSR-HM	Spliced straight GFRP rebar	450 mm	83 MPa Mortar
GSR-UH	Spliced straight GFRP rebar	450 mm	165 MPa UHPC
GCH-HM	Concrete Headed GRFP rebar	225 mm	83 MPa Mortar
GCH-UH	Concrete Headed GRFP rebar	225 mm	165 MPa UHPC
GLC-HM	Non-concrete Headed GRFP rebar	225 mm	83 MPa Mortar
GLC-UH	Non-concrete Headed GRFP rebar	225 mm	165 MPa UHPC

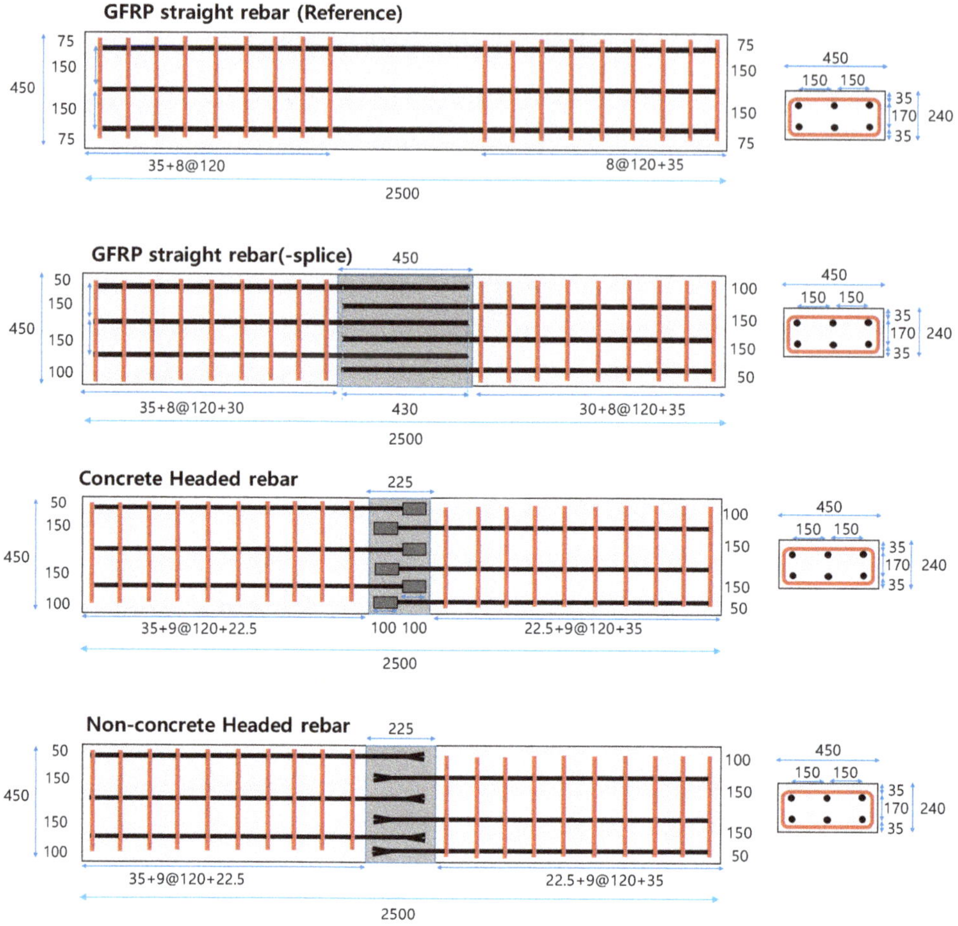

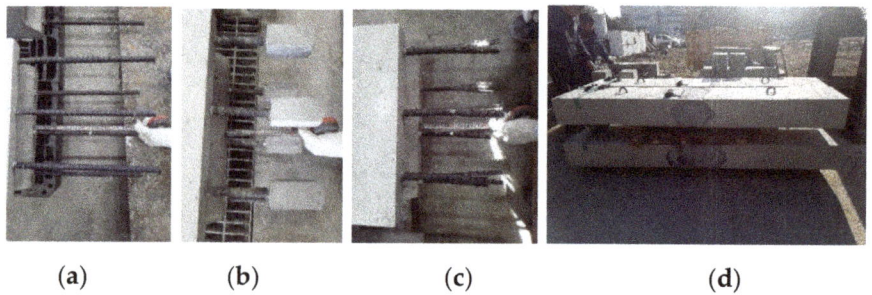

Figure 10. Details of the connected precast concrete deck.

Figure 11. Photographs of the specimens: (**a**) a spliced straight rebar, (**b**) a concrete headed rebar, (**c**) a non-concrete headed rebar, and (**d**) a connected precast concrete deck.

2.2.2. Testing Method and Setup

All specimens were tested with a four-point loading condition, as depicted in Figure 12. The length between the loading points is 600 mm. The decks are simply supported with a span length of 2300 mm. All specimens were loaded monotonically until failure. This loading configuration resulted in the connection region being subjected to a constant moment without shear. The vertical displacements were measured by using LVDTs at the loading points and the center of the deck (Figure 13b).

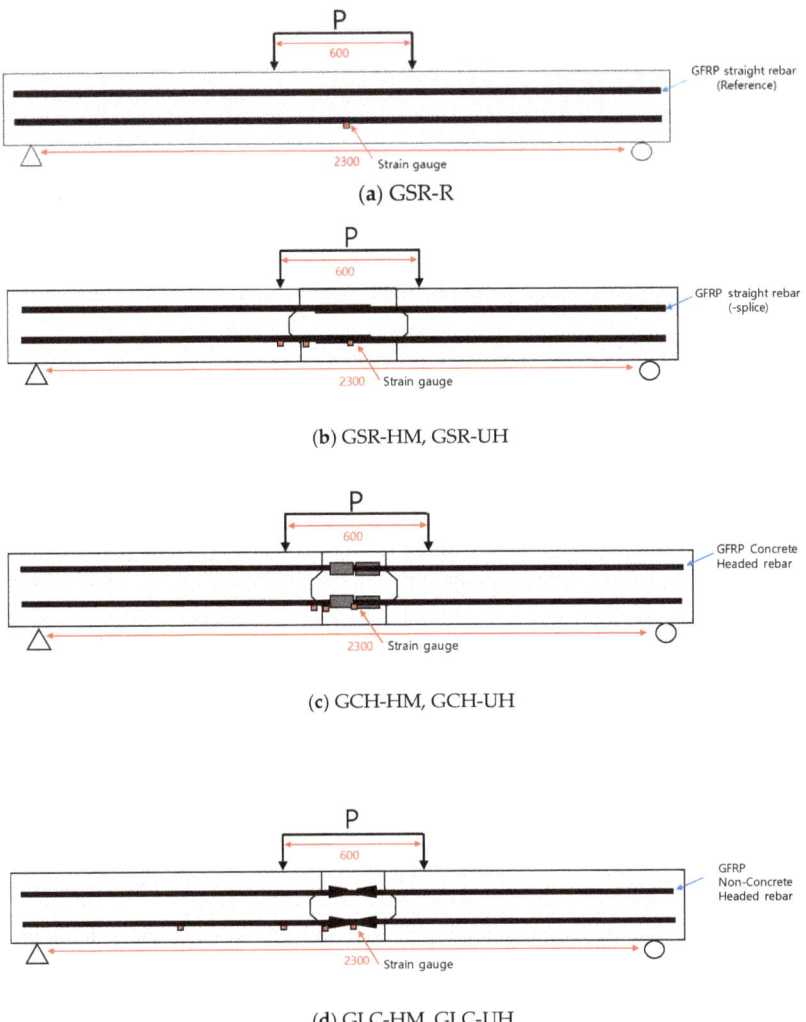

Figure 12. Schematic diagrams of test setup for flexural test. (**a**) GSR-R. (**b**) GSR-HM, GSR-UH. (**c**) GCH-HM, GCH-UH. (**d**) GLC-HM, GLC-UH.

Figure 13. Test setup: (**a**) a photograph of the test setup and (**b**) a photograph of the measurement sensor (LVDTs).

3. Test Results

3.1. Pull-Out Properties of GFRP Rebars

As stated in ACI 440.1R, pull-out and splitting are two dominant failure modes expected with GFRP rebars in concrete [6]. For D16-NC35-N, all specimens failed in the pullout failure mode. The bond strength–slip curves of the representative specimens are illustrated in Figure 14. The result of the pull-out test of these specimens is also summarized in Table 5. In this test, the average bond strength was 10.84 MPa and the average end slip was 1.66 mm.

Figures 15–17 depict the comparison of failure strengths among the specimens, and Table 6 summarizes the test results of the specimens made of concrete-headed GFRP rebars. On analyzing the data, a few of the data exhibiting inconsistencies were excluded. Consequently, the specimens made with LC and CH failed at a similar failure load. The D16 series exhibited a higher failure load than that of the D19 series. This is consistent with the observations of the previous studies [4,10], which demonstrated that as the diameter of the rebar increased, the bond strength decreased. Even the compressive strength of HM was much higher than the strength of NC, and the lower failure loads were measured in the HM specimens. It is important to note that the 200 mm cube made of UHPC exhibited a significantly higher failure load in comparison with the other specimens. Consequently, the improved structural performance can be expected when UHPC is used for the filler material that is cast in the connection of the precast deck.

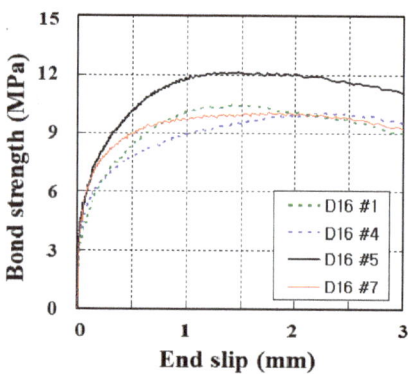

Figure 14. Bond strength–slip curves of GFRP rebars.

Table 5. Test results of the pull-out test (D16 GFRP rebars, normal concrete).

Specimens	Max. Load (kN)	Bond Strength (MPa)	End Slip (mm)	Failure Mode
D16-1	41.92	10.42	1.410	pullout
D16-2	42.71	10.62	1.335	pullout
D16-3	41.91	10.42	1.840	pullout
D16-4	40.49	10.07	2.265	pullout
D16-5	48.78	12.13	1.465	pullout
D16-6	40.46	10.06	2.305	pullout
D16-7d	40.45	10.06	1.755	pullout
D16-8	45.59	11.34	1.420	pullout
D16-9	45.36	11.28	1.410	pullout
D16-10	48.19	11.98	1.365	pullout
Ave.	43.58	10.84	1.66	-
STDEV	3.18	0.79	0.37	-
C.O.V.	7.3%	7.3%	22.3%	-

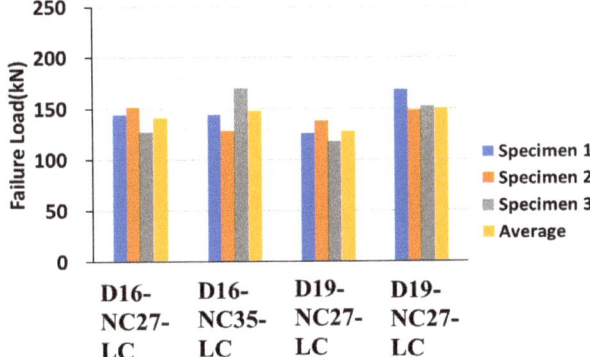

Figure 15. Pull-out test results of 16 and 19 mm GFRP rebars (27 MPa, 35 MPa,).

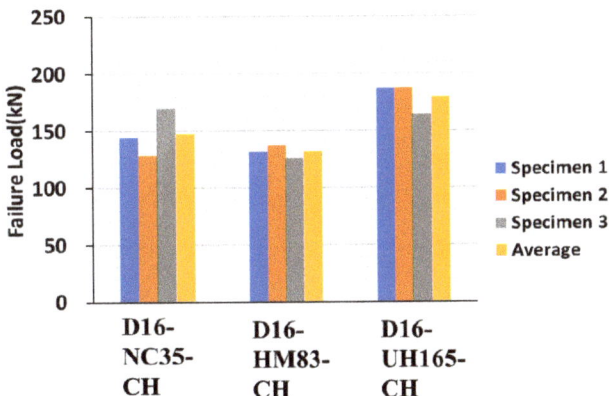

Figure 16. Pull-out test results of 16 mm GFRP rebars (normal concrete, mortar, UHPC).

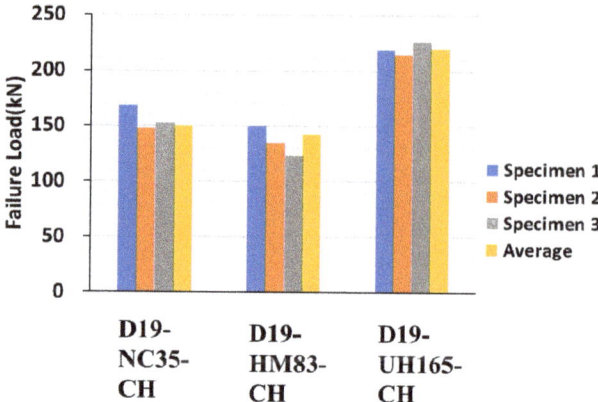

Figure 17. Pull-out test results of 19 mm GFRP rebars (normal concrete, mortar, UHPC).

Table 6. Test results of the pull-out test (D16 and D19 headed GFRP rebars, normal concrete, mortar, UHPC).

Specimen	No.	Pull Out Displacement (mm)	Max. Load (kN)
D16-NC27-CH	1	632	144
	2	644	151
	3	643	127
D16-NC35-CH	1	534	144
	2	639	128
	3	636	170
D16-HM83-CH	1	651	131
	2	654	137
	3	659	126
D16-UH165-CH	1	657	187
	2	656	187
	3	665	164
D19-NC27-CH	1	755	126
	2	753	138
	3	752	118
D19-NC35-CH	1	758	168 *
	2	754	148
	3	754	152
D19-HM83-CH	1	778	150
	2	778	135
	3	778	123 *
D19-UH165-CH	1	774	219
	2	778	215
	3	777	226

* inconsistent data excluded in mean values.

3.2. Flexural Strength of Connected Precast Concrete Deck

All the specimens were loaded monotonically for failure to estimate the ultimate flexural capacity of each connection type. Figure 18 illustrates the load–displacement curves of the connected precast concrete deck. Figure 18a depicts the impropriety of the connection that was connected with the spliced straight GFRP rebar and high strength mortar. Similarly, the specimens made of the concrete headed GFRP rebar and high-strength mortar failed under a significantly low load of less than 100 kN, as depicted in Figure 18c. Only the specimen made with the LC GFRP rebar or GLC-HM exhibited similar behavior to the reference specimen among the specimens connected with the high-strength mortar.

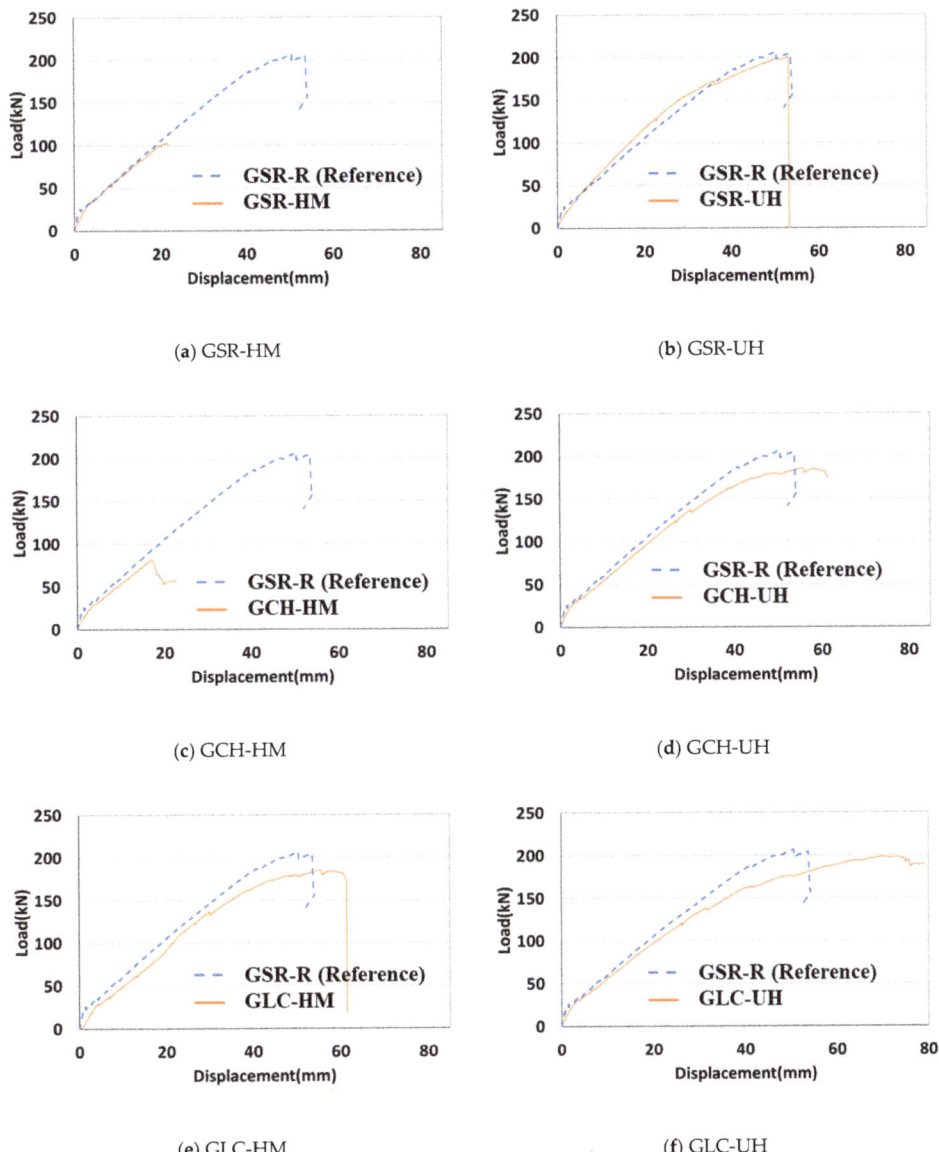

Figure 18. Load–displacement curves of a connected precast concrete deck. (**a**) GSR-HM. (**b**) GSR-UH. (**c**) GCH-HM. (**d**) GCH-UH. (**e**) GLC-HM. (**f**) GLC-UH

In Figure 18b,d,f, all the specimens using UHPC exhibited significantly excellent flexural performance. These experimental results indicate that the use of UHPC as a filler at the connection is appropriate. From the results, the length of the connection in the precast concrete deck can be decreased when the headed GFRP rebars are used, except in the case of GCH-HM. Figure 19 depicts the flexural behavior of GSR-R (Reference) and GLC-UH.

Figure 19. Flexural behavior of (**a**) GSR-R (Reference) and (**b**) GLC-UH.

4. Conclusions

The pull-out behavior of GFRP rebars in concrete was estimated considering the rebar diameter, concrete strength, and head type. Additionally, flexural tests were conducted on the precast concrete decks that were connected with various types of GFRP rebars. From the above discussions, the following conclusions are drawn:

1. In this study, two types of headed GFRP rebars were developed and tested, and their basic pull-out mechanical properties were evaluated. Additionally, by conducting the flexural test on the connected precast concrete deck, the practical effectiveness of the headed GFRP rebar was confirmed.
2. The pull-out test results confirmed the tendency of rebars with larger diameters to have a lower failure strength. In particular, it is important to note that the specimen made with UHPC exhibited a significantly higher failure load in comparison with the other specimens. Consequently, improved structural performance can be expected when UHPC is used as the filler material at the connection between precast deck slabs.
3. The length of the connection in the precast concrete deck can be decreased with the use of headed GFRP rebars. The flexural test results verified that the headed GFRP rebars can provide effective anchorage performance. The LC and CH types did not exhibit significant differences in the structural performance at the connection zone of the precast decks. It is interesting to note that all the specimens using UHPC exhibited excellent flexural performance.

Author Contributions: Conceptualization, W.J.C., J.-Y.L., and Y.-S.Y.; investigation, Y.H.P. and J.-R.C.; writing—original draft preparation, W.J.C. All authors have read and agree to the published version of the manuscript.

Funding: This research was supported by a grant from the Strategic Research Project "Development of safety improvement technology for bridge weak connection and seismic equipment to secure evacuation and recovery route in earthquake" funded by Korea Institute of Civil engineering and building Technology (KICT) and the Smart Civil Infrastructure Research Program (13SCIPA01) funded by the Ministry of Land, Infrastructure and Transport (MOLIT) of the Korean government and the Korea Agency for Infrastructure Technology Advancement (KAIA), Korea.

Conflicts of Interest: The authors declare no conflict of interest.

References

1. You, Y.-J.; Kim, J.-H.J.; Park, Y.-H.; Choi, J.-H. Fatigue Performance of Bridge Deck Reinforced with Cost-to-Performance Optimized GFRP rebar with 900 MPa Guaranteed Tensile Strength. *J. Adv. Concr. Technol.* **2015**, *13*, 252–262. [CrossRef]

2. Benmokrane, B.; Mohamed, H.M.; Manalo, A.; Cousin, P. Evaluation of Physical and Durability Characteristics of New Headed Glass Fiber–Reinforced Polymer Bars for Concrete Structures. *J. Compos. Constr.* **2017**, *21*, 04016081. [CrossRef]
3. Okelo, R.; Yuan, R.L. Bond Strength of Fiber Reinforced Polymer Rebars in Normal Strength Concrete. *J. Compos. Constr.* **2005**, *9*, 203–213. [CrossRef]
4. Cho, J.-R.; Park, Y.-H.; Park, S.-Y.; Kim, S.-T. Pullout behavior of concrete-headed GFRP rebars. In Proceedings of the ASEM17, the 2017 World Congress, Seoul, Korea, 28 August–1 September 2017.
5. Hao, Q.; Wang, Y.; He, Z.; Ou, J. Bond strength of glass fiber reinforced polymer ribbed rebars in normal strength concrete. *Constr. Build. Mater.* **2009**, *23*, 865–871. [CrossRef]
6. Committee ACI 440. *Guide for the Design and Construction of Concrete Reinforced with FRP Bars (ACI 440.1 R-06)*; American Concrete Institute: Detroit, MI, USA, 2006.
7. American Society for Testing and Materials. *ASTM C39: Standard Test Method for Compressive Strength of Cylindrical Concrete Specimens*; ASTM International: West Conshohocken, PA, USA, 2012.
8. Zadeh, H.J.; Nanni, A. Design of RC Columns Using Glass FRP Reinforcement. *J. Compos. Constr.* **2013**, *17*, 294–304. [CrossRef]
9. Porter, S.D.; Julander, J.L.; Halling, M.W.; Barr, P.J.; Boyle, H.; Xing, S. Flexural Testing of Precast Bridge Deck Panel Connections. *J. Bridg. Eng.* **2011**, *16*, 422–430. [CrossRef]
10. Maranan, G.; Manalo, A.; Karunasena, W.M.; Benmokrane, B. Pullout behaviour of GFRP bars with anchor head in geopolymer concrete. *Compos. Struct.* **2015**, *132*, 1–25. [CrossRef]
11. AASHTO. *Bridge Design Guide Specifications for GFRP—Reinforced Concrete Bridge Decks and Traffic Railings*; American Association of State Highway and Transportation Officials: Washington, DC, USA, 2009.
12. Korea Concrete Institute. *Design Guideline of FRP-Reinforced Concrete Structures*; Korea Concrete Institute: Seoul, Korea, 2019.

© 2020 by the authors. Licensee MDPI, Basel, Switzerland. This article is an open access article distributed under the terms and conditions of the Creative Commons Attribution (CC BY) license (http://creativecommons.org/licenses/by/4.0/).

Article

Applications of Sustainable Polymer-Based Phase Change Materials in Mortars Composed by Different Binders

Mariaenrica Frigione [1,*], **Mariateresa Lettieri** [2], **Antonella Sarcinella** [1] and
José Luìs Barroso de Aguiar [3]

1. Innovation Engineering Department, University of Salento, Prov.le Lecce-Monteroni, 73100 Lecce, Italy; antonella.sarcinella@unisalento.it
2. Institute of Archaeological Heritage-Monuments and Sites, CNR-IBAM, Prov.le Lecce-Monteroni, 73100 Lecce, Italy; mariateresa.lettieri@cnr.it
3. Civil Engineering Department, University of Minho, Campus de Azurém, 4800-058 Guimarães, Portugal; aguiar@civil.uminho.pt
* Correspondence: mariaenrica.frigione@unisalento.it; Tel.: +39-0832-297215

Received: 2 October 2019; Accepted: 23 October 2019; Published: 25 October 2019

Abstract: Eco-sustainable, low toxic and low flammable poly-ethylene glycol (PEG) was forced into flakes of the porous Lecce stone (LS), collected as stone cutting wastes, employing a very simple cheap method, to produce a "form-stable" phase change material (PCM). The experimental PCM was included in mortars based on different binders (hydraulic lime, gypsum and cement) in two compositions. The main thermal and mechanical characteristics of the produced mortars were evaluated in order to assess the effects due to the incorporation of the PEG-based PCM. The mortars containing the PEG-based PCM were found to be suitable as thermal energy storage systems, still displaying the characteristics melting and crystallization peaks of PEG polymer, even if the related enthalpies measured on the mortars were appreciably reduced respect to pure PEG. The general reduction in mechanical properties (in flexural and compressive mode) measured on all the mortars, brought about by the presence of PEG-based PCM, was overcome by producing mortars possessing a greater amount of binder. The proposed LS/PEG composite can be considered, therefore, as a promising PCM system for the different mortars analyzed, provided that an optimal composition is identified for each binder.

Keywords: cement; gypsum; hydraulic lime; mechanical properties; mortars; phase-change materials (PCM); sustainable materials for buildings; thermal energy storage

1. Introduction

At the present time, the worldwide research is strongly oriented to identify innovative routes to reduce the global consumption of energy. The concerns about the climate change, from one side, and the rapid depletion of natural resources, from the other, are pushing the international policy to invest on research activities that can address the growing pressing environmental issues, for instance, by improving energy efficiency of buildings. Buildings, in fact, are one of the biggest energy consumers, due to the internal heating/cooling requirements. In addition, the principal source of energy used in buildings comes from non-renewable fossil fuels, whose combustion develops carbon dioxide, with a consequent strong negative impact on the environment [1].

In ancient or old residential buildings low levels of thermal efficiency are frequently registered, since they have been constructed in absence of any regulation or rule from this point of view, neither employing proper insulation materials or devices [2]. Such buildings, therefore, need to be renewed in order to improve the thermal comfort conditions for occupants and to reduce the consumption of energy for heating and cooling; this task can be realized using innovative smart materials [3]. Among the recent techniques devised to improve energy efficiency in buildings, a relevant position is covered by the latent heat thermal energy storage (LHTES) systems, involving the storage of energy in a so-called phase change material (PCM) incorporated in construction materials.

A PCM is able to change its physical status, i.e., from liquid to solid to liquid again, as a consequence of the fluctuations in the external temperature [4]. When the environmental temperature is high (during daytime, for instance), a PCM is able to melt and store the melting enthalpy. In contrast, when the external temperature decreases, the PCM is capable to release the previously stored energy, solidifying again [5]. Due to this novel technology, the temperature inside a building can be maintained fairly constant, with a consequent decrease in heating/cooling energy expenses. The use of a PCM system in construction materials is able to supply several additional advantages, in terms of: reduced gaps between peak and off-peak thermal loads; cut in energy costs; improved interior thermal comfort in buildings and reductions of CO_2 developed in atmosphere [6,7].

The incorporation of a suitable PCM into construction materials, through the passive building concept, has been recognized as the most effective solution. Different elements in a building (i.e., wallboard, floors, bricks, roof and concrete) can be combined with PCMs to increase their thermal energy storage capacity [8–10]. Other solutions can be also used in constructions with the same aim [11,12].

However, according to recent literature, the most feasible solution to include a PCM is based on its introduction inside the building envelope. In this way, the phase change material will be able to absorb and release heat during the hours of daylight [6,13–17].

Mortars, based on different binders, are considered as suitable mediums for PCMs. The incorporation of a PCM in a mortar represents a valuable solution due to large heat exchange area surfaces where mortars are applied. In addition, the PCM material included in the mortar can be shaped in a wide variety of forms and sizes, for each specific need [18]. The first experimental researches on mortars containing PCMs were largely focused on cement and gypsum binders, due to their initial good mechanical performance and thermal properties. However, when PCMs are added to such binders, substantial reductions in mechanical properties are generally registered [19–21].

Starting from these unsatisfactory results, further research moved towards investigation focused on mortars to be used as renders and coatings: these materials, in fact, do not require elevated values of mechanical strength. These mortars can be realized using different binders, such as aerial or hydraulic lime and, in some case, geopolymers [2,22,23]. Lime-based mortars, in addition, can be employed for building retrofitting, where render compatibility must be assured [24].

The form-stable is one of the easiest methods to incorporate an active PCM component into a porous inert support material [7,19]. The PCM composite can be, in fact, obtained by immersing the inert support in the liquid PCM; a vacuum pump can be employed to force the impregnation process.

In the first part of the research [25,26], a novel eco-sustainable form-stable polymeric PCM has been prepared, starting from small pieces of Lecce stone (LS), as support matrix, and low toxic and low flammable PEG (poly-ethylene glycol) as active phase change material. LS was obtained by waste product from a quarry sited in the same region where University of Salento is located. PEG possesses suitable phase change temperatures [26]; its large melting/crystallization enthalpy further supports the selection of this material. PEG displays also high long-term thermal/chemical stability and resistance to corrosion, with a limited volume change during solid–liquid phase transformation [27]. The originality of the designed PCM resides in the use of a waste natural material (LS) with the addition of an eco-sustainable one (PEG); the simplicity of the procedure used to obtain the PCM composite and the low cost of the resulting composite system represent additional valuable benefits.

In a first paper [26], the obtained LS/PEG form-stable PCM system was added as aggregate to an aerial lime, measuring different physical and mechanical properties of the resulting mortars. In this second paper, the same PCM composite was included to hydraulic lime, gypsum and cement-based mortar formulations. Taking into account the aim of the wide research project, i.e., the assessment of the thermal efficiency of the proposed novel PCM to manufacture mortars based on different binders, the influence of the PCM inclusion on some properties of the fresh and hardened mortars, such as workability, compressive and flexural strengths, was investigated.

2. Materials and Methods

2.1. Materials: LS/PEG Composite

Starting from our previous work [25,26], Lecce stone (LS), a biocalcarenite typical of Salento area (South Italy), was chosen as a porous support to realize form—stable PCM composites, to be added to different mortars. LS was specially selected for its characteristic high open porosity [28]; in addition, LS can be readily available as a waste product of the extraction and working of the stone from quarries. In this study, Lecce stone, supplied in the form of flakes, was further reduced in small pieces and sieved up to a granulometry ranging between 1.6 and 2.0 mm, as illustrated in Figure 1a.

The PCM selected in this study was poly(ethylene glycol). It was supplied in solid form (Sigma–Aldrich company, Germany) with the trade name PEG 1000. According to the data sheet, the density of PEG 1000 at 20 °C was 1.2 g/cm^3. The purchased product is illustrated in Figure 1b. The motivation with which PEG 1000 was selected in this research mainly relied on its favorable melting characteristics (a melting point ranging between 37 °C and 40 °C, and a heat of melting of about 129 J/g), which render this material a potentially optimal phase change material for mortars to be employed in Mediterranean regions [26,29,30]. PEG, in addition, displays a series of positive properties, such as: its cheapness, the low environmental impact and toxicity, and a low flammability, all features highly appreciated in the construction industry.

Figure 1. (a) Milled and sieved Lecce stone and (b) PEG 1000 in solid form.

To prepare the form—stable LS/PEG composite (as shown in Figure 2)—a vacuum impregnation process was employed, a cheap and simple method that can be easily realized in a small scale laboratory as well as at the industrial level. The detailed procedure employed, identified as the best one after several trials, has been reported in our previous work [26]; the percentage of PEG that is absorbed in LS, in these specific conditions, is 23% by weight. In the same paper, it was demonstrated that the sustainable LS/PEG stable-form PCM, produced following the optimized procedure, displayed appropriate LHTES properties and, therefore, it is a promising candidate to produce mortars able to improve the thermal efficiency of the buildings, increasing the comfort conditions of occupants.

Figure 2. Form–stable LS/PEG composite obtained through the vacuum impregnation process.

2.2. Materials: Mortars and Their Manufacture

Different binders were employed in this study to produce mortars containing form-stable LS/PEG, i.e., hydraulic lime, gypsum and cement. In a previous paper, aerial lime-based mortar formulations containing this PCM have been already produced and investigated [26]. It was found, however, that the addition of the PEG-based PCM caused an unsuitable reduction of compressive and flexural strength values of the aerial lime-based mortar. The ongoing research project, therefore, continues the investigations in order to identify a different mortar, with an appropriate composition, that can take advantage from the addition of LS/PEG in terms of energy saving, still displaying adequate mechanical properties for the intended applications.

Different Portuguese companies supplied the following binders: a natural hydraulic lime (NHL) with a density of 2700 kg/m^3, was supplied by CIMPOR (Lisbon, Portugal); a conventional gypsum, with high fineness and density of 2960 kg/m^3, was provided by SIVAL (Souto da Carpalhosa, Leira, Portugal); finally, a CEM I 42.5 R cement, with a density of 3030 kg/m^3, was supplied by SECIL (Lisbon, Portugal). The chemical composition of the cement was SiO_2, Al_2O_3, Fe_2O_3, CaO, MgO, SO_3, K_2O and Na_2O, with a specific surface area of 4007 cm^2/g.

A superplasticizer (SP), i.e., a polyacrylate (MasterGlenium SKY 627, by BASF company), was always added to each mortar composition in the same quantity, in order to reduce the amount of water required for the mixing. The density of the SP is 1050 kg/m^3. In Table 1, the composition of all the mortars realized and analyzed, produced according to the European Norm EN 998-1 [31], are reported.

Table 1. Mortar compositions (reported as kg/m^3 of produced mortar).

System	Binder/Content	Aggregates		SP	Water Saturation	Water	Water/Binder
		LS	PEG Content				
HL_{500}_LS	Hydraulic Lime/500	1480	0	15	370	300	0.60
HL_{500}_LS/PEG	Hydraulic Lime/500	1678	386	15	0	350	0.70
G_{500}_LS	Gypsum/500	1454	0	15	366	325	0.65
G_{500}_LS/PEG	Gypsum/500	1645	378	15	0	375	0.75
C_{500}_LS	Cement/500	1392	0	15	350	340	0.68
C_{500}_LS/PEG	Cement/500	1790	412	15	0	350	0.70
HL_{800}_LS	Hydraulic Lime/800	1092	0	15	275	320	0.40
HL_{800}_LS/PEG	Hydraulic Lime/800	1729	398	15	0	375	0.47
G_{800}_LS	Gypsum/800	1169	0	15	294	320	0.40
G_{800}_LS/PEG	Gypsum/800	1472	339	15	0	340	0.43
C_{800}_LS	Cement/800	1070	0	15	269	296	0.37
C_{800}_LS/PEG	Cement/800	1347	310	15	0	360	0.45

A total of twelve compositions were developed: six of them were produced by adding different percentages of LS/PEG to the binders, in order to evaluate the thermal properties of the single mortars as a function of the binder and of the PCM contents. For comparison purposes, six control formulations were prepared by introducing LS alone as aggregate. The indication "water saturation" in Table 1 accounts for the water used to saturate the LS aggregates, possessing a high porosity, to prevent them from absorbing the required water for the mortars manufacture. This additional water was not required when LS/PEG composite was added, since PEG was able to (almost) completely saturate the pores of Lecce stone.

Referring to the compositions selected to manufacture the mortars, it is well known that a certain reduction in mechanical properties of the mortars containing the PCM composite can be expected [32]. Starting from this consideration, some mortar compositions possessing a high amount of binder were also produced. The aim of the present study was, in fact, to identify the most convenient composition for each binder able to produce mortars with a good thermal efficiency and, at the same time, good mechanical properties.

2.3. Methods and Test Procedures

The workability in the fresh state of the produced mortars (summarized in Table 1) was first assessed. To this aim, the flow table method was employed, according to the European code EN 1015-3 [33]. The test was repeated twice, at least, on each produced formulation and the results averaged.

Then, the mechanical properties of the 28-days cured mortars (cast in iron molds, de-molded after 2 days and left for 26 additional days in standard conditions of 25 °C, R.H. of 50%), were measured in both flexural and compressive mode, following the standard EN 1015-11 recommendations [34]. For each composition of the different mortars systems, three prismatic specimens ($40 \times 40 \times 160$ mm^3) were tested using a Lloyd dynamometer machine (LR50K Plus by Ametek Company), with a load cell of 50 kN, and the results averaged. The speeds employed to perform mechanical tests were 6 μm/s for flexural tests and 12 μm/s for compressive ones, respectively.

Calorimetry was employed to measure the phase change processes taking place in each mortar formulation containing the PEG-based PCM [25]. With this purpose, a DSC1 (Stare System, Mettler Toledo) instrument was employed to analyze small samples of each mortar if subjected to heating–cooling thermal cycles, performed at 10 °C/min under nitrogen atmosphere (flow rate: 60 mL min^{-1}): the first from −10° to 100 °C, the second from 100° to −10 °C. For each mortar formulation, three specimens were analyzed, averaging the results. For comparison purposes, samples of the pure PEG were analyzed in DSC, using the same procedure previously described.

3. Results and Discussion

3.1. Workability of Mortars

Firstly, the workability of the produced mortars was assessed; the values measured on each formulation are reported in Table 2. Keeping in mind that an appropriate value of workability for all the mortars produced (based on hydraulic lime, gypsum and cement) should lie in the range 160–180 mm [32], from the observation of data in Table 2 it is concluded that all the studied mortar formulations display an adequate value of workability.

Sometimes, in order to reach a good workability, it is necessary to increase the amount of water of the mortar composition and, according to Figure 3, it was observed that the amount of water increase when the LS/PEG aggregates were used.

The incorporation of PCM led to a rise in water content of about 15% for the hydraulic lime based mortar and for the gypsum; while for the cement based mortar was less, especially when it was used a higher content of binder. On the other hand, as reported in the literature, the incorporation of PCM causes an increase in the amount of water [21,35]. It is well known that a proper content of water is necessary to assure a good workability of the mortar formulation and that this amount should not be excessive: it can lead, in fact, to a high microporosity in the mortar and, as a consequence, to unsuitable mechanical properties [36]. Therefore, superplasticizers (SP) are commonly used as they improve the workability of the mortars even using limited amounts of water.

Table 2. Values of workability of the mortars possessing different compositions.

System	Workability (mm)
HL_{500}_LS	170 ± 2
HL_{500}_LS/PEG	177 ± 3
G_{500}_LS	160 ± 1
G_{500}_LS/PEG	180 ± 3
C_{500}_LS	160 ± 1
C_{500}_LS/PEG	160 ± 1
HL_{800}_LS	165 ± 2
HL_{800}_LS/PEG	175 ± 2
G_{800}_LS	160 ± 1
G_{800}_LS/PEG	160 ± 1
C_{800}_LS	160 ± 1
C_{800}_LS/PEG	178 ± 3

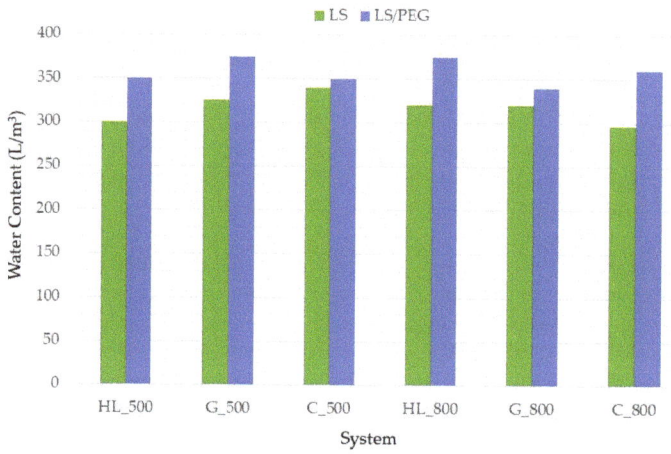

Figure 3. Variation of the water content for each mortar composition with the addition of the phase change material (PCM).

3.2. Results of DSC Analysis Performed on Mortars

The DSC trace obtained from calorimetric analysis performed on pure PEG is shown in Figure 4.

As previously reported [26], the used PEG polymer exhibits an endothermic (melting) peak approximately at 43 °C during the heating stage and an exothermic (crystallization) peak around 23 °C when the temperature is reduced down. Melting/crystallization enthalpy of about 129 J/g was calculated from DSC measurements, in accordance with results of different studies performed on the same material [27,37]. As already underlined, the thermal characteristics displayed by PEG 1000, i.e., the phase change temperatures and enthalpies, are appropriate for developing a form-stable PCM to be used as thermal energy storage material. In fact, also the prepared LS/PEG composite exhibited thermal properties good enough to include this composite as an effective PCM in indoor mortars, especially for applications in buildings located in warm (for instance Mediterranean) regions [26].

Then, in order to assess if LS/PEG aggregates can display suitable phase change characteristics even when incorporated into different mortars, DSC analyses were performed on small pieces of cured mortars containing the LS/PEG composite. The DSC traces recorded for mortar specimens containing LS/PEG composite are shown in Figure 5. The results obtained from the DSC experiments are summarized in Table 3.

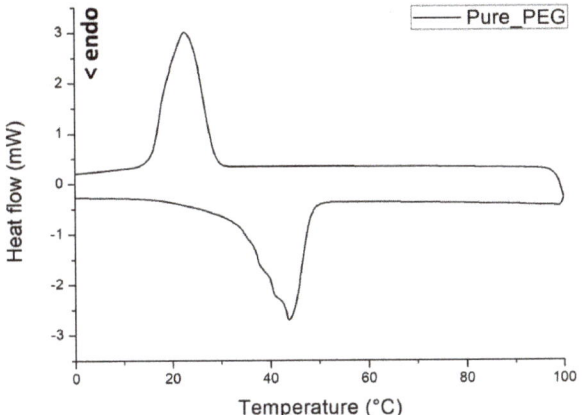

Figure 4. DSC thermograms recorded of pure PEG.

The theoretical enthalpy (ΔH_{Theor}) was also determined using Equation (1) [38].

$$\Delta H_{Theor} = (PEG\% \times \Delta H_{PEG})/100 \qquad (1)$$

where: PEG% is the PEG content in percentage and ΔH_{PEG} denotes the latent heat of the pristine PEG.

On the DSC thermograms of the mortars based on different binders and containing PEG added to Lecce stone (Figure 5), endothermic and exothermic peaks appeared during heating and cooling cycles, respectively, representing the melting and crystallization processes taking place in the PEG component. This observation confirms that phase transitions occurred in the mortar formulations containing the PCM under analysis, even if a low amount of the "active" component of the PCM is present [39]. Generally speaking, the higher the PCM content, the better the heat storage capacity of the mortar [40,41]. As expected, the specimens of mortars containing only LS did not display any melting/crystallization phenomena in the investigated range of temperatures (i.e., up to 100 °C).

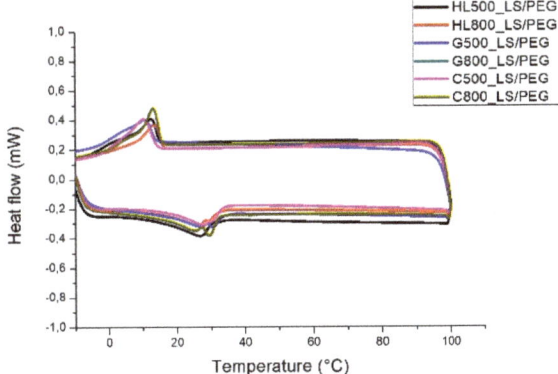

Figure 5. DSC thermograms of mortars containing LS/PEG PCM.

Table 3. Characteristic (initial, end and peak) temperatures and enthalpy measured during heating stage (melting) and subsequent cooling stage (crystallization) on pure PEG, on LS/PEG composite and on mortars, based on different binders, containing LS/PEG composite. For each investigated system, the content in PEG as a percentage and the theoretical enthalpies (as defined in Equation (1)), for both melting (ΔHm_{Theor}) and crystallization (ΔHc_{Theor}), are reported.

	System	PEG Content (%)	Onset (°C)	Endset (°C)	Tm (°C)	ΔHm (J/g)	ΔHm Theor (J/g)
	PEG	100	36.4 ± 0.6	50.2 ± 0.3	42.8 ± 1.1	129.3 ± 1.2	—
	LS/PEG	23	12.4 ± 0.5	50.2 ± 0.8	39.3 ± 0.7	27.7 ± 0.9	29.7
	HL_{500}_LS/PEG	15.2	9.7 ± 0.4	37.9 ± 0.6	26.9 ± 0.7	6.8 ± 1.1	19.7
Heating Stage	HL_{800}_LS/PEG	13.6	2.2 ± 1.0	41.8 ± 1.1	26.0 ± 0.8	7.9 ± 0.9	17.6
	G_{500}_LS/PEG	14.9	3.5 ± 0.4	42.8 ± 0.2	30.4 ± 0.4	8.5 ± 0.9	19.3
	G_{800}_LS/PEG	12.9	2.7 ± 0.5	35.9 ± 0.7	28.9 ± 1.0	7.8 ± 1.2	16.7
	C_{500}_LS/PEG	15.5	4.3 ± 1.1	38.4 ± 0.7	27.6 ± 0.3	9.0 ± 1.3	20.0
	C_{800}_LS/PEG	12.3	3.4 ± 0.1	36.3 ± 0.1	30.0 ± 0.3	7.7 ± 0.2	15.9
	System	PEG Content (%)	Endset (°C)	Onset (°C)	Tc (°C)	ΔHc (J/g)	ΔHc Theor (J/g)
	PEG	100	18.3 ± 1.1	26.7 ± 1.1	23.6 ± 1.2	129.8 ± 0.8	—
	LS/PEG	23	3.7 ± 0.1	28.8 ± 0.3	19.4 ± 0.2	28.6 ± 0.1	38.6
	HL_{500}_LS/PEG	15.2	0.4 ± 0.5	17.6 ± 0.3	12.5 ± 0.4	6.2 ± 1.1	25.5
Cooling Stage	HL_{800}_LS/PEG	13.6	1.6 ± 0.4	18.9 ± 0.8	13.5 ± 0.2	6.0 ± 0.7	22.8
	G_{500}_LS/PEG	14.9	−1.1 ± 0.6	19.2 ± 0.7	10.6 ± 0.8	7.4 ± 0.9	25.0
	G_{800}_LS/PEG	12.9	0.6 ± 0.9	16.9 ± 1.0	10.8 ± 0.7	7.5 ± 1.2	21.7
	C_{500}_LS/PEG	15.5	0.2 ± 0.6	19.2 ± 0.7	10.3 ± 0.3	8.8 ± 1.2	26.0
	C_{800}_LS/PEG	12.3	−1.1 ± 0.2	16.7 ± 0.2	10.8 ± 0.3	8.7 ± 0.4	20.6

The melting and crystallization peak temperatures calculated for the mortar specimens containing LS/PEG aggregates are in the range 27–30 °C and 10–13 °C, respectively, irrespective of the kind of binder. These temperatures are reported to be favorable to obtain a PCM-based mortar to be employed as thermal energy storage system included in the exterior and/or in indoor walls of buildings located in warm regions [35].

From the observation of the data reported in Table 3, it is noticed that both the melting and crystallization processes taking place in the PEG component contained in the mortars occurred at lower temperatures then those of pure PEG, being the decrease in the temperature peak approximately of 13–17 °C for melting process and of 10–13°C for the crystallization one. The shift in melting and crystallization processes toward lower temperatures has already found in different similar studies [25,38,42–48]. It has been attributed to physical surface interactions (such as capillary forces, hydrogen bonds and surface adsorption) between PEG and the other different components of each mortar.

Referring to the melting/crystallization enthalpies, the observed results, denoting a drastic decrease of the values (around 7–9 J/g) measured for the mortars containing PCM in comparison to the enthalpies of pure PEG (129 J/g), are mainly due to the low amount of PEG (that is the only crystallizable phase) present into the mortars samples. However, peak temperatures and enthalpy values lower than those expected (i.e., the theoretical values, normalized to the PEG content), were always measured. The presumed reason for these results may be that most of the PEG chains embedded in the stone pores could experience the phase changes only to a limited extent [49]. Phase transition temperatures and enthalpies decline, until the peak disappearance, as a consequence of confinement [50–52]; this occurs since the change from the crystalline to the melting state (and vice versa) is hindered. In the case of the investigated mortars, it can be hypothesized that, not only the store pore confinement influenced the thermal properties, but also the binder surrounding the LS/PEG aggregates can limit the PEG movements, thus further reducing the phase transition temperatures and enthalpies [53,54].

Nevertheless, the presence of well-defined and measurable melting/crystallization peaks also in the mortars containing the proposed PCM testifies that LS/PEG composite has a potential to act as an efficient phase change material.

3.3. Mechanical Properties of Mortars

The results of flexural and compressive mechanical tests performed on the different mortar formulations under study are reported in Table 4.

Table 4. Mechanical properties of the cured mortars measured in flexural and compressive mode.

System	Flexural Strength (MPa)	Compressive Strength (MPa)
HL_{500}_LS	1.1 ± 0.3	2.8 ± 0.8 [CSII] [a]
HL_{500}_LS/PEG	0.1 ± 0.1	0.4 ± 0.1 [CSI] [a]
G_{500}_LS	3.2 ± 0.2	4.8 ± 0.2 [CSII] [a]
G_{500}_LS/PEG	0.5 ± 0.1	0.4 ± 0.2 [CSI] [a]
C_{500}_LS	5.8 ± 0.3	20.5 ± 0.4 [CSIV] [a]
C_{500}_LS/PEG	1.0 ± 0.2	1.1 ± 0.2 [CSI] [a]
HL_{800}_LS	2.8 ± 0.5	17.0 ± 0.2 [CSIV] [a]
HL_{800}_LS/PEG	0.4 ± 0.1	1.5 ± 0.1 [CSI/CSII] [a]
G_{800}_LS	4.1 ± 0.2	16.4 ± 0.6 [CSIV] [a]
G_{800}_LS/PEG	1.6 ± 0.2	3.3 ± 0.3 [CSII] [a]
C_{800}_LS	9.2 ± 0.9	26.3 ± 0.4 [CSIV] [a]
C_{800}_LS/PEG	1.9 ± 0.3	3.4 ± 0.8 [CSII] [a]

[a] Category of the mechanical resistance of the mortar according to the standard NP EN 998-1.

As observed from the data reported in Table 4, an appreciable decrease in both flexural and compressive strength values is related to the introduction of LS/PEG composite and, most likely, to the greater content of water necessary when the PCM is added to the mortar compositions, irrespective to the kind of binder. It is reported, in fact, that the increase of water content in mortars upon addition of PCMs determines an increase in their microporosity, leading the latter to reductions in their mechanical strength [55]. In fact, the efforts of researchers moved towards the study of mortars basically employed for interior and/or exterior coatings, since a very high value of mechanical strength is not mandatory in such applications. It could be also hypothesized, however, that the decrease of the flexural and compressive strength values can be, at least partly, attributed to a loss of adhesion between the PCM aggregate and the binder paste. Analytical studies are in progress to prove, or exclude, this hypothesis. Finally, as expected, by increasing the percentage of binder, it is possible to increase the mechanical characteristics (flexural and compressive mode) even in the mortars containing LS/PEG composite [56].

The reductions in flexural strength brought about by the addition of PEG go from 62%, registered for G_{800}_LS/PEG system, to 95%, in the case of HL_{500}_LS/PEG. Referring to the compressive strength values, hydraulic lime and gypsum with a binder content of 500 kg/m^3 present a decrease of one strength class when PEG is added to Lecce stone aggregate for the mortars, i.e., from CSII to CSI.

The reduction in compressive strength was much more appreciable in the case of cement-based mortar containing a binder content of 500 kg/m^3 (about 94%), with a drop in the strength class classification from CSIV to CSI. By increasing the binder content up to 800 kg/m^3, hydraulic lime presented a noticeable decrease in compressive strength by the incorporation of PEG, with a consequent strong reduction in the strength class classification, from CSIV to CSI/CSII. Referring to gypsum, a high decrease in compressive strength was again recorded (nearly 80%), with a consequent fall from CSIV to CSII classification. Similar results were found for cement-based mortars, with a higher decrease in compressive strength (about 87%) and a drop from CSIV to CSII strength class classification.

Keeping in mind that, to be successfully employed in the construction field, the mortars should respect the recommendations reported in the standard NP EN 998-1, thus they should have a minimum classification of CS II, the systems G_{800}_LS/PEG and C_{800}_LS/PEG are both fully respecting this requisite, while the formulation HL_{800}_LS/PEG is close to the strength target value. The mortars containing the

PCM based on LS/PEG with a content of binder 500 kg/m^3, on the other hand, display compressive characteristics not fully adequate (i.e., falling within CS I type).

Hydraulic lime is generally used to produce plastic mortars easily to apply and capable to set and harden in extreme conditions, including underwater, making this material appropriate for applications located close to the sea, lakes, rivers, etc. They are typically used in applications characterized by not excessive loads and to realize foundation mortars as well as renders and plasters for conservation, restoration and new build construction. An advantage is represented by its permeability to water vapor, i.e., it does not trap moisture in the walls allowing buildings to "breathe".

Despite the fact that most of the research on mortars containing PCMs is focused on gypsum and cement formulations, due to the outstanding thermal characteristics and mechanical properties of these binders, some papers dealing with the use of phase change materials in hydraulic lime appeared in the last years [22,55–58]. Referring to the mechanical characteristics displayed by these modified-mortars, the addition of 20% and 40% microencapsulated PCM to hydraulic lime (500 kg/m^3 content) determined a decrease in compressive strength of about 47% and 52% (the latter from 5.37 MPa for the hydraulic lime control system to 2.58 MPa), respectively [55,56]. While, the decrease in flexural strength of the hydraulic lime mortar including 20% or 40% of PCM microcapsules was found around 20% and 30%, respectively. Similar results were found in a different study again focused on hydraulic lime mortar, again with the addition of a 40% microencapsulated PCM, by the same authors, with a decrease in flexural strength of about 27% (from 2.2 MPa for control mortar to 1.6 MPa for composite formulation) [58].

In conclusion, the results of mechanical characteristics (flexural and compressive strength) found in the present study for hydraulic lime mortar containing LS/PEG composite, when compared with the same mortar modified with different PCM systems, appeared to be fairly satisfactory for the proposed applications.

Gypsum is one of the more widely used construction materials, mainly in interior designing. It is mainly used as surface materials, being its application prominent for finishing wall and ceiling in form of plaster. Gypsum is light, long-lasting and presents an intrinsic very high fire resistance.

There are several papers published in the last years on gypsum-based mortars containing different PCM systems [55,56,58–61]. Even for this kind of binder, the good mechanical properties of gypsum were found to be severely compromised by the addition of a PCM system. In fact, it was found that the introduction of 20% and 40% of PCM microcapsules determined a decrease in the flexural strength of gypsum (500 kg/m^3 content) of around 40% and up to 80%, respectively, and in compressive strength of about 50% and 64%, respectively [55,56,58]. The reductions in strength found in our research for PCM-modified gypsum, therefore, are more or less in line with those reported in previous literature, especially if we refer to the gypsum mortar containing the highest content of binder, i.e., 800 kg/m^3. It is confirmed, therefore, that we can consider this system suitable for the proposed applications.

The use of cement-based mortars in constructions is extremely wide: from restoration and repairing of damaged concrete to the patching or filling concrete, rendering and floor leveling; it can also be used to produce precast products. These mortars display a great mechanical resistance and a low porosity, with consequent impermeability to water vapor.

The reductions in flexural strength values calculated on mortars based on cement (CEM I type) and containing a PCM system achieved up to 50% depending on the kind and amount of PCM system [56]. The reductions in compressive strength are even more severe (up to 70%) at the highest contents of encapsulated PCM [20,21,56,62]. Our cement-based mortars containing LS/PEG PCM, therefore, followed the same trend, once again referring to the mortar containing the highest content of cement, i.e., 800 kg/m^3.

From the results of mechanical (flexural and compressive) tests, and the relative discussion, it can be concluded that all the produced mortar compositions display values of flexural and compressive strengths that are included in the standard classification for not structural mortars. On the other hand, not all of them fall into the CSII type based on the compressive strength, as requested by the standard

NP EN 998-1. Adequate mechanical properties for the intended purpose are those recorded for the gypsum- and cement-based mortars containing a greater amount of binder (800 kg/m^3); close to the strength target value is also the formulation HL$_{800}$_LS/PEG. None of the mortars containing the PCM based on LS/PEG with a content of binder 500 kg/m^3 falls within the CS I type.

Experiments are in progress on some of the produced formulations using the largest amount of binder (800 kg/m^3), in order to verify the thermal properties of such formulations and, in turn, to assess the effective efficiency as phase change material of LS/PEG composite.

4. Conclusions

A wide experimental investigation was carried out in order to evaluate the effect of the incorporation of a PEG-based PCM on the mechanical and thermal properties of mortars based on different binders. The selected PCM was composed by eco-sustainable PEG simply included in a waste of natural stone, i.e., Lecce stone. The low cost of the method used to produce the PCM composite, low toxicity and low flammability of both components constitute advantageous additional characteristics for applications in the construction industry.

The obtained results showed that the addition of a phase change material in mortars employed for the thermal isolation of building walls (for indoor/outdoor applications) caused significant changes in their properties, either in the fresh and hardened state. As expected, the incorporation of LS/PEG aggregates required an increase in the amount of water to reach a suitable workability in all mortars. It was verified that the thermal properties of the LS/PEG composite decreased when it was incorporated as an aggregate in mortar mixes. However, well-defined and measurable melting/crystallization peaks were still observed in the mortars containing the proposed PCM. The melting and crystallization peak temperatures allowed for obtaining PCM-based mortars suitable as thermal energy storage systems in both exterior and indoor walls of buildings, especially those located in warm regions. The mechanical properties (in flexural and compressive mode) decreased when LS/PEG aggregates were included in the mortar compositions. This phenomenon, however, is common for any mortar containing a PCM system, as reported in the literature. Nevertheless, by increasing the amount of binder, it was possible to achieve mortars that can be classified at least as CSII, except for the system based on hydraulic lime, i.e., HL800LS/PEG. Even this system, however, can be classified in between CSI and CSII categories.

In order to assess if the produced PCM-mortars are effectively able to produce an improvement in the thermal performance of building walls realized with them, and consequently to limit the energy consumption for cooling and heating, experiments are in progress to evaluate the thermal performance of elements produced with such novel mortars.

Author Contributions: Conceptualization, M.F.; experimental work and material characterization, A.S.; analysis of the data, A.S.; data curation, M.F., J.L.B.d.A., M.L.; writing—original draft preparation, A.S.; writing—review and editing, M.F., J.L.B.d.A., M.L.; supervision, M.F., J.L.B.d.A.

Funding: This research received no external funding.

Acknowledgments: The Authors wish to thank: L. Pascali and the Staff of S.I.PRE. S.r.l. (Cutrofiano, Lecce, Italy) for the technical support; Tecnoprove Company (Ostuni, Brindisi, Italy), and in particular V. Parisi for his technical support; Pitardi Cavamonti Company (Melpignano, Lecce, Italy) for supplying the flakes of Lecce Stone.

Conflicts of Interest: The authors declare no conflict of interest.

References

1. Bilgen, S. Structure and Environmental Impact of Global Energy Consumption. *Renew. Sustain. Energy Rev.* **2014**, *38*, 890–902. [CrossRef]
2. Cunha, S.; Aguiar, J.; Ferreira, V. Mortars with Incorporation of Phase Change Materials for Thermal Rehabilitation. *Int. J. Arch. Herit.* **2016**, 1–10. [CrossRef]
3. Munarim, U.; Ghisi, E. Environmental Feasibility of Heritage Buildings Rehabilitation. *Renew. Sustain. Energy Rev.* **2016**, *58*, 235–249. [CrossRef]

4. Janarthanan, B.; Sagadevan, S. Thermal Energy Storage Using Phase Change Materials and Their Applications: A Review. *Int. J. ChemTech Rse.* **2015**, *8*, 250–256.
5. Pomianowski, M.; Heiselberg, P.; Zhang, Y. Review of Thermal Energy Storage Technologies Based on PCM Application in Buildings. *Energy Build.* **2013**, *67*, 56–69. [CrossRef]
6. Kalnæs, S.E.; Jelle, B.P. Phase Change Materials and Products for Building Applications: A State-Of-The-Art Review and Future Research Opportunities. *Energy Build.* **2015**, *94*, 150–176. [CrossRef]
7. Zhang, P.; Xiao, X.; Ma, Z. A Review of the Composite Phase Change Materials: Fabrication, Characterization, Mathematical Modeling and Application to Performance Enhancement. *Appl. Energy* **2016**, *165*, 472–510. [CrossRef]
8. Cui, Y.Q.; Riffat, S. Review on Phase Change Materials for Building Applications. *Appl. Mech. Mater.* **2011**, *71*, 1958–1962. [CrossRef]
9. Madessa, H.B. A Review of the Performance of Buildings Integrated with Phase Change Material: Opportunities for Application in Cold Climate. *Energy Procedia* **2014**, *62*, 318–328. [CrossRef]
10. Nkwetta, D.N.; Haghighat, F. Thermal Energy Storage with Phase Change Material—A State-Of-The Art Review. *Sustain. Cities Soc.* **2014**, *10*, 87–100. [CrossRef]
11. Ogrodnik, P.; Zegardlo, B.; Szeląg, M. The Use of Heat-Resistant Concrete Made with Ceramic Sanitary Ware Waste for a Thermal Energy Storage. *Appl. Sci.* **2017**, *7*, 1303. [CrossRef]
12. John, E.; Hale, M.; Selvam, P. Concrete as a Thermal Energy Storage Medium for Thermocline Solar Energy Storage Systems. *Sol. Energy* **2013**, *96*, 194–204. [CrossRef]
13. Lu, S.; Li, Y.; Kong, X.; Pang, B.; Chen, Y.; Zheng, S.; Sun, L. A Review of PCM Energy Storage Technology Used in Buildings for the Global Warming Solution. In *Energy Solutions to Combat Global Warming*; Zhang, X., Dincer, I., Eds.; Springer International Publishing: Cham, Switzerland; Berlin, Germany, 2017; Volume 33, pp. 611–644, ISBN 978-3-319-26948-1.
14. Memon, S.A. Phase Change Materials Integrated in Building Walls: A State of the Art Review. *Renew. Sustain. Energy Rev.* **2014**, *31*, 870–906. [CrossRef]
15. Huang, X.; Alva, G.; Jia, Y.; Fang, G. Morphological Characterization and Applications of Phase Change Materials in Thermal Energy Storage: A Review. *Renew. Sustain. Energy Rev.* **2017**, *72*, 128–145. [CrossRef]
16. Song, M.; Niu, F.; Mao, N.; Hu, Y.; Deng, S. Review on Building Energy Performance Improvement Using Phase Change Materials. *Energy Build.* **2018**, *158*, 776–793. [CrossRef]
17. Kusama, Y.; Ishidoya, Y. Thermal Effects of a Novel Phase Change Material (PCM) Plaster Under Different Insulation and Heating Scenarios. *Energy Build.* **2017**, *141*, 226–237. [CrossRef]
18. Frigione, M.; Lettieri, M.; Sarcinella, A. Phase Change Materials for Energy Efficiency in Buildings and Their Use in Mortars. *Materials* **2019**, *12*, 1260. [CrossRef]
19. Aguayo, M.; Das, S.; Maroli, A.; Kabay, N.; Mertens, J.C.; Rajan, S.D.; Sant, G.; Chawla, N.; Neithalath, N. The Influence of Microencapsulated Phase Change Material (PCM) Characteristics on the Microstructure and Strength of Cementitious Composites: Experiments and Finite Element Simulations. *Cem. Concr. Compos.* **2016**, *73*, 29–41. [CrossRef]
20. Coppola, L.; Coffetti, D.; Lorenzi, S. Cement-Based Renders Manufactured with Phase-Change Materials: Applications and Feasibility. *Adv. Mater. Sci. Eng.* **2016**, *2016*, 1–6. [CrossRef]
21. Lecompte, T.; Le Bideau, P.; Glouannec, P.; Nortershäuser, D.; Le Masson, S. Mechanical and Thermo-Physical Behaviour of Concretes and Mortars Containing Phase Change Material. *Energy Build.* **2015**, *94*, 52–60. [CrossRef]
22. Pavlik, Z.; Trnik, A.; Keppert, M.; Pavlikova, M.; Zumar, J.; Cerny, R. Experimental Investigation of the Properties of Lime-Based Plaster-Containing PCM for Enhancing the Heat-Storage Capacity of Building Envelopes. *Int. J. Thermophys.* **2014**, *35*, 767–782. [CrossRef]
23. Kheradmand, M.; Abdollahnejad, Z.; Torgal, F.P. Mechanical Performance of Fly Ash Geopolymeric Mortars Containing Phase Change Materials. In *Proceedings of the International Congress on Polymers in Concrete (ICPIC 2018)*; Taha, M.M.R., Ed.; Springer International Publishing: Berlin, Germany, 2018; pp. 211–216.
24. Pacheco-Torgal, F.; Faria, J.; Jalali, S. Some Considerations about the Use of Lime–Cement Mortars for Building Conservation Purposes in Portugal: A Reprehensible Option or a Lesser Evil. *Constr. Build. Mater.* **2012**, *30*, 488–494. [CrossRef]

25. Frigione, M.; Lettieri, M.; Sarcinella, A.; De Aguiar, J.B. Mortars with Phase Change Materials (PCM) and Stone Waste to Improve Energy Efficiency in Buildings. In *International Congress on Polymers in Concrete (ICPIC 2018)*; Springer Science and Business Media LLC: Berlin, Germany, 2018; pp. 195–201.
26. Frigione, M.; Lettieri, M.; Sarcinella, A.; De Aguiar, J.B. Sustainable Polymer-Based Phase Change Materials for Energy Efficiency in Buildings and Their Application in Aerial Lime Mortars. *Constr. Build. Mater.* **2020**, *231*, 117149. [CrossRef]
27. Kou, Y.; Wang, S.; Luo, J.; Sun, K.; Zhang, J.; Tan, Z.; Shi, Q. Thermal Analysis and Heat Capacity Study of Polyethylene Glycol (PEG) Phase Change Materials for Thermal Energy Storage Applications. *J. Chem. Thermodyn.* **2019**, *128*, 259–274. [CrossRef]
28. Andriani, G.F.; Walsh, N. Petrophysical and Mechanical Properties of Soft and Porous Building Rocks Used in Apulian Monuments (South Italy). *Geol. Soc. London Spec. Publ.* **2010**, *333*, 129–141. [CrossRef]
29. Whiffen, T.R.; Riffat, S.B. A Review of PCM Technology for Thermal Energy Storage in the Built Environment: Part I. *Int. J. Low Carbon Technol.* **2013**, *8*, 147–158. [CrossRef]
30. Kamali, S. Review of Free Cooling System Using Phase Change Material for Building. *Energy Build.* **2014**, *80*, 131–136. [CrossRef]
31. European Committee for Standardization. *EN 998-1: 2010: Specification for Mortar for Masonry-Part 1: Rendering and Plastering Mortar*; CEN: Brussels, Belgium, 2010.
32. Cunha, S.; Aguiar, J.; Ferreira, V.; Tadeu, A. Mortars Based in Different Binders with Incorporation of Phase-Change Materials: Physical and Mechanical Properties. *Eur. J. Environ. Civ. Eng.* **2015**, *19*, 1–18. [CrossRef]
33. European Committee for Standardization (CEN). *EN 1015-3 (2004), Methods of Test for Mortar for Mansonry-Part 3: Determination of Consistence of Fresh Mortar (by flow table)*; CEN: Brussels, Belgium, 2004.
34. European Committee for Standardization (CEN). *EN 1015-11 (1999), Methods of Test for Mortar for Masonry-Part 11: Determination of Flexural and Compressive Strength of Hardened Mortar*; CEN: Brussels, Belgium, 1999.
35. Snoeck, D.; Priem, B.; Dubruel, P.; De Belie, N. Encapsulated Phase-Change Materials as Additives in Cementitious Materials to Promote Thermal Comfort in Concrete Constructions. *Mater. Struct.* **2016**, *49*, 225–239. [CrossRef]
36. Cunha, S.; Lucas, S.; Aguiar, J.; Ferreira, V.; Bragança, L. Influence of Incorporation Phase Change Materials, PCM, Granulates on Workability, Mechanical Strenght and Aesthetical Appearance of Lime and Gypsum Mortars. *Archit. Civ. Eng. Environ.* **2013**, *6*, 39–48.
37. Karaman, S.; Karaipekli, A.; Sarı, A.; Biçer, A. Polyethylene Glycol (PEG)/Diatomite Composite as a Novel Form-Stable Phase Change Material for Thermal Energy Storage. *Sol. Energy Mater. Sol. Cells* **2011**, *95*, 1647–1653. [CrossRef]
38. Qian, T.; Li, J.; Min, X.; Deng, Y.; Guan, W.; Ma, H. Polyethylene Glycol/Mesoporous Calcium Silicate Shape-Stabilized Composite Phase Change Material: Preparation, Characterization, and Adjustable Thermal Property. *Energy* **2015**, *82*, 333–340. [CrossRef]
39. Zhang, Z.; Shi, G.; Wang, S.; Fang, X.; Liu, X. Thermal Energy Storage Cement Mortar Containing n-Octadecane/Expanded Graphite Composite Phase Change Material. *Renew. Energy* **2013**, *50*, 670–675. [CrossRef]
40. Li, M.; Wu, Z.; Tan, J. Heat Storage Properties of the Cement Mortar Incorporated with Composite Phase Change Material. *Appl. Energy* **2013**, *103*, 393–399. [CrossRef]
41. Shadnia, R.; Zhang, L.; Li, P. Experimental Study of Geopolymer Mortar with Incorporated PCM. *Constr. Build. Mater.* **2015**, *84*, 95–102. [CrossRef]
42. Min, X.; Fang, M.; Huang, Z.; Liu, Y.; Huang, Y.; Wen, R.; Qian, T.; Wu, X. Enhanced Thermal Properties of Novel Shape-Stabilized PEG Composite Phase Change Materials with Radial Mesoporous Silica Sphere for Thermal Energy Storage. *Sci. Rep.* **2015**, *5*, 12964. [CrossRef]
43. Deng, Y.; He, M.; Li, J.; Yang, Z. Polyethylene Glycol-Carbon Nanotubes/Expanded Vermiculite Form-Stable Composite Phase Change Materials: Simultaneously Enhanced Latent Heat and Heat Transfer. *Polymers* **2018**, *10*, 889. [CrossRef]
44. Wang, Z.; Su, H.; Zhao, S.; Zhao, N. Influence of Phase Change Material on Mechanical and Thermal Properties of Clay Geopolymer Mortar. *Constr. Build. Mater.* **2016**, *120*, 329–334. [CrossRef]
45. Wang, P.; Li, N.; Zhao, C.S.; Wu, L.Y.; Han, G.B. A Phase Change Storage Material that May be Used in the Fire Resistance of Building Structure. *Procedia Eng.* **2014**, *71*, 261–264. [CrossRef]

46. Ramakrishnan, S.; Wang, X.; Sanjayan, J.; Wilson, J. Thermal Performance Assessment of Phase Change Material Integrated Cementitious Composites in Buildings: Experimental and Numerical Approach. *Appl. Energy* **2017**, *207*, 654–664. [CrossRef]
47. Liu, F.; Wang, J.; Qian, X. Integrating Phase Change Materials into Concrete Through Microencapsulation Using Cenospheres. *Cem. Concr. Compos.* **2017**, *80*, 317–325. [CrossRef]
48. Li, T.; Yuan, Y.; Zhang, N. Thermal Properties of Phase Change Cement Board with Capric Acid/Expanded Perlite Form-Stable Phase Change Material. *Adv. Mech. Eng.* **2017**, *9*. [CrossRef]
49. Yang, H.; Feng, L.; Wang, C.; Zhao, W.; Li, X. Confinement Effect of SiO_2 Framework on Phase Change of PEG in Shape-Stabilized PEG/SiO_2 Composites. *Eur. Polym. J.* **2012**, *48*, 803–810. [CrossRef]
50. Gao, C.F.; Wang, L.P.; Li, Q.F.; Wang, C.; Nan, Z.D.; Lan, X.Z. Tuning Thermal Properties of Latent Heat Storage Material Through Confinement in Porous Media: The Case of $(1-C_nH_{2n+1}NH_3)_2ZnCl_4$ (n = 10 and 12). *Sol. Energy Mater. Sol. Cells* **2014**, *128*, 221–230. [CrossRef]
51. Wang, C.; Li, Q.; Wang, L.; Lan, X. Phase Transition of Neopentyl Glycol in Nanopores for Thermal Energy Storage. *Thermochim. Acta* **2016**, *632*, 10–17. [CrossRef]
52. Deng, S.; Wang, D.; Wang, X.; Wei, Y.; Waterhouse, G.I.; Lan, X.Z. Effect of Nanopore Confinement on the Thermal and Structural Properties of Heneicosan. *Thermochim. Acta* **2018**, *664*, 57–63. [CrossRef]
53. Sarı, A. Thermal Energy Storage Characteristics of Bentonite-Based Composite PCMs with Enhanced Thermal Conductivity as Novel Thermal Storage Building Materials. *Energy Convers. Manag.* **2016**, *117*, 132–141. [CrossRef]
54. Radhakrishnan, R.; Gubbins, K.E. Free Energy Studies of Freezing in Slit Pores: An Order-Parameter Approach Using Monte Carlo Simulation. *Mol. Phys.* **1999**, *96*, 1249–1267. [CrossRef]
55. Cunha, S.; Aguiar, J.; Pacheco-Torgal, F. Effect of Temperature on Mortars with Incorporation of Phase Change Materials. *Constr. Build. Mater.* **2015**, *98*, 89–101. [CrossRef]
56. Cunha, S.; Aguiar, J.; Ferreira, V.; Tadeu, A.; Garbacz, A. Mortars with Phase Change Materials-Part I: Physical and Mechanical Characterization. *Key Eng. Mater.* **2014**, *634*, 22–32. [CrossRef]
57. Pavlik, Z.; Trnik, A.; Ondruska, J.; Keppert, M.; Pavlikova, M.; Volfova, P.; Kaulich, V.; Cerny, R. Apparent Thermal Properties of Phase-Change Materials: An Analysis Using Differential Scanning Calorimetry and Impulse Method. *Int. J. Thermophys.* **2013**, *34*, 851–864. [CrossRef]
58. Cunha, S.; Aguiar, J.; Zalegowski, K.; Garbacz, A.; Soares, P.; Azevedo, J.; Ferreira, V.; Tadeu, A. Sustainable Mortars with Incorporation of Microencapsulated Phase Change Materials. *Adv. Mater. Res.* **2015**, *1129*, 621–628. [CrossRef]
59. Silva, N.; Aguiar, J.B.; Bragança, L.M.; Freire, T.; Cardoso, I. Properties of Gypsum-PCM Based Mortars for Interior Plastering of Construction Systems. *Mater. Sci. Forum* **2008**, *587*, 913–917. [CrossRef]
60. Jaworski, M.; Abeid, S. Thermal Conductivity of Gypsum with Incorporated Phase Change Material (PCM) for Building Applications. *J. Power Technol.* **2011**, *91*, 49–53.
61. Cunha, S.; Aguiar, J.B.; Ferreira, V.M.; Tadeu, A. Influence of the Type of Phase Change Materials Microcapsules on the Properties of Lime-Gypsum Thermal Mortars: Influence of the Type of Phase Change Materials Microcapsules. *Adv. Eng. Mater.* **2014**, *16*, 433–441. [CrossRef]
62. Djamai, Z.I.; Salvatore, F.; Larbi, A.S.; Cai, G.; El Mankibi, M. Multiphysics Analysis of Effects of Encapsulated Phase Change Materials (PCMs) in Cement Mortars. *Cem. Concr. Res.* **2019**, *119*, 51–63. [CrossRef]

© 2019 by the authors. Licensee MDPI, Basel, Switzerland. This article is an open access article distributed under the terms and conditions of the Creative Commons Attribution (CC BY) license (http://creativecommons.org/licenses/by/4.0/).

Article

Experimental and Numerical Analysis of a Composite Thin-Walled Cylindrical Structures with Different Variants of Stiffeners, Subjected to Torsion

Tomasz Kopecki *, Przemysław Mazurek and Tomasz Lis

Faculty of Mechanical Engineering and Aeronautics, Rzeszów University of Technology, 35-959 Rzeszów, Poland; pmazurek@prz.edu.pl (P.M.); list@prz.edu.pl (T.L.)
* Correspondence: tkopecki@prz.edu.pl; Tel.: +48-17-865-1923

Received: 23 August 2019; Accepted: 28 September 2019; Published: 2 October 2019

Abstract: The aim of the study was to determine the impact of the use of isogrid stiffeners on the stress and displacement distribution of a thin-walled cylindrical shell made of layered composites subjected to torsion. It also strives to define criteria for assessing the results of non-linear numerical analysis of models of the examined structures by comparing them with the results of the model experiment. The study contains the results of experimental research using models made of glass–epoxy composites and the results of numerical analyses in non-linear terms. The experiment was carried out using a special test stand. The research involved two types of considered structures. The results of the research allowed to create the concept of an adequate numerical model in terms of the finite element method, allowing to determine the distribution of stress and strain in the components of the studied structures. Simultaneously, the obtained conformity between the results of non-linear numerical analyses and the experiment allows to consider the results of analyses of the modified model in order to determine the properties of different stiffening variants as reliable. The presented research allows to determine the nature of the deformation of composite thin-walled structures in which local loss of stability of the covering is acceptable in the area of post-critical loads.

Keywords: isogrid; aircraft load-bearing structures; finite elements method; nonlinear numerical analyses; stability; equilibrium path

1. Introduction

Research devoted to the issue of loss of stability of systems that are elements of load-bearing structures used in technology, generally focuses on problems related to determining the value of critical loads. Analyses of post-critical conditions of structures become much more rarely the subjects thereof. This is due to the fact that in the vast majority of technical fields, the moment of loss of stability by the structure is identified with its destruction [1,2].

In aviation technology, due to the very specific nature of the objects under consideration, specific standards affecting design processes and operational assumptions have also been established. One of the principles, referring to the most commonly used in aviation metal structures, allows for post-critical deformations of selected types of systems, in the scope of operational loads [3,4].

In the general case, due to the need to minimize the mass of the object, loss of stability of the covering under operating conditions is allowed, if this phenomenon is elastic and occurs locally, i.e., within the shell segment limited by skeleton elements. The exceptions are coverings, e.g., of wing torsion box and other parts of the structure responsible for ensuring its appropriate torsional stiffness, as well as fragments of coverings, where large deformations are not desirable due to the need to maintain the aerodynamic properties [5,6].

Many years of research on aircraft structures, initiated by the Junkers construction office, have shown that limiting the area of post-critical deformations can be realized not only by increasing the number of skeleton elements. In many cases, an equally effective way to ensure the local nature of the phenomena has been the use of various forms of integral stiffeners.

Although light metal systems are still the basic components of most of the aircraft load-bearing structures in operation, a clear tendency has emerged in recent years to increase the use of different types of composites. Layer composites are the most commonly used in aviation, based on glass, carbon and aramid fabrics, as well as polymeric resins [7].

Due to insufficient knowledge about the overall changes in the mechanical properties of composites caused by their long-term exploitation, the bearing structures based on them for many years were designed and implemented as shell-like, using the spacers to prevent loss of stability by bearing coverings. At present, in the pursuit of meeting increasingly strict operational and economic criteria, the design doctrine allowing the local loss of stability of some fragments of composite coverings is considered, e.g., in the case of metal. This type of assumption allows the use of semi-monocoque structures characterized by more favorable mechanical properties in relation to the mass than the layered monocoque structures [8,9].

The permissibility of loss of stability of composite coverings causes similar structural problems to appear as in the case of metal. One of them is the necessity to reduce this phenomenon, with as little weight increase as possible. Achieving this goal is possible through the use of stiffeners of an integral or "quasi-integral" nature.

The forms of integral stiffeners, which allow to obtain a significant increase in the stiffness of the covering, as well as relatively high values of critical loads, are grid structures (Figure 1).

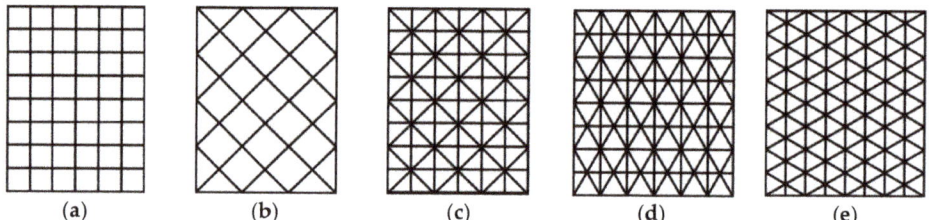

Figure 1. Schemes of grid stiffening structures: (a) ortho-grid, (b) x-grid, (c) bi-grid, (d) iso-grid type I, (e) iso-grid type II

In the case of metal structures, the use of this type of solution is due to the need for precision machining, quite problematic (the need for precision machining). It is much simpler to realize it in the case of composite structures. As results from the published research results, the interest in constructors is focused mainly on isogrid structures [10,11]. Also, in the field of their applications, a number of experiments and numerical calculations have been made in composite constructions [12,13]. It should be emphasized, however, that in most cases the subject matter of the publication is limited to analyses of cylindrical shells subjected to compression. In the case of load-bearing systems used in aviation, the reason for the loss of stability of thin-walled systems is primarily torsion. This study focuses on the problem of local loss of stability of twisted shells, under the conditions of permissible loads and on post-buckling analysis.

2. Materials and Methods

The aim of the research presented in this study was to perform a comparative analysis of two types of structural solutions of the aircraft structure fragment, represented by a thin-walled cylindrical structure with a composite covering, subject to post-critical deformations under operating conditions. The subject of the study were constitutions of equal dimensions (Figure 2), differing in structural solutions of the skeleton. The first of them was a reference structure with a stiffening corresponding to a classic semi-monocoque structure, consisting of four stringers and three ribs. External ribs were made of plywood; the mechanical properties are obtained with the constants defined in the aviation standards: E1 = 8500 MPa, E2 = 7500 MPa, G12 = 1000 MPa, ν12 = 0.34.

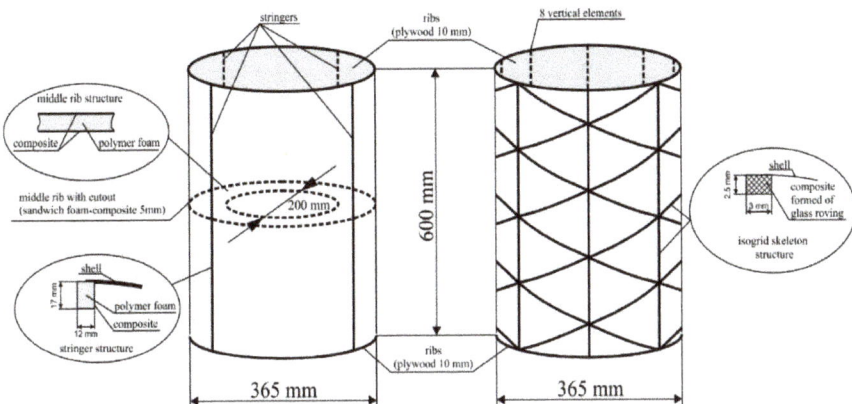

Figure 2. Schemes of models for experimental research: reference structure (on the left), structure with isogrid stiffener (on the right).

The second system had Type 2 isogrid stiffening. In both cases the shell of the model for experimental research was made as a composite structure which consisted of two layers of glass fabrics: 50 and 163 g/m^2. Stringers of the reference structures were made as a closed circuit formed of two layers of glass fabric with a weight of 163 g/m^2. The middle of every circuit was filled with polymer foam. The symmetric glass fabrics Interglass 02037 and 92110 were used to build the models. The matrix was a filling mixture based on epoxy resin MGS L285/H286 with known mechanical properties. Mechanical properties of the composite were obtained with the measured solid constants: E_{11} = 22,000 MPa, E_{22} = 22,000 MPa, ν12 = 0.11, G12 = 4600 MPa. Models were made using the contact method, with a 50/50 reinforcement ratio. The main directions of the composite orthotropy were oriented at 45 degrees to the direction of the axis of the cylindrical structure.

As it was proved during laboratory tests of some types of glass fabrics and epoxy resins applied in aviation, the physical constants for the single layer composites are almost identical in case of different fabrics with different weights [14]. So, the measured constants characterizes the behavior of each of layers.

The isogrid skeleton was formed by means of glass roving fibers in a polymeric mold, using the aforementioned filling mixture.

The aim of the comparative analyses was to examine the differences between the character of post-critical deformations in both types of structures and the preliminary estimation of the impact of the form and size of deformations on the operational durability of the tested systems.

The results of numerical analyses were evaluated by accepting the criterion of satisfactory similarity of the nature of post-critical deformations and representative equilibrium paths with reference to the results of the experiment. As a result, it became possible to determine stress distributions, based on the principle of unambiguity of solutions, according to which for an elastic system the deformation of the structure is responsible for one and only one variant of stress distribution.

For the experiment, a special test stand was used, with high stiffness, whose own deformations can be considered negligibly small (Figure 3). Loads were carried out in a gravitational manner. The ATOS optical scanner and micrometer sensors were used for structure deformation measurements, on the basis of which the total torsion angle of the structure was determined. Due to the load application method, subsequent measurements were made for determined deformation states of the structure.

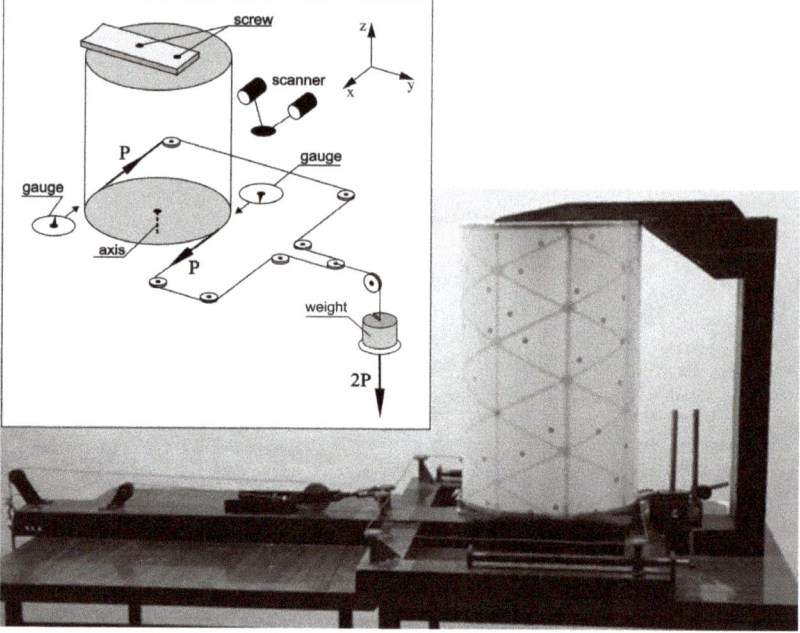

Figure 3. Stand for experimental research, load and deformation measurement diagram.

The results of experimental research constituted the material allowing us to obtain information about stress distributions in the tested systems by developing effective, adequate computational models in terms of the finite element method.

Numerical modelling of the analyzed structures was based on the commercial MSC PATRAN/MARC software (version 2012, MSC Software, Newport Beach, CA, USA), which proved its effectiveness in the case of post-critical deformations analyses of coverings made of isotropic materials [15]. In the case of layered composites, the key element of the software is an algorithm whose task is to determine the properties of the laminate, based on sets of constants corresponding to individual layers. In the case of commercial software used, this algorithm is an integral preprocessor procedure and does not allow the user to intervene.

The feature of composite structures which makes the creation of numerical mappings very difficult, is their heterogeneity, resulting not only from the conditions of lamination of individual layers, but also as a result of assembly operations, i.e., the presence of local surplus resin and varied thickness of glue joint. These factors may result in local changes in the stiffness of the covering and affect the form of post-critical deformations. Even small errors in the selection of geometric parameters of the numerical model, introducing a deviation from the actual boundary conditions of the shell segment, generate significant errors during non-linear analysis.

The basic relationship in a non-linear problem, defining the relationship between the state of the structure and the load is the so-called equilibrium path of the system, in general, a hypersurface in state hyperspace [16–18]. It is a relation that satisfies the matrix equation of residual forces:

$$r(u, \Lambda) = 0 \tag{1}$$

in which *u* is a state vector, containing the displacement components of the structure nodes corresponding to its current geometric configuration, Λ is a matrix containing control parameters corresponding to the current load level, while *r* is a residual vector containing unbalanced force components related to the current state of system deformation.

In general, excluding the singularities resulting from the shape of the equilibrium path, it can be assumed that there are continuous relationships between the values contained in the above equation. For Clapeyron systems, the vector r for a fixed parameter value Λ is defined as a gradient of total potential energy $\Pi\,(u, \Lambda)$ of the system:

$$\mathbf{r} = \frac{\partial \Pi}{\partial \mathbf{u}} \tag{2}$$

which expresses that the condition of the static equilibrium of the system under consideration is a zero increase in potential energy.

Equation (1) can also be presented in the form of relationship:

$$\mathbf{p}(\mathbf{u}) = \mathbf{f}(\mathbf{u}, \Lambda) \tag{3}$$

where **p** is a matrix containing internal forces corresponding to the current state of deformation, while f is a vector of external forces, which may also depend on the current state of deformation, which can be presented in the form of equations:

$$\mathbf{p} = \frac{\partial U}{\partial \mathbf{u}}, \ \mathbf{f} = \frac{\partial P}{\partial \mathbf{u}} \tag{4}$$

where **p** and **u** are respectively elastic strain energy and the work of external loads. The total potential energy of the system is expressed by the equation:

$$\Pi = U - P \tag{5}$$

Stiffness matrix **K** of the system corresponding to the temporary, current configuration of the system is defined as a derivative of the residual vector r relative to the components of the state vector **u**:

$$\mathbf{K} = \frac{\partial \mathbf{r}}{\partial \mathbf{u}} \tag{6}$$

The $\mathbf{K}^{-1}$ inverse matrix is the system flexibility matrix. Excluding singularities corresponding to the characteristic points of the equilibrium path, both matrices are symmetric matrices.

By determining the derivative of the residual vector r relative to the control parameters, a control matrix, also called a load matrix, can be determined:

$$\mathbf{Q} = -\frac{\partial \mathbf{r}}{\partial \Lambda} \tag{7}$$

The concept of stepwise changes in the configuration of the structure corresponding to the staged increase in load results in the possibility of binding the matrix **u** and Λ with a dimensionless parameter determining the degree of task completion, called the pseudo-time parameter:

$$\mathbf{u} = \mathbf{u}(t), \ \Lambda = \Lambda(t) \tag{8}$$

The derivative of the residual vector component r in relation to the pseudo-time—t has the form:

$$\overset{\circ}{r_i} = \frac{\partial r_i}{\partial u_j} \cdot \overset{\circ}{u_j} + \frac{\partial r_i}{\partial \Lambda_i} \cdot \overset{\circ}{\Lambda_j} \tag{9}$$

where:
$$\mathring{r} = \frac{\partial r}{\partial t} \tag{10}$$

From the above compound and from dependences 6 and 7 the matrix equation follows:
$$\mathring{\mathbf{r}} = \mathbf{K} \cdot \mathring{\mathbf{u}} - \mathbf{Q} \cdot \mathring{\mathbf{\Lambda}} \tag{11}$$

By determining the second derivative of the residual vector r relative to the pseudo-time parameter, we obtain:
$$\mathring{\mathring{\mathbf{r}}} = \mathbf{K} \cdot \mathring{\mathring{\mathbf{u}}} + \mathring{\mathbf{K}} \cdot \mathring{\mathbf{u}} - \mathbf{Q} \cdot \mathring{\mathring{\mathbf{\Lambda}}} - \mathring{\mathbf{Q}} \cdot \mathring{\mathbf{\Lambda}} \tag{12}$$

where $\mathring{\mathbf{K}}$ and $\mathring{\mathbf{Q}}$ are matrices:
$$\mathring{\mathbf{K}} = \frac{\partial \mathbf{K}}{\partial t}, \mathring{\mathbf{Q}} = \frac{\partial \mathbf{Q}}{\partial t} \tag{13}$$

In numerical algorithms for non-linear problems, all components of the matrix is expressed as functions a single parameter λ, called the state control parameter. This parameter is a measure of the increase in the associated load, directly or indirectly, with the pseudo-time parameter—t. Thus, the equation of state 1 can be written in the form:
$$\mathbf{r}(\mathbf{u}, \lambda) = \mathbf{0} \tag{14}$$

called the monoparametric equation of residual forces. The corresponding derivatives in relation to the pseudo-time can be written as follows:
$$\mathring{\mathbf{r}} = \mathbf{K} \cdot \mathring{\mathbf{u}} - \mathbf{q} \cdot \mathring{\lambda} \tag{15}$$

$$\mathring{\mathring{\mathbf{r}}} = \mathbf{K} \cdot \mathring{\mathring{\mathbf{u}}} + \mathring{\mathbf{K}} \cdot \mathring{\mathbf{u}} - \mathbf{q} \cdot \mathring{\mathring{\lambda}} - \mathring{\mathbf{q}} \cdot \mathring{\lambda} \tag{16}$$

where $\mathbf{K} = \frac{\partial \mathbf{r}}{\partial \mathbf{u}}$ is the defined by Equation (6) system stiffness matrix, also called the *tangent matrix* to the equilibrium path, while:
$$\mathbf{q} = -\frac{\partial \mathbf{r}}{\partial \lambda} \tag{17}$$

is a *vector of load increase*.

Because at each stage of the solution a static equilibrium of the system is assumed, the vector r in each stage assumes zero values and does not change in relation to pseudo-time. The following relationships result:
$$\mathring{\mathbf{r}} = \mathbf{0} \Rightarrow \mathbf{K} \cdot \mathring{\mathbf{u}} = \mathbf{q} \cdot \mathring{\lambda} \tag{18}$$

$$\mathring{\mathring{\mathbf{r}}} = \mathbf{0} \Rightarrow \mathbf{K} \cdot \mathring{\mathring{\mathbf{u}}} + \mathring{\mathbf{K}} \cdot \mathring{\mathbf{u}} = \mathbf{q} \cdot \mathring{\mathring{\lambda}} + \mathring{\mathbf{q}} \cdot \mathring{\lambda} \tag{19}$$

Thus, for all points of the equilibrium path (for which K is a non-singular matrix), the relationship resulting from the Equation (18) can be used:
$$\mathring{\mathbf{u}} = \mathbf{K}^{-1} \cdot \mathbf{q} \cdot \mathring{\lambda} = \mathbf{v} \cdot \mathring{\lambda} \tag{20}$$

where in
$$\mathbf{v} = \mathbf{K}^{-1} \cdot \mathbf{q} = \frac{\partial \mathbf{u}}{\partial \lambda} \tag{21}$$

is a *vector of velocity of load increase*.

The prediction-correction methods of determining the consecutive points of the equilibrium path used in modern programs also include the correction phase based on the fulfillment by the system of an additional equation, called the increment control equation or the constraint Equation (18):

$$c(\Delta u_n, \Delta \lambda_n) = 0 \qquad (22)$$

where the increases:

$$\Delta u_n = u_{n+1} - u_n \text{ and } \Delta \lambda_n = \lambda_{n+1} - \lambda_n \qquad (23)$$

correspond to the transition from state n to state n + 1.

As in the case of the experiment, since in the case of systems with the number of freedom greater than 2, it is difficult to interpret the equilibrium path in a clear graph form, in practice, for comparative purposes, representative equilibrium paths are used, which are the relationships between the chosen parameter characterizing deformation the system and a single control parameter related to the load. As a confirmation of the reliability of the results of non-linear numerical analyses in terms of FEM, it is considered that satisfactory convergence between representative equilibrium paths: determined during the experiment and obtained on the numerical way. It is also necessary to converge the forms of deformations that are the effects of calculations with the result of the experiment. Based on the aforementioned principle of unambiguity of solutions, the distributions of effective stress in the deformed shell can also be considered reliable.

The geometric structure of numerical models was based mainly on surface objects. In the case of the reference model, three-dimensional objects were also used, to model the stringers (Figure 4). It has to be emphasized that pictures below does not present complete models. Their present a half of each numerical model, for the better visualization of the structural details.

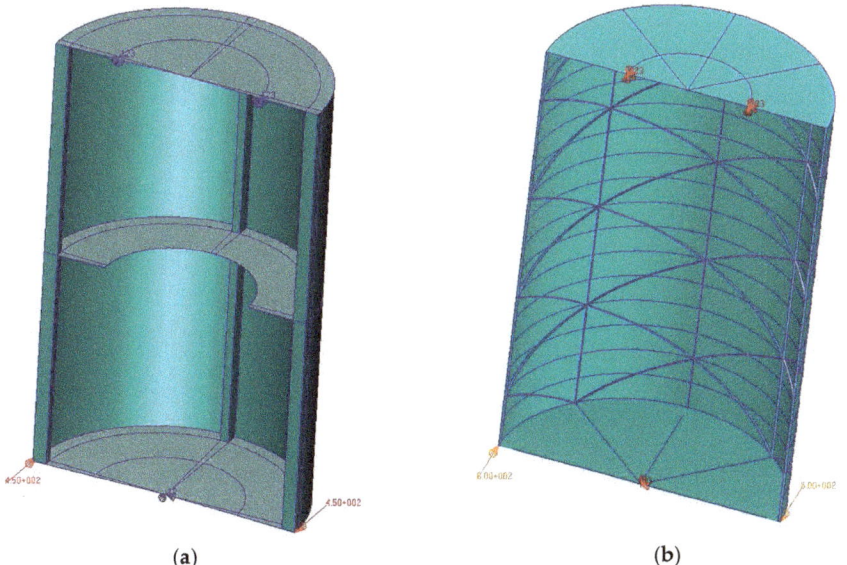

Figure 4. Fragments of geometric models, with visible methods of attachment and loads: (**a**) reference structure, (**b**) structure with isogrid.

The use of this kind of solution resulted from the desire to map the actual proportions between the dimensions of the coverage segments.

The non-linear numerical analysis is an iterative process, aimed at determining subsequent equilibrium states, so its correctness is largely determined by the correct selection of the prediction method, correction strategy and a whole range of control parameters. In the described case, the Newton–Raphson method was used, related to the Crisfield hypersferrical correction [16,17].

After the series of numerical tests in the scope of choosing the topology of the model, the mesh consisting of about 3000 bilinear, four-node shell elements were used for the reference structure. The number of used elements was the result of analyses executed using various versions of the numerical models and it was a minimum providing the nonlinear analysis convergence and the compliance of results with the experiment. The necessity to a bilinear element resulted from the fact that other types of them, contained in the MSC MARC software library, which can be assigned to the properties of layered composites, do not have the ability to map geometrically complex objects, due to the type and number of degrees of freedom.

For the modeling of the stringers, in the case of the reference structure, a total of 120 three-dimensional, eight-node elements were used. The structure model stiffened integrally was based entirely on surface elements, most of which were 4–node ones, the total number of which was 8700.

The material models were made taking into account the mechanical properties of composites based on the components used during the experimental phase, with the constants given above.

The process of nonlinear numerical calculations was multistage and a lot of mesh variants were tested. The goal of this procedure was to obtain the solution reproducing the results of experiment, however, under this procedure a quality of the mesh was also tested and analyzed. The convergence of the mesh was verified first, before the comparison with the experimental results were carried out, to obtain an assurance that in case of any mesh refinement the results do not change significantly. Presented and described results were obtained by means of mesh variants, which were considered as verified.

3. Results

3.1. Experimental Research

Measurements of total torsion angles of examined structures formed the basis for determining and comparing representative equilibrium paths of the studied systems (Figure 5). As the representative equilibrium path, the relationship between the said total torsion angle, characterizing the state of the structure and the control parameter for which the torsional moment was obtained, was assumed. For selected states, corresponding to the torsional moment values 160 Nm for the reference structure and 220 Nm for the structure with isogrid, the deformed surfaces of the structures were also scanned (Figure 6).

The shape of representative equilibrium paths indicates a completely different course of the phenomenon in both analyzed objects. In the case of a reference structure containing areas of coverage characterized by a high ratio of curvature to their surface area, the post-critical deformations occurred with pronounced skips, which was reflected in the occurrence of a linear section and refraction of the characteristic and a further significant increase in the torsion angle of the structure with relatively small increase of torsional moment. This corresponded to the emergence and deepening of large, double folds within each of the coverage segments. Although such deformations meet theoretical acceptability criteria, they cause a significant decrease in structural stiffness and deterioration of its operational properties. In the case of composite coverings, there is a serious concern that with such large deformations there may be local damage to the structure, e.g., in the form of delamination, resulting in a significant reduction in the operating durability of the system.

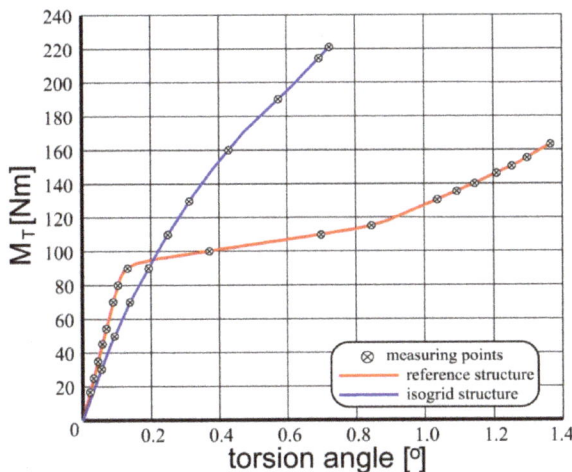

Figure 5. Representative equilibrium paths of the examined structures.

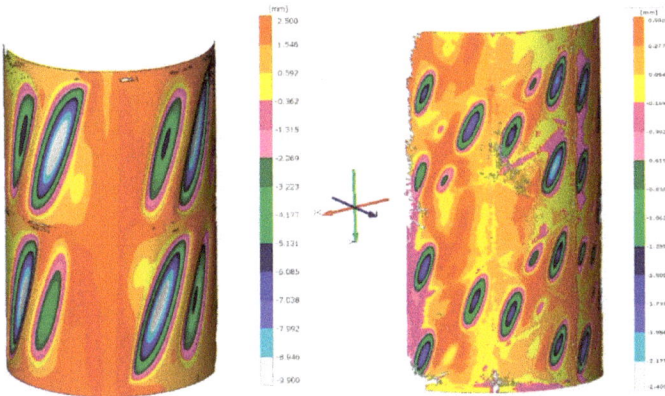

Figure 6. Distributions of deformations of the examined structures obtained as a result of scanning.

In the case of an integrally stiffened structure, the coverage segments between the frame elements constitute surfaces close to flat ones, as a result of which a loss of stability within the majority of them occurs at a relatively low load value. However, the structure quickly achieves a state of post-critical equilibrium, gaining high stiffness, thanks to which a representative equilibrium path maintains a linear-like character even with high load values. The depth of folds formed within the coverage segments is small, which allows to conclude that the structure has a much higher durability compared to the reference system.

It should also be emphasized that the masses of both structures are very similar (1849 g, the reference structure, 1861 g, the structure stiffened integrally), while the stiffness of the structure with the isogrid type is almost 70% higher than the stiffness of the reference structure.

3.2. Nonlinear Numerical Analyzes

As a result of conducted non-linear numerical analyses, representative equilibrium paths were determined and their comparison was made with the appropriate characteristics obtained during the experiment (Figure 7).

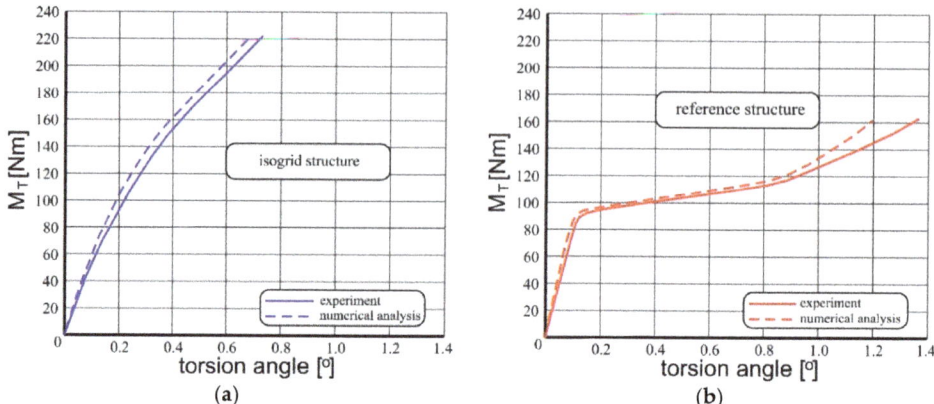

Figure 7. List of representative equilibrium paths obtained in numerical way and as a result of the experiment: (**a**) structure stiffened with isogrid, (**b**) reference structure.

In both cases, the stiffens of the numerical models were slightly higher stiff then their experimental counterparts. It results from the simplifications, which must be applied due to limited effectiveness and efficiency of the numerical procedures applied in commercial FEM software.

Both numerical models were characterized by slightly higher stiffness than experimental models, although the deformation of some parts of the covering in both cases turned out to be greater than in the case of the latter (Figure 8). This is a consequence of the fact that the state of the structure depends on the proportion between many parameters defining it. As a result, different combinations of state parameters in different subsets may result in a similar value of a representative parameter which in the described case, is the total torsion angle.

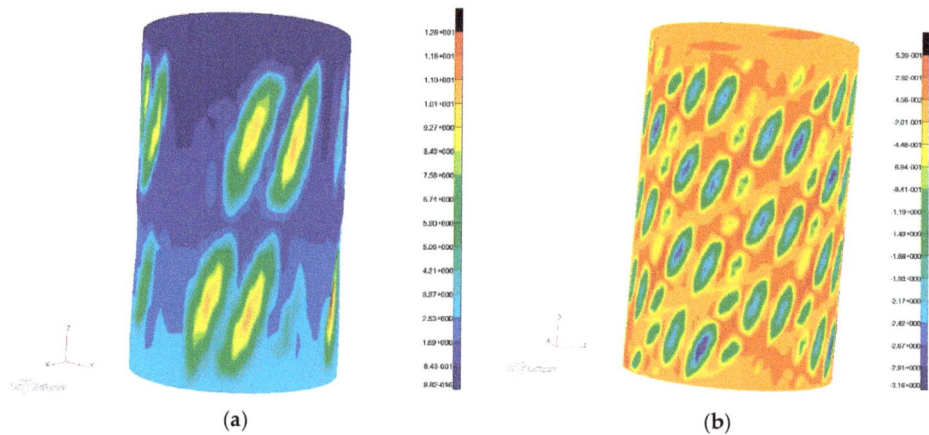

Figure 8. Distributions of the radial deformation component at the maximum load value: (**a**) reference structure, (**b**) structure with isogrid.

However, despite these divergences, the results of non-linear numerical analyzes can be considered satisfactory. In particular, in the case of the isogrid structure, the inconsistency of the representative equilibrium paths did not exceed 7%. On the other hand, in the case of the reference structure, this inconsistency for most of the range of post-critical deformations did not exceed 15%.

Satisfactory similarity of the form of post-critical deformations was also achieved, and based on the above-mentioned principle of unambiguity of solutions, the distribution of effective stress based on the hypothesis of the highest tensile stress can be considered reliable (Figure 9). However, stress levels in the cylindrical part of the structure stiffened with isogrid were much lower than in the reference structure, for better visual comparison of the two cases of stress distribution, similar maximum scales were applied.

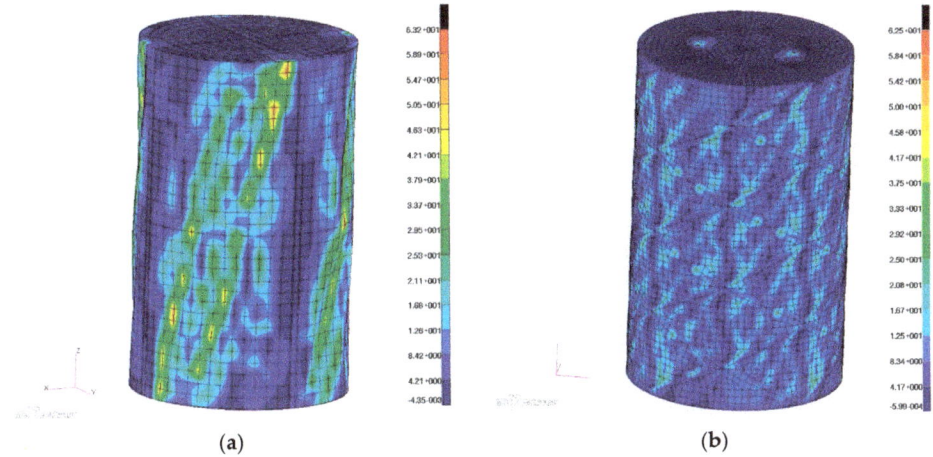

Figure 9. Distribution of effective stress according to the hypothesis of the highest tensile stress: (**a**) reference structure, (**b**) structure with isogrid.

4. Discussion

The analysis of the nature of post-critical deformations of the considered systems and the comparison of their masses, reveals a number of advantages of a structural solution based on an isogrid integral stiffener. The semi-monocoque reference structure used for the study was an example of the use of a minimum number of skeleton elements, as a result of which relatively large covering segments with significant curvature underwent deformation. In this case, the nature of the deformation may cause a significant reduction in the reliability of the structure, resulting from the possibility of local damage of the composite. Traditional solutions used in aviation technology rely on increasing the number of frames and stringers. Due to such a solution the covering segments gain smaller dimensions, and the relations between them and the shell curvature become more advantageous. However, increasing the number of skeleton elements leads to a significant increase in the mass of the structure. Therefore, if additional frames or stringers are used, the semi-monocoque structure would have to have a larger mass than the structure with isogrid type stiffeners.

It should be emphasized that inference about the operational lifetime of a structure, regardless of the material used, requires knowledge about its behavior under cyclic loading conditions. However, even in the absence of such information, in the case of static studies, it can be assumed that local damage to the structure may appear primarily in the areas of high stress gradients. From this point of view, the structure with isogrid type stiffeners has a great advantage, characterized by a very regular distribution of stresses that is devoid of explicit concentrations.

The results of experimental studies allowed the development of adequate numerical models based on the finite element method, using commercial software. This allows to conclude that the processes of designing thin-walled composite structures subjected to post-critical deformations under the conditions of permissible loads can be numerically assisted, however the results of numerical analyses should be verified experimentally, even using a simplified model experiment. The numerical models presented in this study were subjected to repeated tests and in a number of them, erroneous results were obtained, characterized by incompatibility with the actual deformations and the deformation of the analysed systems. Relatively simple and cheap experiment allowed to correct errors and selection of the most appropriate sets of numerical procedures, as well as parameters controlling the course of computational processes. This confirms the validity of the thesis that relying on unverified results of numerical analyses may lead to the appearance of significant construction errors.

5. Conclusions

The presented study was limited to the analysis of one type of isogrid stiffening, due to the inconvenient and time-consuming process of creating a model for experimental research. The search for the most appropriate solution, from the point of view of distribution of stresses and strains, would have to include the analysis of a whole range of stiffening variants and their impact on the aforementioned size and total mass of the structure. In this context, the methodology of gradual modifications of the numerical model verified by the experiment, aimed at testing successive versions of structural solutions, seems to be useful. However, it should be emphasized that the greater the deviation of the numerical model from its verified form, the greater the probability of obtaining incorrect results. Therefore, the final stage of the analysis seems to be the final experimental verification of the solution considered to be the most effective.

The presented results should be evaluated in the context of the expected broader scope of research, aimed at determining the best possible proportions between the dimensions of the structure and the number and form of integral stiffening elements. It should be emphasized that it is also planned to subject selected representative systems to cyclic loads and to attempt to develop an effective method of detecting emerging damages.

Author Contributions: Conceptualization, T.K.; methodology, T.K. and T.L.; software, T.K. and P.M.; validation, T.K., P.M. and T.L.; formal analysis, T.K.; investigation, T.K., P.M. and T.L.; resources, T.L.; data curation, P.M.; visualization, T.K. and P.M.; supervision, T.K.; project administration, T.K.

Funding: This research received no external funding.

Conflicts of Interest: The authors declare no conflict of interest.

References

1. Dębski, H.; Sadowski, T. Modelling of microcracks initiation and evolution along interfaces of the WC/Co composite by the finite element method. *Comput. Mater. Sci.* **2014**, *83*, 403–411. [CrossRef]
2. Dębski, H.; Teter, A.; Kubiak, T. Numerical and experimental studies of compressed composite columns. *Compos. Struct.* **2014**, *118*, 28–36. [CrossRef]
3. Arborcz, J. Post-buckling behavior of structures. Numerical techniques for more complicated structures. *Lect. Notes Phys.* **1985**, *288*, 83–142.
4. Kopecki, T.; Bakunowicz, J.; Lis, T. Post-critical deformation states of composite thin-walled aircraft load-bearing structures. *J. Theor. Appl. Mech.* **2016**, *54*, 195–204. [CrossRef]
5. Goraj, Z. Load composite structure in aeronautical engineering. *Trans. Inst. Aviat. Warszawa* **2007**, *191*, 13–32.
6. Lynch, C.A. Finite Element Study of the Post Buckling Behavior of a Typical Aircraft Fuselage Panel. Ph.D. Thesis, Queen's University Belfast, Belfast, UK, 2000.
7. Teter, A.; Debski, H.; Samborski, S. On buckling collapse and failure analysis of thin-walled composite lipped-channel columns subjected to uniaxial compression. *Thinn-Walled Struct.* **2014**, *85*, 324–331. [CrossRef]
8. Paschero, M.; Hyer, M.W. Axial buckling of an orthotropic circular cylinder: Application to orthogrid concept. *Int. J. Solids Struct.* **2009**, *46*, 2151–2171. [CrossRef]
9. Riks, E. An incremental approach to the solution of snapping and buckling problems. *Int. J. Solid Struct.* **1979**, *15*, 529–551. [CrossRef]
10. Huybrechts, S.; Tsai, S.W. Analysis and behavior of grid structures. *Compos. Sci. Technol.* **1996**, *56*, 1001–1015. [CrossRef]

11. Kim, T.D. Fabrication and testing of composite isogrid stiffened cylinder. *Compos. Struct.* **1999**, *45*, 1–6. [CrossRef]
12. Goetzendorf-Grabowski, T.; Mieloszyk, J. Common Computational Model for coupling panel method with finite element method. *Aircr. Eng. Aerosp. Technol.* **2017**, *89*, 654–662. [CrossRef]
13. Liang, H.; Sheikh, A.; Ching-Tai, N.G.; Griffith, M.C. An efficient finite element model for buckling analysis of grid stiffened laminated composite plates. *Compos. Struct.* **2015**, *122*, 41–50.
14. Stafiej, W. *Calculations Used in Sailplanes Designs*; Warsaw university of Technology: Warsaw, Poland, 2000.
15. Kopecki, T. Numerical-experimental analysis of the post-buckling state of a multi-segment multi-member thin-walled structure subjected to torsion. *J. Theor. Appl. Mech.* **2011**, *49*, 227–242.
16. de Borst, R.; Crisfield, M.A.; Remmers, J.J.; Verhoose, C.V. *Non-Linear Finite Element Analysis of Solid and Structures*, 2nd ed.; John Wiley & Sons: Hoboken, NJ, USA, 2012.
17. Doyle, J.F. *Nonlinear Analysis of Thin-Walled Structures*; Springer: Luxemburg, 2001.
18. Felippa, C.A.; Crivelli, L.A.; Haugen, B. A survey of the core-congruential formulation for nonlinear finite element. *Arch. Comput. Methods Eng.* **1994**, *1*, 1–48. [CrossRef]

© 2019 by the authors. Licensee MDPI, Basel, Switzerland. This article is an open access article distributed under the terms and conditions of the Creative Commons Attribution (CC BY) license (http://creativecommons.org/licenses/by/4.0/).

Review

Insights into the Current Trends in the Utilization of Bacteria for Microbially Induced Calcium Carbonate Precipitation

Sing Chuong Chuo [1,2], Sarajul Fikri Mohamed [2,*], Siti Hamidah Mohd Setapar [1,3,*], Akil Ahmad [1,3,*], Mohammad Jawaid [4,*], Waseem A. Wani [5], Asim Ali Yaqoob [6] and Mohamad Nasir Mohamad Ibrahim [6]

1. Centre of Lipids Engineering and Applied Research, Universiti Teknologi Malaysia, Skudai 81310 UTM, Johor, Malaysia; scchuo2@yahoo.com.my
2. Department of Quantity Surveying, Faculty of Built Environment, Universiti Teknologi Malaysia, Skudai 81310 UTM, Johor, Malaysia
3. Malaysia-Japan International Institute of Technology, Jalan Sultan Yahya Petra, Universiti Teknologi, Malaysia, Kuala Lumpur 54100, Malaysia
4. Laboratory of Biocomposite Technology, Institute of Tropical Forestry and Forest Products (INTROP), Universiti Putra Malaysia, Serdang 43400 UPM, Selangor, Malaysia
5. Department of Chemistry, Govt. Degree College Tral, Kashmir J&K-192123, India; waseemorg@gmail.com
6. School of Chemical Sciences, Universiti Sains Malaysia, Penang 11800, Malaysia; asimchem4@gmail.com (A.A.Y.); mnm@usm.my (M.N.M.I.)

* Correspondence: sarajul@utm.my (S.F.M.); siti-h@utm.my (S.H.M.S.); akilchem@yahoo.com (A.A.); jawaid_md@yahoo.co.in (M.J.); Tel.: +60-75535496 (S.H.M.S.); Fax: +60-75581463 (S.H.M.S.)

Received: 8 September 2020; Accepted: 14 October 2020; Published: 5 November 2020

Abstract: Nowadays, microbially induced calcium carbonate precipitation (MICP) has received great attention for its potential in construction and geotechnical applications. This technique has been used in biocementation of sand, consolidation of soil, production of self-healing concrete or mortar, and removal of heavy metal ions from water. The products of MICP often have enhanced strength, durability, and self-healing ability. Utilization of the MICP technique can also increase sustainability, especially in the construction industry where a huge portion of the materials used is not sustainable. The presence of bacteria is essential for MICP to occur. Bacteria promote the conversion of suitable compounds into carbonate ions, change the microenvironment to favor precipitation of calcium carbonate, and act as precipitation sites for calcium carbonate crystals. Many bacteria have been discovered and tested for MICP potential. This paper reviews the bacteria used for MICP in some of the most recent studies. Bacteria that can cause MICP include ureolytic bacteria, non-ureolytic bacteria, cyanobacteria, nitrate reducing bacteria, and sulfate reducing bacteria. The most studied bacterium for MICP over the years is *Sporosarcina pasteurii*. Other bacteria from Bacillus species are also frequently investigated. Several factors that affect MICP performance are bacterial strain, bacterial concentration, nutrient concentration, calcium source concentration, addition of other substances, and methods to distribute bacteria. Several suggestions for future studies such as CO_2 sequestration through MICP, cost reduction by using plant or animal wastes as media, and genetic modification of bacteria to enhance MICP have been put forward.

Keywords: bacteria; biocementation; construction; microbially induced calcium carbonate precipitation

1. Introduction

Microbially induced calcium carbonate precipitation (MICP) is a process that occurs when microorganisms, especially bacteria, are provided with appropriate substrates and thus induce the formation of calcium carbonate ($CaCO_3$) crystals. The $CaCO_3$ formed is very useful in coating surfaces and binding different particles together [1–3]. MICP can occur under atmospheric pressure and other mild conditions. In fact, it happens in nature all around the world. This process has coated the surfaces of various natural structures and left hints about past ages for researchers to discover. Formation of $CaCO_3$ by microorganisms has been studied through biomimetic approach and then applied in various fields such as construction, environment, geo-techniques, and nanotechnology [4,5]. An example of the MICP process is shown in Figure 1.

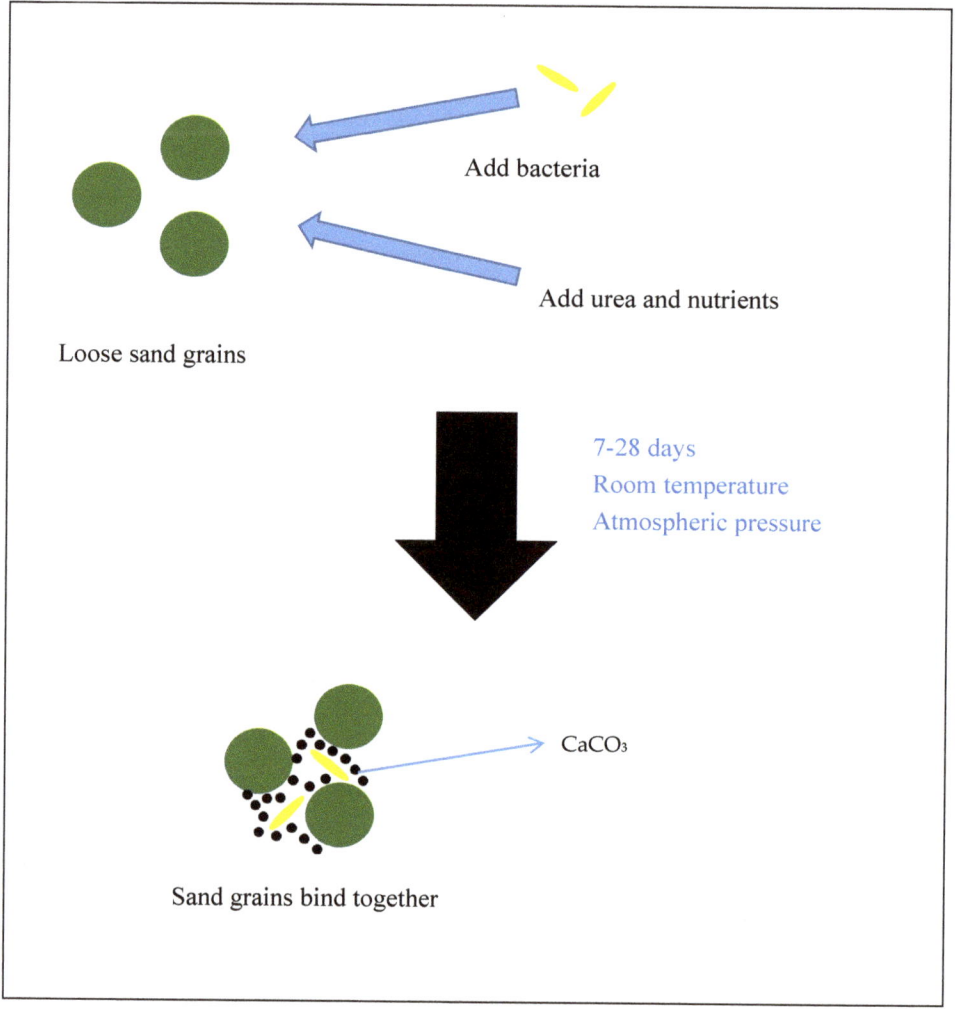

Figure 1. Schematic diagram of the microbially induced calcium carbonate precipitation (MICP) process for biocementation of sand.

In recent years, there are increasing interests in MICP among researchers worldwide. A search in SciFinder with the keywords 'MICP' and 'biocement' showed increasing number of studies about MICP from the year 2010 to 2020. The increasing interest of researchers towards MICP and biocement may be due to increasing awareness on sustainability globally. A lot of studies focus on finding or developing sustainable materials and processes to replace conventional non-sustainable ones.

The construction industry is one of the fastest growing fields due to rapid urbanization [6]. Large amounts of building materials are being consumed every day to build, maintain, and renew various structures. These building materials, especially ordinary Portland cement (OPC) and concrete, are not sustainable. Construction consumes a lot of natural resources and energy, while at the same time contributing 50% of CO_2 emission worldwide [7,8]. Therefore, it is desirable to change current building materials into sustainable ones. In addition, biocementation through MICP can be used to reduce usage of OPC.

Construction costs are rising over the years and are expected to stay high for future times ahead. Producing building materials that are more durable and longer lasting can help to reduce maintenance costs. MICP was reported by various researchers to enhance strength and durability of building materials. Development of self-healing building materials also helps to reduce resources spent on routine repairs [9,10]. By including appropriate bacteria into cement or concrete, the formation of cracks will be stopped and sealed by the bacteria due to $CaCO_3$ precipitation. Figure 2 shows the schematic diagram of self-healing process. Bacterial solution can also be applied from outside to seal a cracked surface on old building materials.

Conducting an MICP process requires knowledge from different fields including biotechnology, geotechniques, civil engineering, material engineering, and nanotechnology [11]. Current trends of technology integration encouraged researchers to investigate applications of MICP for various purposes. Most recent studies on MICP focus on consolidation of sand and soil, self-healing concrete and crack sealing [12,13], and removal of heavy metals/ions from water [14,15]. Technology integration in construction and other industries will make the applications of MICP easier.

The presence of bacteria is crucial for MICP to occur. The bacteria produce necessary enzymes such as urease and carbonic anhydrase to convert appropriate compounds into carbonate ions [16]. These activities change the microenvironment to favor precipitation of $CaCO_3$ in the presence of calcium ions. Surface charges of bacterial cells attract calcium ions and then the cells serve as precipitation sites for $CaCO_3$ crystals. Some bacteria also produce extracellular polymeric substances (EPS) that can enhance the MICP process. Different bacteria have been studied for their MICP potential. The bacteria must have high cell availability and high enzyme activity because they are often placed in harsh environments associated with high alkalinity, lack of nutrients, and high compressive force [17]. Bacterial strain and medium composition will affect $CaCO_3$ crystal morphologies (calcite, vaterite and aragonite); thus, affecting the stability and strength of structures formed [18]. Therefore, careful considerations are required when choosing bacteria and its medium to obtain desired products.

The purpose of this paper was to review the bacteria used in some of the most recent studies of MICP. By learning and comparing the behavior of the bacteria and the results of MICP processes, insights on choosing suitable bacteria for certain applications have been proposed. This paper will thus help in future studies to further improve the MICP processes.

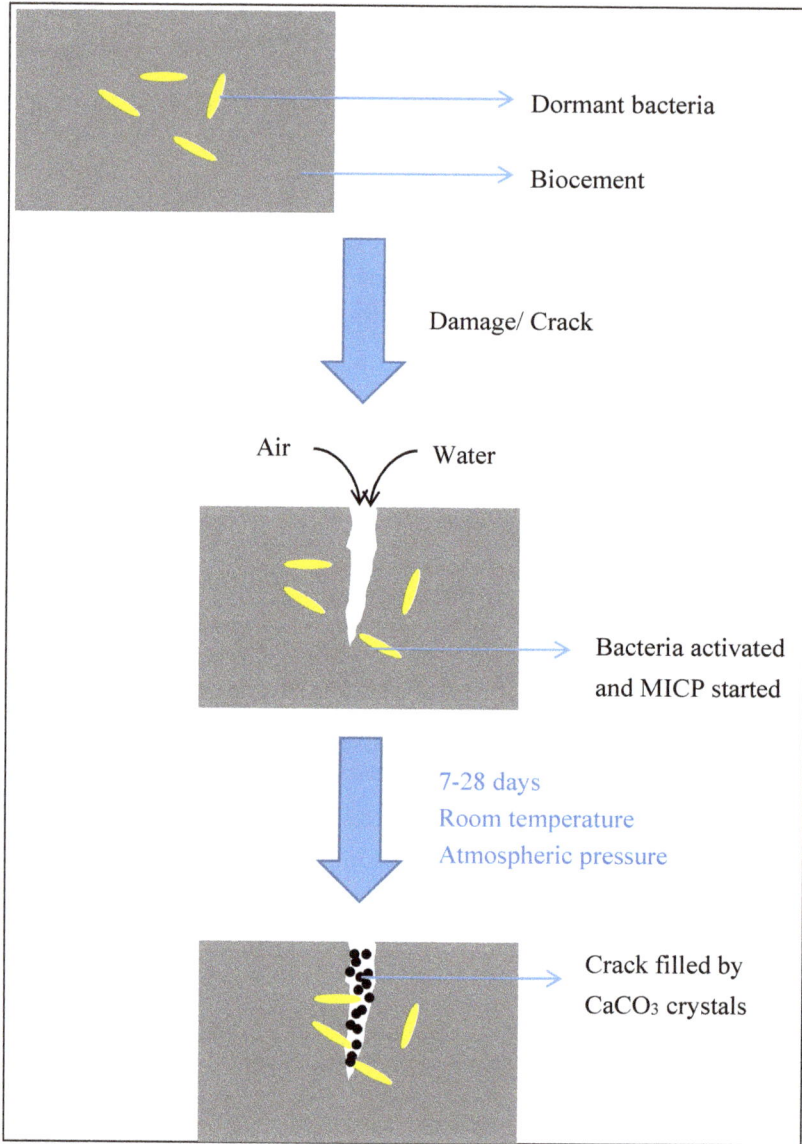

Figure 2. Schematic diagram for self-healing of biocement.

2. Bacteria Used in MICP

Globally, different bacterial strains have been tried by different researchers for MICP. Some of the commonly used bacterial strains that were successfully employed for MICP are discussed in the succeeding sections.

2.1. Sporosarcina pasteurii

Sporosarcina pasteurii, previously known as *Bacillus pasteurii* is the most commonly used bacterium for studying MICP due to its high urease activity. It is a non-pathogenic bacterial strain. Urease catalyzes the hydrolysis of urea to form ammonia and carbonic acid as shown in Equations (1) and (2).

$$CO(NH_2)_2 + H_2O \longrightarrow NH_2COOH + NH_3 \tag{1}$$

$$NH_2COOH + H_2O \longleftrightarrow NH_3 + H_2CO_3 \tag{2}$$

Ammonia then forms ammonium and hydroxide ions in water (Equation (3)).

$$NH_3 + H_2O \longleftrightarrow NH_4^+ + OH^- \tag{3}$$

Carbonic acid also forms bicarbonate and hydrogen ions in water (Equation (4)).

$$H_2CO_3 \longleftrightarrow HCO_3^- + H^+ \tag{4}$$

Formation of hydroxide ions causes pH to increase and shifts the bicarbonate equilibrium. This causes the formation of carbonate ions. The overall equation becomes as shown below (Equation (5)).

$$HCO_3^- + H^+ + 2NH_4^+ + 2OH^- \longleftrightarrow CO_3^{2-} + 2NH_4^+ + 2H_2O \tag{5}$$

In the presence of calcium ions, calcium carbonate crystals can be precipitated as shown in Equation (6).

$$Ca^{2+} + CO_3^{2-} \longleftrightarrow CaCO_3 \tag{6}$$

Table 1 shows some studies on sand and soil improvements using MICP with *Sporosarcina pasteurii*. In laboratory scale experiments, researchers often test the biocementation potential of MICP by *Sporosarcina pasteurii* on sand columns composed of various types of sands and additives. Strength and durability of the sands are often enhanced after the MICP process. The research group of Cardoso et al. [19] used *Sporosarcina pasteurii* for biocementation of sand columns and showed that compressibility and tensile strength increased while permeability was decreased after MICP. They also showed that addition of clay in the sand column further enhanced its properties. Bu et al. [20] studied biocementation of sand using *Sporosarcina pasteurii* and compared with sands treated with normal cement or lime. Sands treated with MICP had higher unconfined compressive strength (UCS) than sands with 10% cement and flexure strength similar to sands with 20–25% cement. Lime treated sands were the weakest among those samples. Porter et al. [21] investigated the combinations of MICP by *Sporosarcina pasteurii*, OPC, and metakaolin to treat sand columns. Combination of *Sporosarcina pasteurii* and OPC had better performance in terms of UCS and water absorption as compared to other combinations or single treatment. Analysis revealed that MICP enhanced bridges formed by OPC or metakaolin between sand particles. A total of 93% of $CaCO_3$ crystals formed at the bridging zones in sand columns treated by *Sporosarcina pasteurii* and OPC. This study suggested that there is synergistic relationship between chemical and microbial cementation process. Choi et al. [22] used *Sporosarcina pasteurii* for biocementation of PVA fiber reinforced sand columns. They found that at fixed $CaCO_3$ concentration, increasing PVA fiber content further increased UCS and splitting tensile strength and decreased permeability of the sand columns. They also found that brittleness of sand column was greatly reduced with addition of fiber. They suggested that addition of fiber enhanced biocementation by filling more pores. Xiao et al. [23] showed that MICP by *Sporosarcina pasteurii* enhanced cyclic shear resistance of calcareous sand. The authors found that increasing biocementation solution can further reduce degree of sand deformation and increase liquefaction resistance due to more and bigger $CaCO_3$ crystals filling voids. The $CaCO_3$ crystals made the surface rougher and that enhanced bonding between sand particles. Sasaki and Kuwano [24] used *Sporosarcina pasteurii* to consolidate sands with different non-plastic fines content. They found that presence of fines content greatly reduced liquefaction resistance after MICP due to smaller void ratio, formation of smaller and unevenly distributed $CaCO_3$. Higher concentration of biocementation solution or more cycles of treatment were suggested to enhance sand liquefaction resistance. Salifu et al . [25] treated sloped sand with MICP by *Sporosarcina pasteurii* and then tested it with tidal cycles. Sand slope angle was maintained after 30 simulated tidal cycles. This indicated that MICP process successfully stabilized and reduced erosion of slope surface significantly.

A critical analysis of these research reports indicates that *Sporosarcina pasteurii* has been effectively used by different research groups for compressibility and tensile strength enhancement, biocementation, UCS, and water absorption. The utilization of *Sporosarcina pasteurii* has been cost effective too, in addition to being effective.

There are several studies that have discussed the factors affecting MICP in sand columns. Tang et al. [26] stated that $CaCO_3$ content from MICP in sand columns is affected by flow rate and hold time of biocementation solution. $CaCO_3$ content in sand columns decreased at high flow rate and its distribution depended on hold time. They reported that sand columns treated with MICP by *Sporosarcina pasteurii* can achieve compressive strength of 3.29 MPa at 3 h hold time and 0.5 mol/L biocementation solution. Omoregie et al . [27] used different strains of *Sporosarcina pasteurii* to treat sand columns. They optimized the temperature, initial pH, incubation time, and urea concentration for MICP process. Their results indicated that final enhancement varied according to the strains of bacteria used. Duo et al . [28] used *Sporosarcina pasteurii* for biocementation of desert aeolian sand. They studied the effects of urea-$CaCl_2$ concentration on $CaCO_3$ amount, dry density, permeability, and UCS of the sand columns after MICP. All properties increased with higher urea-$CaCl_2$ concentration, which showed that the formation of $CaCO_3$ crystals consolidated the sand column. USC of sand columns was greatly increased when more than 14% $CaCO_3$ formed. Their study showed that $CaCO_3$ crystals mostly formed at sand particle surfaces and pores between sand particle when the concentration of urea-$CaCl_2$ used was 0.5–1.0 mol/L. Solidifying and connecting properties became more significant at 1.5–2.5 mol/L urea-$CaCl_2$. Methods to apply bacteria and cementation solutions into sand columns affect the MICP performance. Similar study was also carried out by Sharaky et al. and applied biocementation solution including *Sporosarcina pasteurii* for consolidation of sand columns [29]. Minto et al. [30] used *Sporosarcina pasteurii* for MICP on marble grains in columns and found that only the top to middle portion of the columns were solid enough. Rate of MICP increased while porosity and permeability decreased towards top of columns. This is due to the formation of $CaCO_3$ during MICP that blocked the path and affected flow pattern of biocementation solution. Rowshanbakht et al. [31] used *Sporosarcina pasteurii* to enhance sand columns through a two phase-injection method. Bacteria retention, optical density (OD), and urease activity were optimized at 2/3 injection pore volume and 85% sand relative density. Maximum UCS was achieved at 1/3 injection pore volume and 85% sand relative density. The author found that the portion of the sand column near the injection point had a lower UCS than the other end. Permeability was found to decrease with increasing injection pore volume and sand relative density. More calcite formed near the injection point and calcite content throughout the sand column varied from 4.5% to 8%. Kakelar et al. [32] replaced yeast extract using sodium acetate in ratio for MICP of *Sporosarcina pasteurii* and concluded that cost can be reduced by using this technique. Minto et al . [33] applied MICP on sandstone cores through continuous injection of *Sporosarcina pasteurii* and nutrients. After that, they tested the sample with acidic fluids. Less than 1% permeability drop was reported due to $CaCO_3$ blocking preferential flow paths and also buffered the acidic fluids. Tobler et al . [34] studied the transportation of *Sporosarcina pasteurii* in sandstone. The bacteria cells were found easily trapped in sandstone. Higher injection rate can enhance cells transportation. Although initial injection can distribute the cells uniformly, they were easily trapped on $CaCO_3$ after MICP started. The authors stated that transport behavior must be determined for each bacteria strain individually.

A critical analysis this section indicates that injection method is superior to mixing method for the consolidation of sand columns using *Sporosarcina pasteurii*. Various factors (injection rate, injection pore volume, medium, etc.) need to be considered while designing an MICP process using *Sporosarcina pasteurii*.

Consolidation of soils through MICP is also frequently investigated. Grabiec et al. [35] mixed *Sporosarcina pasteurii* in silty soils to make cylinder samples. They found that soil shear strength and rigidity increased while soil deformation under stress reduced after the MICP process. This method further ensured soil lithification. The authors also demonstrated that high mechanical pressure involved during sample making may reduce bacterial survival rate, although their study showed that a number of compaction strokes had insignificant effect on bacterial survival in soil. Canakci et al. [36] used *Sporosarcina pasteurii* to consolidate organic soil and found that MICP was able to improve soil shear strength due to enhanced cohesion and internal friction. However, it was limited by strength of organic particles in the soil. Around 20% $CaCO_3$ formed after MICP process. This is less than the amount of $CaCO_3$ formed in sandy soil from other studies and therefore organic matter may have inhibited growth of $CaCO_3$ crystals. Feng et al. [37] used simulation to study biocementation of soil through MICP by *Sporosarcina pasteurii* and OPC. They claimed that mechanical properties of bio-cemented sand can be predicted through careful calibration.

In addition to consolidation of sands and soils, researchers also showed great interest in making bacterial-based bricks, concretes, and mortars [38]. Some works are shown in Table 2. Bernardi et al. [39] made bio-bricks with silica rich masonry sand through MICP by using *Sporosarcina pasteurii*. They recorded the transition of bio-bricks from ductile to brittle within 28 days curing time. Up to 2.2 MPa compressive strength was achieved for their bio-bricks. Cuzman et al. [40] made bio-blocks with sand though MICP by *Sporosarcina pasteurii* with cement kiln dust as the calcium source. Ground granulated blast furnace slug was also added as a mean for solid waste recycling. Addition of solid wastes was shown to reduce urease activity due to high alkalinity and inhibitory effects. Nevertheless, this study suggested that it is a possible method to reduce construction costs and environmental pollution. There are many factors that need to be considered when making bacteria-based materials. Okyay and Frigi Rodrigues [41] attempted to optimize MICP by *Sporosarcina pasteurii* through a center composite design by varying the concentrations of urea, $CaCl_2$, and nickel nitrate. Concentrations of urea and $CaCl_2$ were identified as the significant factors. High urea to $CaCl_2$ ratio enhanced the MICP process. No clear relation between bacterial growth and rate of MICP was observed. Zhang et al. [42] studied the effects of calcium source on MICP by *Sporosarcina pasteurii* in mortar samples. The calcium sources tested were calcium acetate, calcium chloride, and calcium nitrate. The amount of $CaCO_3$ formed and water adsorption of sand column was not affected by type of calcium sources tested. However, samples using calcium acetate had the highest UCS and tensile strength. Those samples also had smaller and more uniformly distributed pore structure. This is related to the formation of 88% aragonite and 12% calcite in samples using calcium acetate, while only calcite formed in samples using other two calcium salts.

Many studies were conducted to enhance MICP by *Sporosarcina pasteurii* so that better materials can be produced at lower costs. Amiri and Bundur [43] compared the effects of different nutrients and calcium salts on MICP by *Sporosarcina pasteurii* to make mortar samples. Similar bacterial growth was observed in corn steep liquor (CSL) and yeast extract, but cells in CSL had lower surface charge. Both nutrients caused the setting time to increase but CSL caused it less than yeast extract. CSL also produced more $CaCO_3$ than yeast extract after 28 days. However, compressive strength of yeast extract sample was higher than CSL sample. The authors also stated that calcium salts affect $CaCO_3$ crystal morphology due to their different solubility.

Since the cost of the nutrient source can be up to 60% of total costs, it is obvious that the use of cheaper alternatives can reduce the cost of biocement production. In this direction, the research group of Yoosathaporn et al. [44] used chicken manure effluent as an alternative nutrient source for *Sporosarcina pasteurii* to make biocement cubes. The biocement had 30.27% higher compressive strength, 5.38% higher density, 3.2% more voids, and slightly higher water adsorption than normal cement. It was also documented that calcite and vaterite were formed. The authors reported that chicken manure effluent enabled more than two times urease production than commonly used nutrient broth, and was 88.2% cheaper. Critically speaking, it is pertinent to mention that chicken manure effluent–urea medium may create tiny air bubbles within cement that can weaken the cement structure and therefore more studies should be conducted to find low cost nutrient sources for biocement production.

Harsh conditions in cement and concrete often lead to low viability of bacterial cells. Williams et al. [45] simulated the harsh conditions in concrete and studied the effects on *Sporosarcina pasteurii*. Cell viability was greatly decreased at high temperature and high alkaline conditions. Urease activity was halted in high alkaline condition and greatly reduced at temperature higher than 45 °C. However, urease activity was not solely affected by cell viability. In order to ensure sufficient cells for MICP, carriers can be used to protect the cells from direct contact with its surroundings. This technique is often used by researchers to make self-healing samples. Amiri et al. [46] studied the effects of encapsulating *Sporosarcina pasteurii* in Air Entraining Agents (AEA) surfactant in cement mortar. They found that AEA has insignificant effects on cells zeta potential and in-vitro MICP but cell viability was greatly reduced. MICP process in mortar did not differ much with the addition of AEA. The author stated that AEA may encapsulate bacteria and the surfactant tails prevented water and nutrient from reaching the bacteria, thus killing them. This encapsulation method should be tested on bacteria endospores.

Looking at the research reports discussed above, it is very important to consider the effects of the process conditions (temperature, alkaline and acidic conditions, and carriers such as surfactants) on the urease activity of *Sporosarcina pasteurii*.

MICP can be applied on various surfaces by immersion or spraying method. Usually, the purpose of the processes is to heal cracks on the surface and to increase durability of the material. Choi et al . [47] studied the potential of *Sporosarcina pasteurii* bacterial solution to treat cracks on mortar samples. The authors applied treatment cycle once per day and found that seven cycles sealed most small cracks less than 0.52 mm and 21 cycles seal all cracks up to 1.64 mm. A portion of water permeability was recovered and only 8% tensile strength was recovered. This is due to the fact that all the voids in the cracks were not filled by $CaCO_3$ crystals, and therefore adhesion and bridging effects were weak. Analysis revealed 1–2 mm $CaCO_3$ layer at crack surface with a mix of calcite and vaterite. Balam et al. [48] compared MICP by *Sporosarcina pasteurii* and *Bacillus subtilis* to reduce water adsorption of various types of concrete aggregates categorized by their weight. They found that *Sporosarcina pasteurii* is better than *Bacillus subtilis* in reducing water adsorption of aggregates by up to 20%. The percentage of water adsorption reduction that was achieved varied from 0.6% to 28.2% depending on type of aggregates due to different microstructure and pore distribution. Bacteria concentration also affected the result. Generally, water adsorption reduction by MICP was more effective on lightweight aggregates compared to normal weight aggregates. Nosouhian et al . [49] showed that MICP by *Sporosarcina pasteurii* for surface treatment of concrete can help to enhance durability of concrete exposed to sulphate condition.

Some structures have very different environments such as those at subsurface. Verba et al. [50] used *Sporosarcina pasteurii* to make biocement-sandstone. The experiment environment was adjusted to mimic subsurface conditions with brine, 10 MPa high pressure, and supercritical CO_2. Bacteria growth was greatly reduced by the presence of brine, although MICP still occurred in this condition. High pressure (10 MPa) and CO_2 concentration did not have significant effects on *Sporosarcina pasteurii*. However, temperature at 40 °C greatly reduced the bacterial density. Calcite, vaterite, and aragonite were observed after MICP process. The result from this study is beneficial for subsurface MICP applications such as wellbore sealing. Cunningham et al. [51] also conducted similar studies on sandstone core for mitigation of wellbore leakage. MICP by *Sporosarcina pasteurii* greatly reduced sample pore size at 75.8 bar. It also greatly reduced permeability with the presence of brine but delayed sealing was observed. Usage of excess chemicals was suggested for field applications. Cunningham et al . [51] suggested using less expensive nutrient sources to reduce cost from $2.34 per liter to $0.28 per liter. Their field application was successful to reduce permeability and enhance wellbore integrity of a well in Alabama within four days. Phillips et al. [52] conducted a field scale study on wellbore cement sealing using MICP by *Sporosarcina pasteurii*. They reported that after four days of treatment, injectivity and pressure falloff were greatly reduced while solid content was greatly increased.

During formation of $CaCO_3$ crystals, some other metal ions can also bind together and onto the crystals. This phenomenon has been exploited to remove metal contaminants. Mugwar and Harbottle [14] tested the potential *Sporosarcina pasteurii* to remove various heavy metals through MICP. They reported the following findings: complete removal of up to 0.5 mM Zinc in 7 days; near complete removal of up to 1.5 mM cadmium in 3 days; almost complete removal of up to 5 mM lead in 1 day; and almost complete removal of up to 0.01 mM copper in 1 day. The author stated that removal of heavy metals may be due to sorption or co-precipitation of the metals on or within $CaCO_3$ crystals during MICP process.

Carbonates other than $CaCO_3$ can also be precipitated by using appropriate sources. Yu et al. [53] used *Sporosarcina pasteurii* to treat loose quartz sand in columns through injection method. Magnesium chloride ($MgCl_2$) was used instead of calcium salts; thus, magnesium carbonates were precipitated. The number of injections varied from 2 to 6 injections. They found that hydraulic conductivity, porosity, and maximum defect volume decreased with the number of injections. On the other hand, compressive strength and density increased with number of injections. They also showed that the application of biocementation solution through spraying method only once was sufficient to reduce wind erosion rate of the sand column to zero. Ruan et al. [54] used *Sporosarcina pasteurii* isolated from activated sludge to treat cracks in reactive magnesia cement. Magnesium carbonate was formed on crack surfaces and completely healed cracks wider than 0.15 mm after two cycles of treatment. They noted that urea concentration did not improve the healing process but affected pH and carbonate morphology.

Overall, *Sporosarcina pasteurii* has been extensively explored for the induction of MICP in different kinds of structures in different conditions. Several reports demonstrated the urease producing ability of this bacterial strain that bestows it with the effectivity of inducing MICP in different media and different environments. The optimization of the process conditions forms the corner stone of MICP processes using *Sporosarcina pasteurii*.

Table 1. MICP with *Sporosarcina pasteurii* for sand and soil improvement.

Ingredients	Structure and Properties after MICP	Reference
Sand, clay	Increased tensile strength (40.8 kPa) and compressibility, decreased permeability (0.53×10^{-7} m/s)	[19]
Ottawa silica sand	Unconfined compressive strength (UCS) 1.3 MPa, flexure strength 0.95 MPa	[20]
Sand, metakaolin, OPC	OPC-MICP has best properties with UCS 1.2 MPa, water absorption 8%	[21]
Sand, PVA fiber	Highest UCS 1.6 MPa, highest splitting tensile strength 440 kPa, lowest permeability 1.05×10^{-5} m/s	[22]
Sand	Highest CS 3.29 MPa	[26]
Desert aeolian sand	Highest UCS 18 MPa, lowest permeability 0.92×10^{-8} m/s	[28]
Medium/fine sand	UCS 1.74 MPa, durability and water stability increased	[29]
Poorly graded course sand	UCS 525 kPa	[32]
Poorly graded sandy soil	UCS 400 kPa	[55]
Sandy soil	UCS 625 kPa, permeability 1.8×10^{-7} m/s	[56]
Sandy soil	Highest 6.4 MPa after 4 treatments, permeability 1.0×10^{-5} m/s	[57]

Table 2. Construction materials made by MICP with *Sporosarcina pasteurii*.

Materials	Structure and Properties after MICP	Reference
Bio-brick from silica rich masonry sand	Highest CS 2.2 MPa	[37]
Red brick (treatment)	CS 7.54 MPa, reduce water absorption by 49% after treatment	[39]
Concrete with light weight aggregates	Highest CS 40 MPa, lowest water absorption 5%	[58]
Bio-mortar	Highest CS 39.6 MPa, tensile strength 37% higher than normal mortar	[59]
Bio-mortar	Highest UCS 43 MPa, lowest water absorption 2.5%	[42]
Bio-mortar	Highest UCS 44 MPa	[60]
Bio-mortar	Highest CS 54/70 MPa at 7/28 days curing	[43]
Bio-mortar with superplasticizers	Crack width healed 0.35 mm	[44]
Bio-cement	CS 42 MPa, water absorption 21%	[45]
Bio-mortar	Crack width healed 0.41 mm, water adsorption restored 95%, CS restored 84%	[61]
Bio-mortar	Crack width healed 0.27 mm, CS restored 63%	[62]
Bio-mortar with fiber and zeolite as bacteria carriers	CS 70/100 MPa at 7/270 days	[63]
Geopolymer	Self-healing observed in 1 month old sample	[64]

2.2. Bacillus sphaericus

Bacillus sphaericus, now reclassified as *Lysinibacillus sphaericus*, is an aerobic Gram-positive, mesophilic, rod shaped bacterium commonly found in soil and aquatic habitats. This bacterium is often used to produce mosquitocide [65,66]. It is able to produce urease and is tolerant to high alkalinity. Therefore, it is also often used in MICP experiments. Table 3 shows some of the works conducted. Moravej et al. [67] used *Bacillus sphaericus* for biocementation of dispersive soil. They optimized the MICP process at 1.5 bacteria OD, 7.5 g/L $CaCl_2$ concentration, and 28 °C. The authors observed that calcite was formed that connected the soil grains together. Four to five days of treatment was enough to consolidate the soil. The effect of soil pH during treatment was also studied and it was observed to decrease greatly until day four to neutral. The pH change reduced double layer thickness and stabilized exchangeable sodium ions; thus, reducing dispersity in soil. Gupta et al. [68] added *Bacillus sphaericus* immobilized in biochar to make mortar samples. Biochar as carrier protect and distribute bacteria more uniformly throughout the mortar samples. They also added superabsorbent polymer to provide moisture for bacteria in mortar and polypropylene microfiber to reduce crack width and promote better self-healing. Addition of superabsorbent polymer decreased the compressive and flexural strength of mortar. These mortar samples showed more than 90% crack sealing for all crack width observed with 77% reduced water penetration and near 100% strength regained. Analysis showed that calcite was formed from the MICP process. Seifan et al. [69] reported that aeration with sufficient oxygen enhanced MICP of the *Bacillus sphaericus* and *Bacillus licheniformis*. The bacteria were able to tolerate pH up to 12. Higher pH enhanced MICP and caused smaller crystals to form. Their study showed that more vaterite was formed at pH 9–10 while more calcite was formed at pH 11–12. The same research group [70] also optimized MICP of *Bacillus sphaericus* and *Bacillus licheniformis* by adding oxygen releasing compounds. They suggested that the addition of oxygen releasing compounds in self-healing concrete healed cracks deep inside the concrete. In another investigation, the same group [71] investigated the effect of cell immobilization of *Bacillus sphaericus* and *Bacillus licheniformis* on magnetic iron oxide nanoparticles. It was found that when magnetic iron oxide nanoparticles exceeded 150 µg/mL, bacterial growth decreased but $CaCO_3$ precipitation increased. The nanoparticles were absorbed on $CaCO_3$ surface but did not affect crystal morphology.

Thus, it was obvious that the use of various additives such as the nanoparticles may be employed to enhance concrete properties. It was further suggested that more studies are needed to fully investigate the interactions of bacteria with MICP for the production of materials with good strength, durability and self-healing ability. Shirakawa et al. [72] treated fiber reinforced cement with *Bacillus sphaericus* and then left the cement for 22 months exposed outdoors. They reported that MICP treatment with live *Bacillus sphaericus* with calcium acetate, yeast extract, glucose, and urea gave the best biodeterioration resistance, low water absorption, and porosity. The MICP process formed the smallest calcite crystals and more homogenous layer on cement surface to provide better protection.

Table 3. Materials produced using MICP with *Bacillus sphaericus*.

Material	Structure Properties after MICP	Reference
Bio-mortar with biochar, superabsorbent polymer, polypropylene fiber	CS 35–60 MPa, flexure strength 9–12 MPa Crack width healed up to 0.9 mm, water penetration restored 70%	[68]
Bio-concrete with fly ash	Highest CS 32.5 MPa, highest tensile strength 4.1 MPa, highest flexure strength 3.5 MPa	[73]
Bio-concrete with fly ash	CS 30–40 MPa, tensile strength 2.9–5.0 MPa	[74]
Bonding repair mortar	Highest slant shear strength 17 MPa	[75]
Industrial ceramic aggregates (treatment)	Water absorption 6–16%, weight gained 3–7%	[76]

2.3. Bacillus megaterium

Bacillus megaterium is a Gram-positive, rod shaped bacterium. It is the largest of all Bacillus species [77]. It was used as a model organism for various research before *Bacillus subtilis* was introduced. It is also ureolytic, thus being studied for MICP potential. Dhami et al. [78] used *Bacillus megaterium* solution to treat sand columns of different grain sizes from 0.2 mm to 1.5 mm. Cell viability was reported to be lower in bigger grain size samples initially, but the cell viability difference among all grain sizes became smaller later as MICP proceeded. The study showed that initial MICP for smaller grain size samples was higher but rate of increase over time was slower, while the opposite trend was observed for bigger grain size samples. Analysis showed that mostly calcite was formed from the MICP process. The authors also proposed an indirect method to measure the rate of MICP for site applications by effluent chemical analysis.

Although some recent studies showed the potential of *Bacillus megaterium* in making bacteria-based materials, most other studies are focused on its MICP behaviors. Jiang et al. [79] studied the ureolytic activity of *Bacillus megaterium* in oxic and anoxic conditions. They found that anoxic conditions enhanced the ureolytic activity, thus claiming that the bacteria have potential use in sub-seafloor environment with low temperature, high pressure, and anoxic condition. Bains et al. [80] investigated the effects of EPS produced by *Bacillus megaterium* SS3 on its MICP. They found that culture media affected bacteria growth, enzyme production, and EPS production. Calcium consumption was greatly increased at higher EPS concentration because EPS provided more nucleation sites for MICP. Dhami et al. [81] compared the effects of urease and carbonic anhydrase produced by *Bacillus megaterium* SS3 on MICP. During the MICP process, urease was found to maintain alkaline pH. Carbonic anhydrase showed a better MICP rate than urease during the initial 4 h while urease had a better MICP rate afterwards. The best pH and temperature reported for urease were 9 and 35 °C, respectively, while for carbonic anhydrase the values were 8 and 40 °C, respectively. Urease inhibitor caused greater MICP reduction compared to carbonic anhydrase inhibitor indicating that urease may be the main $CaCO_3$ producer. The study showed that urease and carbonic anhydrase work synergistically for the MICP process. The authors in their other study [82] attempted to optimize media for *Bacillus megaterium* SS3 to obtain best MICP performance. They found that glucose and peptone are the best carbon and nitrogen sources for the bacteria. They also found that glucose, urea, and $NaHCO_3$ had significant positive effects on the MICP performance. The optimized media was able to increase $CaCO_3$ production by 70%. Table 4 shows some of the works conducted.

Table 4. Materials produced using MICP with *Bacillus megaterium*.

Material	Structure and Properties after MICP	Reference
Treat sand column with varying grain size	Up to 30% $CaCO_3$ formation	[82]
Bio-concrete with recycled aggregates, nanosilica	Water absorption 5%, void volume 10%	[83]
Bio-mortar	Highest CS 36 MPa, permeability 5×10^{-5} m/s	[84]

Overall, *Bacillus megaterium* has been successful in treating sand columns of different grain sizes. Bacterial concrete has been developed by using *Bacillus megaterium* in combination with recycled aggregates and nanosilica. The bacterium has affected the formation of materials in different environments effectively. However, further research is needed to optimize the parameters for MICP using *Bacillus megaterium*.

2.4. Bacillus subtilis

Bacillus subtilis is an aerobic, Gram-positive, spore forming, and rod-shaped bacterium commonly found in soil, water, and plants. It can produce various metabolites and has great potential in industrial applications [85–88]. It is non-ureolytic but studies have shown that that it has functional urease and can be activated though specific procedures. Although the exact mechanism is unknown, it is able to induce $CaCO_3$ precipitation in appropriate media in the presence of calcium. It can also survive in harsh conditions, and thus has been studied for the preparation of bacteria-based materials. Mondal and Ghosh [89] added *Bacillus subtilis* at 103, 105, or 107 cell/mL cell densities to make mortar samples. The highest compressive strength increase was obtained at 105 cell/mL bacteria and the lowest water adsorption was obtained at 107 cell/mL bacteria. Self-healing potential of the mortars was shown to increase with higher bacteria density. Cracks up to 1.2 mm wide were completely healed within 28 days with 107 cell/mL bacteria. The author stated that higher bacteria concentration caused more $CaCO_3$ crystals to form at surface thus providing better protection and reduce water permeation into inner layer. This condition may have affected cement hydration process and caused lower compressive strength at 107 cell/mL compared to 105 cell/mL bacteria. Mortar properties can be controlled by bacteria type, water to cement ratio, and cement to sand ratio. Perito et al . [90] found that a solution with dead *Bacillus subtilis* cells was able to precipitate $CaCO_3$. The advantage is that dead cells have high heat resistance (up to 100 °C). In addition, calcite is produced only on the dead cell wall; thus, the MICP process can potentially be controlled. The authors used it to treat stones and an Angera Church wall. Water adsorption reduction of 16.7% on laboratory stones and 6.8% on Angera Church after treatment was reported. In addition to $CaCO_3$, *Bacillus subtilis* was reported to produce phosphates when suitable chemicals were added and biosandstone with compressive strength of 2.1 MPa was obtained [91]. Table 5 shows some of the works conducted.

The reports discussed above indicate that *Bacillus subtilis* is activated through specific procedures. Its utilization in MICP processes ensures the formation of concrete with appreciable compressive strength. The MICP processes can also be controlled by tuning certain parameters.

Table 5. Materials produced using MICP with *Bacillus subtilis*.

Material	Structure and Properties after MICP	Reference
Bio-mortar	Highest CS 50 MPa, lowest water absorption 5% Crack width healed up to 1.2 mm	[89]
Bio-concrete	Highest CS 44 MPa Self-healing observed	[90]
Bio-shotcrete	Highest CS 34 MPa, highest tensile strength 3.4 MPa, lowest water absorption 6.2% Self-healing observed	[92]
Sand column (mixture of *B. subtilis* and *S. pasteurii*)	UCS 1.69 MPa, permeability 1.06×10^{-5} m/s	[93]

2.5. Bacillus mucilaginous

Little information about *Bacillus mucilaginous* is available. It is known to produce carbonic anhydrase [94]. Carbonic anhydrase is able to extract CO_2 from air or glucose for $CaCO_3$ precipitation. Dhami et al. [82] compared the performance of bacteria urease and carbonic anhydrase for MICP. Urease has better MICP than carbonic anhydrase. Carbonic anhydrase absorbed CO_2 from air, but the $CaCO_3$ produced was less than what was obtained by using $NaHCO_3$. Range of $CaCO_3$ crystals formed among all samples was 151–189 mg/mL. The authors also reported that $CaCO_3$ morphology was affected by bacterial species and carbon sources. It was also documented that urease mainly produced calcite while carbonic anhydrase produced vaterite. Qian et al. [95] used *Bacillus mucilaginous* to seal cracks on mortar samples and reduce their efflorescence. Surface treatment of mortar was not effective for reducing water adsorption. However, immobilization with agar layer greatly reduced the water adsorption to 14% of control sample. The authors demonstrated that agar significantly strengthened the bonding of deposit layer with mortar surface to form a dense film. The bacterial treatment also reduced mortar surface efflorescence by 42.4%. The advantage of using carbonic anhydrase producing bacteria is that they take CO_2 from air and change it to HCO_3^-, which reacts with $Ca(OH)_2$ to form $CaCO_3$. Chen et al. [96] used ceramsite as carriers to encapsulate Bacillus mucilaginous and nutrients for preparation of biocement. Then the biocement was cracked and self-healing was observed. Cracks up to 0.5 mm wide were healed. The biocement has 0.8×10^{-7} m/s water permeability and flexural strength 3.3 times higher than normal cracked cement after 28 days of healing. Analysis showed that calcite was formed during the self-healing process. The study showed that immobilizing bacteria and nutrients with ceramsite can greatly enhance MICP process by increasing the amount of $CaCO_3$ formed. Wang et al. [97] added *Bacillus mucilaginous* to make steel slag bricks and found that MICP greatly enhanced the bio-bricks. Up to 16.8 MPa compressive strength and 4.2 MPa flexural strength was recorded after three hours of MICP. Pore volume of bio-bricks was also greatly reduced after MICP. Analysis showed that MICP was detected up to 40 mm depth and $CaCO_3$ was denser at surface because it is hard for CO_2 to diffuse into the inner layer. The author suggested using higher pressure and more CO_2 to enhance MICP in the inner layer of bio-bricks.

A critical analysis of this section indicates that *Baccilus mucilaginous* is a promising candidate for the production of self-healing biocement through MICP. The beauty of using *Bacillus mucilaginous* is that plenty of air surrounding the surfaces can be used a source of CO_2, which then acts as the precursor for $CaCO_3$ formation and deposition. However, further studies are warranted to ascertain the essentiality of this bacterium for MICP.

2.6. Cyanobacteria

Cyanobacteria are effective in sequestering atmospheric CO_2. They can be modified to convert CO_2 into valuable products through photosynthesis [98]. Therefore, they have the potential for MICP. Cyanobacteria have been reported to form intracellular $CaCO_3$ [99]. These bacteria take calcium into cells and reduce the amount of calcium in solution, thus inhibiting $CaCO_3$ formation in solution. Zhu et al. [100] studied the MICP of several cyanobacteria such as *Synechocystis* sp. and *Synechococcus* sp. on mortar surfaces. The cyanobacteria were either alive or killed by UV, and the MICP condition was either under light or in dark. All these conditions and bacterial species used affected the concentration of calcium consumed and sizes of $CaCO_3$ crystals formed. The authors stated that the detailed mechanism to different calcification behaviors among cyanobacteria was not clear. Light intensity or UV exposure conditions may affect MICP depending on cyanobacterial species. The study showed that overall performance of *Synechocystis* sp. was better than *Synechococcus* sp. Zhu et al. [101] also studied the MICP of live and UV killed Gloeocapsa PCC73106 under light and dark conditions to treat mortar samples. The best properties among all samples studied were recorded for mortar treated with UV killed Gloeocapsa PCC73106, which had 7.7% higher compressive strength and 10% lower water adsorption compared to untreated mortar. The authors reported that more EPS was produced by UV-killed cells, and EPS protected bacterial cells from calcification, thus promoting cell adherence to mortar surface and also $CaCO_3$ precipitation. Zhu et al. [102] studied the MICP of Synechococcus PCC8806 in cement mixture and on the concrete surface. In cement mixture, the bacteria with $CaCl_2$ greatly enhanced $CaCO_3$ precipitation in terms of size, amount, precipitation time, and $CaCO_3$ morphology. Silicification occurred at cell surface before $CaCO_3$ formation and may enhanced rate of $CaCO_3$ precipitation. On the concrete surface, the bacteria produced 200 μm–270 μm thick $CaCO_3$ layer. Water adsorption of the concrete was 3 g/cm^2 after surface treatment. The $CaCO_3$ layer was resistant to scratching with 4% mass lost after sonication test. Bundeleva et al. [103] studied the MICP behavior of *Gloeocapsa* sp. f-6gl. They found that light is important for its MICP process because there was no biomass increase in dark condition. There was no clear relation between biomass and rate of MICP. Only calcite was detected in almost all samples.

A critical analysis of this section indicates that cyanobacteria are potential microorganisms for MICP. The main merits of using cyanobacteria are that they do not need urea and a carbon source as they simply they take CO_2 from atmosphere. Additionally, they do not produce nitrogen-based byproducts, and the costs of the processes are lower.

2.7. Other Bacteria

Several other bacterial strains have been potentially used for inducing MICP under various conditions in the laboratory setups. *Bacillus cereus* is an ureolytic, aerobic, Gram-positive, and rod-shaped bacterium commonly found in soil and food. Some strains of *Bacillus cereus* are harmful to humans, thus careful selection must be made to ensure safety [104]. Li et al. [105] tried MICP by *Bacillus cereus* NS4 to make mortar with addition of metakaolin. The bio-sample has higher compressive strength and lower permeability than normal mortar. Addition of 25% mass metakaolin gave better compressive strength and permeability compared to 0% or 50% mass metakaolin. Rozenbaum et al. [106] treated tuffeau stone with MICP by *Bacillus cereus* and then investigated the water transfer behavior in the stone through modeling. The authors documented that bio-coating has a limited lifetime, thus needing renewal within some period. Zhu et al. [107] reported that *Bacillus cereus* can be used for large scale nickel removal from soil. Their study showed that nickel concentration was reduced from 400 mg/kg soil to 38 mg/kg soil. Most nickel was bound to carbonate after treatment. The study showed that MICP by *Bacillus cereus* can be a potential alternative for large remediation of metal contaminated soil. Zhang et al. [108] added non-ureolytic and alkaliphilic *Bacillus cohnii* encapsulated in expanded perlite or expanded clay as healing agents into concrete mixture and evaluated self-healing capability. This bacterium utilizes organic compounds such as calcium lactate instead of urea to induce $CaCO_3$ precipitation. The experiment showed that maximum crack width healed within 28 days was 0.79 mm for expanded perlite sample, 0.45 mm for expanded clay sample, 0.39 mm for bacterial samples without carriers, and 0.25 mm for normal concrete. Calcite was detected in all bacterial concretes. Expanded perlite had several advantages over expanded clay as carrier. It has high porosity and water adsorption to contain 12% more bacteria than expanded clay. Volume of expanded perlite used was 89% of expanded clay to contain same amount of *Bacillus cohnii*. The structure of expanded perlite allows it to provide more oxygen and water for MICP process. It is also protected by geopolymer coating. Furthermore, it costs only USD 0.22 per kg. The study suggests that expanded perlite can be good carrier for bacteria in making self-healing materials. Lors et al. [109] used *Bacillus pseudofirmus* solution with calcium lactate as calcium salt, calcium nitrate as inorganic salt, and yeast extract as nutrient to heal autogenously healed mortar that had been left for one year. Addition of calcium nitrate enhanced bacterial growth but did not improve $CaCO_3$ precipitation. Nevertheless, it was able to enhance healing slightly. Calcite was detected from the MICP process. The authors stated that organic calcium salt should have organic part that can be used as nutrient for bacteria while anion of inorganic calcium salts need to be in some part of reaction so that $CaCO_3$ crystals can form. Sharma et al. [110] prepared spores of *Bacillus pseudofirmus* DSM 8715. The spores were used to make mortar and treat cracks on concrete through injection method. This bacteria strain has great spore forming and germination properties but suitable germinants such as alanine, inosine, or NaCl are needed. MICP of this bacterium produced calcite and aragonite. Results showed that cracks were healed and water absorption of concrete was restored to normal value. Helmi et al. [111] studied the MICP of ureolytic bacterium, *Bacillus licheniformis*. They found that media having calcium as a pure source enhanced MICP but calcium acetate inhibited MICP due to pH decrease. Optimum pH was 8 and optimum temperature was 35 °C. Analysis showed that 89% calcite and 11% vaterite were formed. Bhaskar et al. [63] used *Sporosarcina ureae* encapsulated in zeolite to make bacterial mortars. The prepared mortars had better properties than normal mortars, but still not as good as mortars made using *Sporosarcina pasteurii*. Zhan et al. [112] reported that *Paenibacillus mucilaginosus* can absorb CO_2 from air for MICP to bind fugitive dust. Erşan et al. [113] studied the potential of two nitrate reducing bacteria *Pseudomonas aeruginosa* and *Diaphorobacter nitroreducens* to make self-healing mortars with expanded clay or granular activated carbon as carriers. Their results showed that cracks up to 400 μm wide can be healed within 28 days using bacteria in expanded clay and cracks up to 500 μm wide can be healed within 56 days using bacteria in granular activated carbon. Up to 85% water tightness regain was reported for using bacteria in granular activated carbon. Calcite and aragonite were observed from the healing process. This study showed that nitrate reducing bacteria can have similar MICP performance as ureolytic and aerobic bacteria. Bai et al. [114] presented visual observations of MICP in *Pseudomonas aeruginosa* biofilm. They reported different MICP behavior compared to other studies. Lin et al. [115] studied the crystal morphology of $CaCO_3$ formed by sulfate reducing bacterium, *Desulfovibrio bizertensis*. They found that $CaCO_3$ crystal morphology is determined during the nucleation stage. They also found that the presence of phosphate inhibited the formation of aragonite.

Overall, a large diversity of bacteria have been investigated for MICP. Exciting results have been reported in terms of compressive strength of the materials developed and the overall efficiency of the MICP processes. However, more investigations are needed to explore the potential of these bacterial strains as far as MICP is concerned.

2.8. Bacteria Isolated from Various Environments

Researchers have been diligently searching for new bacterial species with the hope of achieving more efficient MICP. Diverse classes of bacteria have been isolated from various environments and tested for their MICP potential. Li et al. [116] isolated urease producing bacteria namely *Acinetobacter* sp. SC4 from Yixing Shanjuan Cave, China and tested its MICP potential to repair cracked masonry cement mortar. The repaired cement mortar regained 97.7% of its original compressive strength and has 42.4% lower water adsorption compared to repaired normal mortar. Analysis showed that calcite was formed through the MICP process. Zhang et al. [117] isolated a *Bacillus* sp. strain H4 from a mangrove conservation area in Shenzhen Bay, China and then developed a self-healing system using the bacteria together with oxygen releasing tablets. The oxygen releasing tablets were made from various peroxides and organic acids. Addition of oxygen releasing tablets was shown to greatly enhance the MICP process. Another study [118] also reported that high concentrations of a nitrate and calcium source inhibited the MICP process. Surrounding pH must be controlled at 9.5–11 because this bacteria strain cannot tolerate high alkalinity. Achal and Pan [119] used *Bacillus* sp. CR2 isolated from mine tailing soil of Urumqi, Xinjiang, China and studied the effects of calcium source on its MICP process. The calcium sources tested were $CaCl_2$, calcium oxide, calcium acetate, and calcium nitrate. Results showed that $CaCl_2$ was the best for enhancing bacteria growth profile, urease activity, and $CaCO_3$ precipitation. Lv et al. [120] studied the stability of vaterite formed by *Lysinibacillus* sp. GW-2 isolated from soil in Nanjing Botanical Garden, China. They observed the formation and transitions of different $CaCO_3$ crystals. They reported that organic matters allowed vaterite to remain stable without transforming to other morphology. *Lysinibacillus* sp. YS11 isolated by Lee et al. [121] is able to form spores, EPS, and biofilms. The bacteria showed MICP only in aerobic conditions with sufficient aeration. Xu et al. [122] tested the MICP potential of *Microbacterium* sp. GM-1 isolated from active sludge. They reported that urea concentration was the most significant factor for the MICP and calcite was the dominant crystals formed. Javadi et al. [123] made bio-blocks with recycled concrete aggregates and natural aggregates through MICP by urease producing bacteria *Staphylococcus pasteurii* isolated from a soil sample. There was less than 10% difference between UCS of bio-blocks made with those aggregates, and the maximum UCS obtained was 10 MPa. UCS of the bio-blocks decreased at higher temperature due to calcination and thermal decomposition of $CaCO_3$ crystals. UCS also decreased after several freeze-thaw cycles with recycled concrete aggregates of bio-blocks having greater lowering due to higher water absorption. UCS could be reduced by increasing $CaCO_3$ content to lower water absorption and ensure the distribution of tensile force in the bio-blocks. Vashisht et al. [124] isolated *Lysinibacillus* sp. from alluvial soil and sewage samples collected from different locations of district Solan, India and then made self-healing concrete with the bacteria. The self-healing concrete had 34.6% higher compressive strength than normal concrete. The authors claimed that their concrete had better self-healing ability than concrete made with *Bacillus megaterium*. Siddique et al. [125] isolated ureolytic bacteria *Bacillus aerius* strain AKKR5 from marble sludge to make bacterial concrete with cement baghouse filter dust replacing up to 30% of ordinary Portland cement. Bacteria concrete without cement baghouse filter dust had 10% higher compressive strength than normal concrete. However, addition of cement baghouse filter dust reduced the overall concrete properties shown by decreased compressive strength, increased water absorption, chloride permeability, and porosity. Analysis revealed that calcite and ettringite were formed from the process. In another study, the same group [126] also used the *Bacillus aerius* strain AKKR5 to make bacterial concrete with rice husk ash replacing up to 20% of ordinary Portland cement. Best bacterial concrete properties were obtained using 10% rice husk ash with 14.7% higher compressive strength than normal concrete, 0.8% water absorption, 1.5% porosity, and very low to moderate chloride permeability. Analysis showed that mainly calcite was formed from the process. Krishnapriya et al. [127] isolated some alkali resistant urease producing bacteria viz. *Bacillus megaterium*, *Bacillus licheniformis*, and *Bacillus flexus* from cement factory soil at Coimbatore, Tamil Nadu, India and used them to make bacterial concrete. They all enhanced concrete compressive strength and self-healing capability but not as good as commercial *Bacillus megaterium* MTCC 1684. The authors stated that enhancement of concrete properties is related to the ability of the isolates to form endospores. The *Bacillus flexus* isolate had limited endospore form, thus its bacterial concrete had lower performance compared to the other isolates. Hao et al. [128] used a *Bacillus* sp. strain isolated from soil sample from Perth, Australia for MICP surface treatment of polypropylene before making fiber reinforced cementitious composites. Compressive strength of the composites was decreased by 6.9% but energy adsorption capacity increased by 69.3%. Surface treatment of polypropylene enhanced the bending strength of the composites especially after cracking occurred. The author found that slight deposition of (0.026 g) was too weak to enhance the composites while heavy deposition of $CaCO_3$ (0.372 g) made the $CaCO_3$ layer too brittle and easily de-bonded from polypropylene. Thus, moderate deposition of $CaCO_3$ (0.094 g) was suggested for surface treatment of polypropylene. Montano-Salazar et al. [129]

isolated some bacteria from buildings in the National University of Colombia and tested their MICP potential. Nine isolates showed MICP potential but $CaCO_3$ crystal morphology obtained was different between bacteria strains. *Psychrobacillus psychrodurans* M414 was identified as the best $CaCO_3$ producer and was used to make bacterial mortar. The study showed that concrete compressive strength was greatly increased through immersion in biocementation solution but only slight increase of compressive strength was observed for direct addition of bacteria into concrete mixture. Mwandira et al. [130] isolated ureolytic bacteria *Pararhodobacter* sp. from soil near beachrock in Okinawa, Japan and then used the bacteria to treat lead contaminated sand columns. The contaminant was completely removed through co-precipitation with calcite or vaterite. Maximum UCS obtained was 1.33 MPa for fine sand sample, 2.87 MPa for coarse sand sample, and 2.80 MPa for mixed sand samples. Erşan et al. [131] isolated *Pseudomonas aeruginosa* and *Diaphorobacter nitroreducens* from soil and tested their MICP potential through denitrification in a minimal nutrient condition. The $CaCO_3$ precipitation recorded was 53–72% of using optimal growth conditions. The author claimed that those bacteria have potential use for soil enhancement due to high $CaCO_3$ precipitation in anoxic and minimum nutrient conditions. The bacteria also have potential use in self-healing concrete because they are concrete compatible and no other additives are needed. The research group of Daskalakis et al. [132] isolated *Bacillus pumilus* from a cave in Paiania, Athens, Greece and tested its potential for vaterite precipitation on stone marbles. Temperature and medium concentration were identified as the significant factors. Stone surface was completely covered by vaterite within 9 days and the vaterite was stable even after 1 year. The authors documented that acetate enhanced vaterite formation while the bacteria maintained vaterite stability. Charpe et al. [133] prepared bacteria solution from soil samples collected from Visvesvaraya National Institute of Technology campus in India without isolating the bacteria and then added the solution into cement mixture to make biocement. The biocement had 47.96 MPa compressive strength and 5.8% water adsorption. Analysis revealed that calcite and aragonite were formed during the process. The author claimed that biocement production costs can be reduced by using soil without isolating the bacteria, using lentil seed powder as the protein source, and sugar as the carbon source. Liu et al. [134] reported that some desert soil bacteria were able to utilize atmospheric CO_2 for MICP. The MICP capability depends on the bacteria species.

A large number of bacterial strains have been isolated from diverse sources with varied conditions. The isolated and collected strains of bacteria have been investigated for MICP processes under different conditions. The bacteria have been successfully utilized in the formation of biocement via the precipitation of $CaCO_3$ crystals. Of course, different methodologies for the precipitation of $CaCO_3$ utilizing the bacterial strains isolated from diverse sources have been developed. Additionally, different MICP parameters have been optimized for enhancing the efficiency of the developed processes. However, further studies are needed to fully optimize the process conditions for the enhancement of MICP efficiency in terms of cost and applicability.

2.9. Unidentified or Unknown Bacteria

Some reports did not reveal the exact species of the bacteria used. Nevertheless, these reports can show some MICP behaviors and applications as references for future studies. Seifan et al. [135] studied the effects of several variables on MICP of various bacteria. They identified that bacteria species, concentration of bacteria, yeast extract, $CaCl_2$, urea, and agitation speed were the significant factors. High bacteria concentration enhanced MICP. Too much yeast extract (more than 3 g/L) greatly reduced MICP. $CaCl_2$ was said to be better than calcium lactate, calcium nitrate, or calcium acetate. Too low or high Ca^{2+} concentration will decrease MICP; thus, it must be controlled carefully. On the other hand, temperature was reported to have insignificant effect on MICP. $CaCO_3$ crystal morphologies observed were only calcite and vaterite. More calcite was formed when using calcium lactate while more vaterite was formed using $CaCl_2$. High medium viscosity also caused more calcite to form. Joshi et al. [136] studied the effects of a urease producing bacterium *Bacillus* sp. CT5 by adding the bacteria into cement mixture or spraying the bacteria on the concrete surface. Addition of bacteria into cement mixture delayed the setting time. Both methods led to the lowest sorptivity coefficient, water penetration, and chloride penetration. However, compressive strength of the addition method was higher. Analysis revealed that $CaCO_3$ in the form of calcite and aragonite mostly precipitated in upper depth (0 mm–10 mm) of the samples but none in middle depth (20 mm–30 mm) and inner depth (40 mm–50 mm). Bacteria were found in middle depth for both methods, but were found in inner depth only for the addition method. Xu and Yao [137] added some non-ureolytic Bacillus genus together with calcium sources into concrete and studied the self-healing capability. Calcium glutamate was better than calcium lactate because calcium glutamate caused thicker transition zone which enhanced the bonding in the concrete. However, they also reported that healing agent was less effective than surface treatment due to different amount of activated bacteria and nutrient supply. Chu et al. [138] used a *Bacillus* sp. VS1 for biocementation of sand columns together with metal ions pretreatment. They found that intact bacteria suspension had better MICP ability than washed suspension of bacteria and supernatant. They also found that protease activity greatly reduced urease activity and therefore must be controlled. The study showed that surface coating of Ca^{2+}, Fe^{3+}, or Al^{3+} on sand enhanced bacteria cell adsorption by 31%. The author established some equations relating compressive strength and permeability to $CaCO_3$ content to estimate time needed to achieve certain compressive strength or permeability. Li et al. [139] exposed a Bacillus genus to UV light and obtained a mutant strain LHUM107. Urea degradation efficiency of the strain greatly increased from 67% to 97% after mutation. The mutant showed potential to enhance the MICP process. Rizwan et al. [140] used two types of effective microorganism consortia containing yeast, lactic acid bacteria, and photosynthetic bacteria to make biocement. Setting time of cement paste increased with addition of the consortia solution. This method also required the addition of super plasticizer. The biocement had lower water adsorption than control sample. Highest compressive strength reported for the biocement was 89 MPa. Analysis revealed that mainly calcite and some wollastonite were formed by the effective microorganisms. Luo et al. [141] used some unknown spore forming alkali resistant bacteria to make self-healing concrete. The self-healing capability was 85% for crack width less than 0.3 mm, 50–70% for 0.3 mm–0.5 mm, and less than 30% for up to 0.8 mm within 20 days. Good healing was observed up to 28 days but then decreased greatly at 60 days to 90 days. They also reported that cracked concrete needed to be immersed in water to achieve good healing as self-healing at atmosphere with 90% relative humidity was quite low. In another study, the same group [142] reported crack healing of up to 0.48 mm wide cracks within 80 days and water permeability reduction up to 96% within 28 days of self-healing. Calcite was detected in all cases. Qian et al. [143] compared the performance of calcite and phosphate formed by bacteria on sheet glass interface. Various tests were conducted and the author concluded that bio-calcite was best among all samples in terms of intensity of interface interactions, strength per mass, and interfacial bonding strength. Mors and Jonkers [144] reported that the addition of a bacterial healing agent has insignificant effect on concrete strength but increased the self-healing capability to three times higher than normal concrete. They also proposed a method for its applications to reduce environmental impact and costs. Gat et al. [145] studied the stability of bacterial $CaCO_3$ crystals in aqueous phase. $CaCO_3$ dissolution was observed starting from 20 days after complete MICP and up to 30% $CaCO_3$ loss was recorded at the end of experiment. This dissolution was caused by ammonia volatilization. Therefore, it was suggested to increase Ca^{2+} or CO_3^{2-} to prevent ammonia volatilization or just remove ammonia after complete MICP. Ammonia volatilization effect was only observed on the surface or near the surface but not several cm into the soil. Liu et al. [146] studied the MICP of bacteria in activated sludge on aerobic granules of different sizes in the reactor for wastewater treatment. Local microenvironment varied due to different mass transfer resistance and thus affected the rate of MICP on granules with different sizes. More $CaCO_3$ was formed on larger granules but very large granules (more than 700 μm) may limit MICP. Acetate metabolism enhanced MICP by increasing CO_3^{2-} concentration and pH. Some bacteria consortia can sequester CO_2 from atmosphere through their MICP process. The CO_2 sequestration ability depends on the bacteria species in the consortia [147]. Wiktor and Jonkers [148] reported the application of MICP to heal cracks in a parking garage. Sodium silicate was added to the healing solution to provide rapid initial crack sealing (weaker sealing) and alkaline pH for MICP (stronger sealing). Mass loss due to freeze-thaw cycle was reduced from 3.6 kg/m^3 to 1.9 kg/m^3. All areas previously with heavy leaking only had a few localized dripping zones or no leaking after sealing. Jroundi et al. [149] used microorganisms obtained from historic gypsum plaster to treat gypsum plaster from 13–15th century. Analysis showed that 95% of the microorganisms were carbonatogenic and 10% produced acids with addition of glucose. The authors stated that bacteria can penetrate deeper into the sample compared to other conventional consolidants. Bacterial treated plaster was reported to have better drilling resistance, slightly decreased porosity, no significant color change, and 1.5–2% mass vaterite precipitated.

3. Remarks and Aspects for Future Studies

In recent years, *Sporosarcina pasteurii* is clearly the most studied bacterium for MICP followed by various Bacillus species including *Bacillus sphaericus*, *Bacillus megaterium*, *Bacillus subtilis*, and *Bacillus mucilaginous*. Other bacteria such as cyanobacteria, nitrate reducing bacteria, and sulfate reducing bacteria are also tested for their MICP potential. Researchers have also been diligently isolating bacteria from various environments in order to obtain new bacterial strains that can effectively cause MICP. The MICP experiments conducted are mostly biocementation of sand columns, consolidation of soil, development of self-healing mortar or concrete, and crack sealing. It is interesting to note that MICP processes have potential in heavy metal/ion removal from water samples [14,150–152].

Table 6 shows the performance of bacteria MICP in consolidation of sand and soil. Higher initial concentration of bacteria usually leads to better MICP performance because there are more cells available to induce $CaCO_3$ precipitation. Higher concentrations of urea and $CaCl_2$ also often lead to better MICP performance. Types of sand or soil greatly affect the strength of final MICP products. This may be due to different bacteria cells retention and penetration as well as distribution of $CaCO_3$ crystals in the sand or soil. MICP seems to have lowest biocementation performance on poorly graded sands. Another factor that affects MICP performance is the method to introduce bacteria and biocementation solution into sand or soil to ensure uniform distribution across all volume. A lot of recent studies about consolidation of sand and soil used *Sporosarcina pasteurii* probably because it is the most established bacteria for MICP over the years. Potential of other types of bacteria should also be investigated.

Table 7 shows the performance of bacterial MICP in making bacterial concrete or mortar and their self-healing potential. A large variety of bacteria have been used to make bacterial concrete and mortar. Type of bacteria definitely affected performance of the final MICP products due to their differences in enzyme activity, size, and reaction pathway to precipitate $CaCO_3$ crystals. Generally, these bacterial concretes or mortars have equal or better strength and durability compared to normal concretes or mortars. Bacterial concrete or mortar also had better self-healing capability, and 0.5 mm–1.0 mm wide cracks can be healed. Encapsulation of bacteria in carriers can enhance the MICP performance due to higher cell survivability. Addition of other substances such as fly ash or rice husk ash may improve or reduce MICP performance, thus they must be chosen wisely.

Many bacteria used to study MICP are ureolytic. They have high urease activity to catalyze the hydrolysis of urea and elevate surrounding pH, which leads to the formation of $CaCO_3$. However, this process is sometimes criticized due to the formation of nitrogenous byproducts especially ammonium that can be harmful to living organisms and environment. These byproducts need to be converted into other harmless forms or completely removed after MICP is completed. Utilization of non-ureolytic bacteria can also solve this problem. Non-ureolytic bacteria consume other organic compounds such as lactate instead of urea to form carbonate ions. Some of them can even capture CO_2 from the atmosphere and convert them into carbonate ions to form $CaCO_3$ with calcium ions. This shows that MICP process can be developed to sequester CO_2 and contribute in reducing greenhouse gas in the atmosphere. More studies can be conducted to explore this potential.

Harsh conditions such as high alkalinity and lack of nutrients greatly affect bacteria cell availability and MICP behavior. Genetic modification of bacteria may increase bacteria survivability and enzyme activity to enhance MICP process. In order to develop mutant bacteria, which are safe to use and good for MICP, integration of knowledge from different fields is required. Nevertheless, this can be a good aspect to be included in future studies. One of the limitations of current MICP technique is that $CaCO_3$ crystals only precipitate on surface and in upper to middle parts of larger samples. MICP does not occur in deeper parts of the samples due to lack of necessary compounds there. Therefore, more studies can be conducted to ensure that $CaCO_3$ crystals precipitate uniformly inside large samples. Application costs of the MICP technique can be reduced by using plant and animal waste materials as nutrients. Researchers from various places should explore the potential of local wastes to be used in MICP process. This can not only reduce costs but also increase sustainability of the process.

Table 6. Biocementation of sand and soil through MICP.

Bacteria (Initial Concentration)	Sand/Soil	Cementation Solution	Performance	Reference
Sporosarcina pasteurii + Bacillus subtilis (OD_{600} = 1.2)	Sandy soil	2 M urea + 1 M $CaCl_2$	UCS = 1.69 MPa Permeability = 1.06×10^{-5} m/s	[93]
Sporosarcina pasteurii (OD_{600} = 0.6)	Ottawa silica sand	0.5 M (urea + $CaCl_2$)	UCS = 1.30 MPa	[20]
Sporosarcina pasteurii (OD_{600} = 1.0)	Commercial sand + white kaolin clay	0.5 M (urea + $CaCl_2$)	Tensile strength = 0.04 MPa Permeability = 0.53×10^{-7} m/s	[19]
Sporosarcina pasteurii (OD_{600} = 1.9–2.4)	Desert aeolian sand	2.5 M (urea + $CaCl_2$)	UCS = 18 MPa Permeability = 0.92×10^{-7} m/s	[28]
Sporosarcina pasteurii (OD_{600} = 2.3)	Natural SiO_2 sand	1.0 M (urea + $CaCl_2$)	UCS 1.74 MPa	[29]
Sporosarcina pasteurii (OD_{600} = 0.6)	Ottawa silica sand	0.75 M (urea + $CaCl_2$)	UCS = 6.4 MPa Permeability = 1.00×10^{-5} m/s	[40]
Sporosarcina pasteurii (Not provided)	Poorly graded medium sand	1.0 M (urea + $CaCl_2$)	Surface strength = 4.83 MPa	[27]
Sporosarcina pasteurii (OD_{600} = 2.0)	Loose sand	0.5 M (urea + $CaCl_2$)	UCS (MICP) = 0.10 MPa UCS (MICP + OPC) = 1.10 MPa Water adsorption (MICP) = 11% Water adsorption (MICP + OPC) = 8%	[21]
Sporosarcina pasteurii (OD_{600} = 3.5)	Standard sand	0.5 M (urea + $CaCl_2$)	UCS = 3.29 MPa	[26]
Sporosarcina pasteurii (OD_{600} = 1.5)	Sandy soil	3 mM urea + 2 mM $CaCl_2$	UCS = 0.63 MPa Permeability = 1.80×10^{-5} m/s	[39]
Sporosarcina pasteurii (1.5 g/L)	Ottawa silica sand + PVA fiber	0.5 M (urea + $CaCl_2$)	UCS = 2.20 MPa Splitting tensile strength = 0.60 MPa Permeability = 4.00×10^{-7} m/s	[22]
Sporosarcina pasteurii (OD_{600} = 2.5)	Poorly graded SiO_2 sand	1.0 M (urea + $CaCl_2$)	UCS = 0.53 MPa	[32]
Sporosarcina pasteurii (OD_{600} > 2)	Poorly graded sandy silica	1.0 M (urea + $CaCl_2$)	UCS = 0.50 MPa Permeability = 0.85×10^{-6} m/s	[31]

Table 7. Performance of MICP by various bacteria in making concrete and mortar.

Bacteria (Initial Concentration)	Other Additives	Performance	Reference
Bacillus sphaericus (10^{10} cell/mL)	Biochar, PP fiber, SAP	Compressive strength = 53.0 MPa Water penetration = 9.0 mm Crack width healed = 0.9 mm	[68]
Bacillus sphaericus (Not provided)	Fly ash	Compressive strength = 32.5 MPa	[73]
Bacillus sp. CT5 (OD_{600} = 0.5)	-	Compressive strength = 46.0 MPa Water penetration = 14.2 mm	[135]
Bacillus subtilis (10^3–10^7 cell/mL)	-	Compressive strength = 54.0 MPa Water adsorption = 4% Crack width healed = 1.2 mm	[89]
Lysinibacillus sp. I13 (Not provided)	Fly ash	Compressive strength = 33.6 MPa *Able to heal cracks but no exact values provided	[123]
Sporosarcina pasteurii (10^9 cell/mL)	Calcium sulpho-aluminate cement, silica fume	Compressive strength = 46.8 MPa Crack width healed = 0.4 mm	[55]
Sporosarcina pasteurii (8×10^8 cfu/mL)	-	Compressive strength = 70.0 MPa	[50]
Sporosarcina pasteurii (10^6 cell/mL)	Zeolite, fiber reinforced	Compressive strength = 84.0 MPa Water penetration = 1.5 mm Crack width healed = 0.1 mm	[57]
Sporosarcina pasteurii (10^8–10^9 cell/mL)	-	Compressive strength = 39.6 MPa	[47]
Bacillus cohnii (5.2×10^8 cell/mL)	Expanded pearlite	Crack width healed = 0.8 mm	[108]
Bacillus sphaericus (10^5 cell/mL)	Fly ash	Compressive strength = 40.4 MPa	[72]
Bacillus cereus (5×10^8 cfu/mL)	Metakaolin	Compressive strength = 40.2 MPa	[105]
Bacillus aerius (10^5 cell/mL)	Cement baghouse filter dust	Compressive strength = 36.3 MPa Water adsorption = 1.2%	[125]
Bacillus aerius (10^5 cell/mL)	Rice husk ash	Compressive strength = 35.0 MPa Water adsorption = 1.1%	[126]
Bacillus mucilaginous (10^8–10^9 cell/mL)	Ceramsite	Crack width healed = 0.5 mm Water permeability = 0.8×10^{-7} m/s	[96]
Bacillus megaterium (OD_{600} = 1.5)	-	Compressive strength = 35.0 MPa	[81]
Pseudomonas aeruginosa Diaphorobacter nitroreducens (Not provided)	Granular activated carbon	Crack width healed = 0.5 mm	[113]
Soil bacteria (OD_{600} = 0.866)	-	Compressive strength = 48.0 MPa Water adsorption = 5.8%	[132]

4. Conclusions

The MICP techniques show promising potential for applications in various fields such as construction, geotechnology, and nanotechnology. MICP can reduce OPC usage and enhance sustainability. MICP performances of various bacteria have been discussed in this paper. Some of the studies have indicated that the bacterial strains can extract carbon dioxide from air for the precipitation of $CaCO_3$. On one hand, reduction of accumulation of greenhouse carbon dioxide is ensured and on the other hand cracks in the cement are sealed and healed. This technique is shown to be commonly used for biocementation of sand, consolidation of soil, and development of self-healing concrete. This technique can also apply for removal of heavy metals. Future studies are expected to further enhance the MICP performance, reduce its application costs, and increase its sustainability.

Author Contributions: S.C.C., S.H.M.S., and S.F.M. designed and wrote the first draft of review. A.A., W.A.W., and M.J. were responsible for drafting and critically revising the manuscript. A.A.Y. and M.N.M.I. participated in the technical check of the manuscript and the drawing of the figures. All authors have read and agreed to the published version of the manuscript.

Funding: This work was supported by Ministry of Higher Education, Malaysia, through the funds (grant number PY/2018/02895). The authors thank the Research Management Center (RMC), Universiti Teknologi Malaysia (UTM) and Universiti Putra Malaysia (UPM).

Conflicts of Interest: The authors declare that they have no conflicts of interest.

References

1. Seifan, M.; Sarabadani, Z.; Berenjian, A. Microbially induced calcium carbonate precipitation to design a new type of bio self-healing dental composite. *Appl. Microbiol. Biotechnol.* **2020**, *104*, 2029–2037. [CrossRef] [PubMed]
2. Seifan, M.; Berenjian, A. Microbially induced calcium carbonate precipitation: A widespread phenomenon in the biological world. *Appl. Microbiol. Biotechnol.* **2019**, *103*, 4693–4708. [CrossRef] [PubMed]
3. Zhu, X.; Wang, J.; De Belie, N.; Boon, N. Complementing urea hydrolysis and nitrate reduction for improved microbially induced calcium carbonate precipitation. *Appl. Microbiol. Biotechnol.* **2019**, *103*, 8825–8838. [CrossRef] [PubMed]
4. Zhang, J.; Xie, L.; Huang, X.; Liang, Z.; Liu, B.; Han, N.; Xing, F.; Deng, X. Enhanced calcite precipitation for crack healing by bacteria isolated under low-nitrogen conditions. *Appl. Microbiol. Biotechnol.* **2019**, *103*, 7971–7982. [CrossRef] [PubMed]
5. Ramachandran, A.L.; Polat, P.; Mukherjee, A.; Dhami, N.K. Understanding and creating biocementing beachrocks via biostimulation of indigenous microbial communities. *Appl. Microbiol. Biotechnol.* **2020**, *104*, 3655–3673. [CrossRef] [PubMed]
6. Achal, V.; Mukherjee, A.; Zhang, Q. Unearthing ecological wisdom from natural habitats and its ramifications on development of biocement and sustainable cities. *Landsc. Urban Plan.* **2016**, *155*, 61–68. [CrossRef]
7. Achal, V.; Mukherjee, A. A review of microbial precipitation for sustainable construction. *Constr. Build. Mater.* **2015**, *93*, 1224–1235. [CrossRef]
8. Yaqoob, A.A.; Ibrahim, M.N.M.; Ahmad, A.; Reddy, A.V.B. Toxicology and Environmental Application of Carbon Nanocomposite. In *Environmental Remediation through Carbon Based Nanocomposites. Green Energy and Technology*; Springer: Singapore, 2020. [CrossRef]
9. Gupta, S.; Pang, S.D.; Kua, H.W. Autonomous healing in concrete by bio-based healing agents—A review. *Constr. Build. Mater.* **2017**, *146*, 419–428. [CrossRef]
10. Muhammad, N.Z.; Shafaghat, A.; Keyvanfar, A.; Majid, M.Z.A.; Ghoshal, S.; Yasouj, S.E.M.; Ganiyu, A.A.; Kouchaksaraei, M.S.; Kamyab, H.; Taheri, M.M.; et al. Tests and methods of evaluating the self-healing efficiency of concrete: A review. *Constr. Build. Mater.* **2016**, *112*, 1123–1132. [CrossRef]
11. Aziz, Z.A.A.; Mohd-Nasir, H.; Ahmad, A.; Setapar, S.H.M.; Peng, W.L.; Chuo, S.C.; Khatoon, A.; Umar, K.; Yaqoob, A.A.; Ibrahim, M.N.M. Role of Nanotechnology for Design and Development of Cosmeceutical: Application in Makeup and Skin Care. *Front. Chem.* **2019**, *7*, 739. [CrossRef]
12. Vijay, K.; Murmu, M.; Deo, S.V. Bacteria based self healing concrete—A review. *Constr. Build. Mater.* **2017**, *152*, 1008–1014. [CrossRef]
13. Al-Salloum, Y.; Hadi, S.; Abbas, H.; Almusallam, T.; Moslem, M. Bio-induction and bioremediation of cementitious composites using microbial mineral precipitation—A review. *Constr. Build. Mater.* **2017**, *154*, 857–876. [CrossRef]
14. Mugwar, A.J.; Harbottle, M. Toxicity effects on metal sequestration by microbially-induced carbonate precipitation. *J. Hazard. Mater.* **2016**, *314*, 237–248. [CrossRef] [PubMed]
15. Yaqoob, A.A.; Parveen, T.; Umar, K.; Ibrahim, M.N.M. Role of Nanomaterials in the Treatment of Wastewater: A Review. *Water* **2020**, *12*, 495. [CrossRef]
16. Wong, L.S. Microbial cementation of ureolytic bacteria from the genus Bacillus: A review of the bacterial application on cement-based materials for cleaner production. *J. Clean. Prod.* **2015**, *93*, 5–17. [CrossRef]

17. Kang, C.-H.; So, J.-S. Heavy metal and antibiotic resistance of ureolytic bacteria and their immobilization of heavy metals. *Ecol. Eng.* **2016**, *97*, 304–312. [CrossRef]
18. Li, Q.; Zhang, B.; Ge, Q.; Yang, X. Calcium carbonate precipitation induced by calcifying bacteria in culture experiments: Influence of the medium on morphology and mineralogy. *Int. Biodeterior. Biodegrad.* **2018**, *134*, 83–92. [CrossRef]
19. Cardoso, R.; Pires, I.; Duarte, S.; Monteiro, G.A. Effects of clay's chemical interactions on biocementation. *Appl. Clay Sci.* **2018**, *156*, 96–103. [CrossRef]
20. Bu, C.; Wen, K.; Liu, S.; Ogbonnaya, U.; Li, L. Development of bio-cemented constructional materials through microbial induced calcite precipitation. *Mater. Struct.* **2018**, *51*, 30. [CrossRef]
21. Porter, H.; Dhami, N.K.; Mukherjee, A. Synergistic chemical and microbial cementation for stabilization of aggregates. *Cem. Concr. Compos.* **2017**, *83*, 160–170. [CrossRef]
22. Choi, S.-G.; Wang, K.; Chu, J. Properties of biocemented, fiber reinforced sand. *Constr. Build. Mater.* **2016**, *120*, 623–629. [CrossRef]
23. Xiao, P.; Liu, H.; Xiao, Y.; Stuedlein, A.W.; Evans, T.M. Liquefaction resistance of bio-cemented calcareous sand. *Soil Dyn. Earthq. Eng.* **2018**, *107*, 9–19. [CrossRef]
24. Sasaki, T.; Kuwano, R. Undrained cyclic triaxial testing on sand with non-plastic fines content cemented with microbially induced $CaCO_3$. *Soils Found.* **2016**, *56*, 485–495. [CrossRef]
25. Salifu, E.; MacLachlan, E.; Iyer, K.R.; Knapp, C.W.; Tarantino, A. Application of microbially induced calcite precipitation in erosion mitigation and stabilisation of sandy soil foreshore slopes: A preliminary investigation. *Eng. Geol.* **2016**, *201*, 96–105. [CrossRef]
26. Tang, Y.; Lian, J.; Xu, G.; Yan, Y.; Xu, H. Effect of Cementation on Calcium Carbonate Precipitation of Loose Sand Resulting from Microbial Treatment. *Trans. Tianjin Univ.* **2017**, *23*, 547–554. [CrossRef]
27. Omoregie, A.I.; Khoshdelnezamiha, G.; Senian, N.; Ong, D.E.L.; Nissom, P.M. Experimental optimisation of various cultural conditions on urease activity for isolated Sporosarcina pasteurii strains and evaluation of their biocement potentials. *Ecol. Eng.* **2017**, *109*, 65–75. [CrossRef]
28. Li, D.; Tian, K.-I.; Zhang, H.-I.; Wu, Y.-Y.; Nie, K.-Y.; Zhang, S.-C. Experimental investigation of solidifying desert aeolian sand using microbially induced calcite precipitation. *Constr. Build. Mater.* **2018**, *172*, 251–262. [CrossRef]
29. Sharaky, A.M.; Mohamed, N.S.; Elmashad, M.E.; Shredah, N.M. Application of microbial biocementation to improve the physico-mechanical properties of sandy soil. *Constr. Build. Mater.* **2018**, *190*, 861–869. [CrossRef]
30. Minto, J.M.; Tan, Q.; Lunn, R.J.; El Mountassir, G.; Guo, H.; Cheng, X. 'Microbial mortar'-restoration of degraded marble structures with microbially induced carbonate precipitation. *Constr. Build. Mater.* **2018**, *180*, 44–54. [CrossRef]
31. Rowshanbakht, K.; Khamehchiyan, M.; Sajedi, R.H.; Nikudel, M.R. Effect of injected bacterial suspension volume and relative density on carbonate precipitation resulting from microbial treatment. *Ecol. Eng.* **2016**, *89*, 49–55. [CrossRef]
32. Kakelar, M.M.; Ebrahimi, S.; Hosseini, M. Improvement in soil grouting by biocementation through injection method. *Asia Pac. J. Chem. Eng.* **2016**, *11*, 930–938. [CrossRef]
33. Minto, J.M.; Hingerl, F.F.; Benson, S.M.; Lunn, R.J. X-ray CT and multiphase flow characterization of a 'bio-grouted' sandstone core: The effect of dissolution on seal longevity. *Int. J. Greenh. Gas Control* **2017**, *64*, 152–162. [CrossRef]
34. Tobler, D.J.; Cuthbert, M.O.; Phoenix, V.R. Transport of Sporosarcina pasteurii in sandstone and its significance for subsurface engineering technologies. *Appl. Geochem.* **2014**, *42*, 38–44. [CrossRef]
35. Grabiec, A.M.; Starzyk, J.; Stefaniak, K.; Wierzbicki, J.; Zawal, D. On possibility of improvement of compacted silty soils using biodeposition method. *Constr. Build. Mater.* **2017**, *138*, 134–140. [CrossRef]
36. Canakci, H.; Sidik, W.; Kılıç, I.H.; Canakci, H. Effect of bacterial calcium carbonate precipitation on compressibility and shear strength of organic soil. *Soils Found.* **2015**, *55*, 1211–1221. [CrossRef]
37. Feng, K.; Montoya, B.M.; Evans, T. Discrete element method simulations of bio-cemented sands. *Comput. Geotech.* **2017**, *85*, 139–150. [CrossRef]
38. Yaqoob, A.A.; Khatoon, A.; Setapar, S.H.M.; Umar, K.; Parveen, T.; Ibrahim, M.N.M.; Ahmad, A.; Rafatullah, M. Outlook on the Role of Microbial Fuel Cells in Remediation of Environmental Pollutants with Electricity Generation. *Catalysts* **2020**, *10*, 819. [CrossRef]
39. Bernardi, D.; DeJong, J.; Montoya, B.; Martinez, B. Bio-bricks: Biologically cemented sandstone bricks. *Constr. Build. Mater.* **2014**, *55*, 462–469. [CrossRef]

40. Cuzman, O.A.; Rescic, S.; Richter, K.; Wittig, L.; Tiano, P. Sporosarcina pasteurii use in extreme alkaline conditions for recycling solid industrial wastes. *J. Biotechnol.* **2015**, *214*, 49–56. [CrossRef]
41. Okyay, T.O.; Rodrigues, D.F. Optimized carbonate micro-particle production by Sporosarcina pasteurii using response surface methodology. *Ecol. Eng.* **2014**, *62*, 168–174. [CrossRef]
42. Zhang, Y.; Guo, H.; Cheng, X. Role of calcium sources in the strength and microstructure of microbial mortar. *Constr. Build. Mater.* **2015**, *77*, 160–167. [CrossRef]
43. Amiri, A.; Bundur, Z.B. Use of corn-steep liquor as an alternative carbon source for biomineralization in cement-based materials and its impact on performance. *Constr. Build. Mater.* **2018**, *165*, 655–662. [CrossRef]
44. Yoosathaporn, S.; Tiangburanatham, P.; Bovonsombut, S.; Chaipanich, A.; Pathom-Aree, W. A cost effective cultivation medium for biocalcification of Bacillus pasteurii KCTC 3558 and its effect on cement cubes properties. *Microbiol. Res.* **2016**, *186*, 132–138. [CrossRef]
45. Williams, S.L.; Kirisits, M.J.; Ferron, R.D. Influence of concrete-related environmental stressors on biomineralizing bacteria used in self-healing concrete. *Constr. Build. Mater.* **2017**, *139*, 611–618. [CrossRef]
46. Amiri, A.; Azima, M.; Bundur, Z.B. Crack remediation in mortar via biomineralization: Effects of chemical admixtures on biogenic calcium carbonate. *Constr. Build. Mater.* **2018**, *190*, 317–325. [CrossRef]
47. Choi, S.-G.; Wang, K.; Wen, Z.; Chu, J. Mortar crack repair using microbial induced calcite precipitation method. *Cem. Concr. Compos.* **2017**, *83*, 209–221. [CrossRef]
48. Balam, N.H.; Mostofinejad, D.; Eftekhar, M. Use of carbonate precipitating bacteria to reduce water absorption of aggregates. *Constr. Build. Mater.* **2017**, *141*, 565–577. [CrossRef]
49. Nosouhian, F.; Mostofinejad, D.; Hasheminejad, H. Influence of biodeposition treatment on concrete durability in a sulphate environment. *Biosyst. Eng.* **2015**, *133*, 141–152. [CrossRef]
50. Verba, C.; Thurber, A.; Alleau, Y.; Koley, D.; Colwell, F.; Torres, M. Mineral changes in cement-sandstone matrices induced by biocementation. *Int. J. Greenh. Gas Control* **2016**, *49*, 312–322. [CrossRef]
51. Cunningham, A.B.; Phillips, A.J.; Troyer, E.; Lauchnor, E.; Hiebert, R.; Gerlach, R.; Spangler, L. Wellbore leakage mitigation using engineered biomineralization. *Energy Procedia* **2014**, *63*, 4612–4619. [CrossRef]
52. Phillips, A.J.; Troyer, E.; Hiebert, R.; Kirkland, C.; Gerlach, R.; Cunningham, A.; Spangler, L.; Kirksey, J.; Rowe, W.; Esposito, R. Enhancing wellbore cement integrity with microbially induced calcite precipitation (MICP): A field scale demonstration. *J. Pet. Sci. Eng.* **2018**, *171*, 1141–1148. [CrossRef]
53. Yu, X.; Qian, C.; Sun, L. The influence of the number of injections of bio-composite cement on the properties of bio-sandstone cemented by bio-composite cement. *Constr. Build. Mater.* **2018**, *164*, 682–687. [CrossRef]
54. Ruan, S.; Qiu, J.; Weng, Y.; Yang, Y.; Yang, E.-H.; Chu, J.; Unluer, C. The use of microbial induced carbonate precipitation in healing cracks within reactive magnesia cement-based blends. *Cem. Concr. Res.* **2019**, *115*, 176–188. [CrossRef]
55. Kalantary, F.; Kahani, M. Evaluation of the Ability to Control Biological Precipitation to Improve Sandy Soils. *Procedia Earth Planet. Sci.* **2015**, *15*, 278–284. [CrossRef]
56. Azadi, M.; Pouri, S. Estimation of Reconstructed Strength of Disturbed Biologically Cemented Sand Under Unconfined Compression Tests. *Arab. J. Sci. Eng.* **2016**, *41*, 4847–4854. [CrossRef]
57. Wen, K.; Li, Y.; Liu, S.; Bu, C.; Li, L. Development of an Improved Immersing Method to Enhance Microbial Induced Calcite Precipitation Treated Sandy Soil through Multiple Treatments in Low Cementation Media Concentration. *Geotech. Geol. Eng.* **2018**, *37*, 1015–1027. [CrossRef]
58. Balam, N.H.; Mostofinejad, D.; Eftekhar, M. Effects of bacterial remediation on compressive strength, water absorption, and chloride permeability of lightweight aggregate concrete. *Constr. Build. Mater.* **2017**, *145*, 107–116. [CrossRef]
59. Al-Salloum, Y.; Abbas, H.; Sheikh, Q.; Hadi, S.; Alsayed, S.; Almusallam, T. Effect of some biotic factors on microbially-induced calcite precipitation in cement mortar. *Saudi J. Biol. Sci.* **2017**, *24*, 286–294. [CrossRef] [PubMed]
60. Bundur, Z.B.; Kirisits, M.J.; Ferron, R.D. Biomineralized cement-based materials: Impact of inoculating vegetative bacterial cells on hydration and strength. *Cem. Concr. Res.* **2015**, *67*, 237–245. [CrossRef]
61. Xu, J.; Wang, X. Self-healing of concrete cracks by use of bacteria-containing low alkali cementitious material. *Constr. Build. Mater.* **2018**, *167*, 1–14. [CrossRef]
62. Xu, J.; Wang, X.; Wang, B. Biochemical process of ureolysis-based microbial CaCO3 precipitation and its application in self-healing concrete. *Appl. Microbiol. Biotechnol.* **2018**, *102*, 3121–3132. [CrossRef] [PubMed]

63. Bhaskar, S.; Hossain, K.; Lachemi, M.; Wolfaardt, G.; Kroukamp, M.O. Effect of self-healing on strength and durability of zeolite-immobilized bacterial cementitious mortar composites. *Cem. Concr. Compos.* **2017**, *82*, 23–33. [CrossRef]
64. Jadhav, U.; Lahoti, M.; Chen, Z.; Qiu, J.; Cao, B.; Yang, E.-H. Viability of bacterial spores and crack healing in bacteria-containing geopolymer. *Constr. Build. Mater.* **2018**, *169*, 716–723. [CrossRef]
65. Berry, C. The bacterium, Lysinibacillus sphaericus, as an insect pathogen. *J. Invertebr. Pathol.* **2012**, *109*, 1–10. [CrossRef] [PubMed]
66. Park, H.-W.; Bideshi, D.K.; Federici, B.A. Properties and applied use of the mosquitocidal bacterium, Bacillus sphaericus. *J. Asia Pac. Entomol.* **2010**, *13*, 159–168. [CrossRef] [PubMed]
67. Moravej, S.; Habibagahi, G.; Nikooee, E.; Niazi, A. Stabilization of dispersive soils by means of biological calcite precipitation. *Geoderma* **2018**, *315*, 130–137. [CrossRef]
68. Gupta, S.; Kua, H.W.; Pang, S.D. Healing cement mortar by immobilization of bacteria in biochar: An integrated approach of self-healing and carbon sequestration. *Cem. Concr. Compos.* **2018**, *86*, 238–254. [CrossRef]
69. Seifan, M.; Samani, A.K.; Berenjian, A. New insights into the role of pH and aeration in the bacterial production of calcium carbonate ($CaCO_3$). *Appl. Microbiol. Biotechnol.* **2017**, *101*, 3131–3142. [CrossRef]
70. Seifan, M.; Samani, A.K.; Berenjian, A. A novel approach to accelerate bacterially induced calcium carbonate precipitation using oxygen releasing compounds (ORCs). *Biocatal. Agric. Biotechnol.* **2017**, *12*, 299–307. [CrossRef]
71. Seifan, M.; Ebrahiminezhad, A.; Ghasemi, Y.; Samani, A.K.; Berenjian, A. The role of magnetic iron oxide nanoparticles in the bacterially induced calcium carbonate precipitation. *Appl. Microbiol. Biotechnol.* **2018**, *102*, 3595–3606. [CrossRef]
72. Shirakawa, M.; John, V.M.; De Belie, N.; Alves, J.; Pinto, J.; Gaylarde, C.C. Susceptibility of biocalcite-modified fiber cement to biodeterioration. *Int. Biodeterior. Biodegrad.* **2015**, *103*, 215–220. [CrossRef]
73. Jagannathan, P.; Narayanan, K.S.; Arunachalam, K.D.; Annamalai, S.K. Studies on the mechanical properties of bacterial concrete with two bacterial species. *Mater. Today: Proc.* **2018**, *5*, 8875–8879. [CrossRef]
74. Kadapure, S.A.; Kulkarni, G.S.; Prakash, K.B. A Laboratory Investigation on the Production of Sustainable Bacteria-Blended Fly Ash Concrete. *Arab. J. Sci. Eng.* **2016**, *42*, 1039–1048. [CrossRef]
75. Snoeck, D.; Wang, J.; Bentz, D.; De Belie, N. Applying a biodeposition layer to increase the bond of a repair mortar on a mortar substrate. *Cem. Concr. Compos.* **2018**, *86*, 30–39. [CrossRef] [PubMed]
76. García-González, J.; Rodríguez-Robles, D.; Wang, J.; De Belie, N.; Del Pozo, J.M.M.; Guerra-Romero, M.I.; Juan-Valdés, A. Quality improvement of mixed and ceramic recycled aggregates by biodeposition of calcium carbonate. *Constr. Build. Mater.* **2017**, *154*, 1015–1023. [CrossRef]
77. Eppinger, M.; Bunk, B.; Johns, M.A.; Edirisinghe, J.N.; Kutumbaka, K.K.; Koenig, S.S.K.; Creasy, H.H.; Rosovitz, M.J.; Riley, D.R.; Daugherty, S.; et al. Genome Sequences of the Biotechnologically Important Bacillus megaterium Strains QM B1551 and DSM319. *J. Bacteriol.* **2011**, *193*, 4199–4213. [CrossRef]
78. Dhami, N.K.; Reddy, M.S.; Mukherjee, A. Significant indicators for biomineralisation in sand of varying grain sizes. *Constr. Build. Mater.* **2016**, *104*, 198–207. [CrossRef]
79. Jiang, N.-J.; Yoshioka, H.; Yamamoto, K.; Soga, K. Ureolytic activities of a urease-producing bacterium and purified urease enzyme in the anoxic condition: Implication for subseafloor sand production control by microbially induced carbonate precipitation (MICP). *Ecol. Eng.* **2016**, *90*, 96–104. [CrossRef]
80. Bains, A.; Dhami, N.K.; Mukherjee, A.; Reddy, M.S. Influence of Exopolymeric Materials on Bacterially Induced Mineralization of Carbonates. *Appl. Biochem. Biotechnol.* **2015**, *175*, 3531–3541. [CrossRef]
81. Dhami, N.K.; Reddy, M.S.; Mukherjee, A. Synergistic Role of Bacterial Urease and Carbonic Anhydrase in Carbonate Mineralization. *Appl. Biochem. Biotechnol.* **2014**, *172*, 2552–2561. [CrossRef]
82. Dhami, N.K.; Mukherjee, A.; Reddy, M.S. Micrographical, mineralogical and nano-mechanical characterisation of microbial carbonates from urease and carbonic anhydrase producing bacteria. *Ecol. Eng.* **2016**, *94*, 443–454. [CrossRef]
83. Singh, L.; Bisht, V.; Aswathy, M.; Chaurasia, L.; Gupta, S. Studies on performance enhancement of recycled aggregate by incorporating bio and nano materials. *Constr. Build. Mater.* **2018**, *181*, 217–226. [CrossRef]
84. Kaur, G.; Dhami, N.K.; Goyal, S.; Mukherjee, A.; Reddy, M.S. Utilization of carbon dioxide as an alternative to urea in biocementation. *Constr. Build. Mater.* **2016**, *123*, 527–533. [CrossRef]
85. Yaqoob, A.A.; Ibrahim, M.N.M.; Rodriguez-Couto, S. Development And modification of materials to build cost-effective Anodes for microbial fuel cells (MFCs): An overview. *Biochem. Eng. J.* **2020**, *3*, 107779. [CrossRef]
86. Yaqoob, A.A.; Ibrahim, M.N.M.; Rafatullah, M.; Chua, Y.S.; Ahmad, A.; Umar, K. Recent Advances in Anodes for Microbial Fuel Cells: An Overview. *Materials* **2020**, *13*, 2078. [CrossRef] [PubMed]

87. Gu, Y.; Xu, X.; Wu, Y.; Niu, T.; Liu, Y.; Li, J.; Du, G.; Liu, L. Advances and prospects of Bacillus subtilis cellular factories: From rational design to industrial applications. *Metab. Eng.* **2018**, *50*, 109–121. [CrossRef]
88. Wang, T.; Liang, Y.; Wu, M.; Chen, Z.; Lin, J.; Yang, L. Natural products from Bacillus subtilis with antimicrobial properties. *Chin. J. Chem. Eng.* **2015**, *23*, 744–754. [CrossRef]
89. Mondal, S.; Ghosh, A. Investigation into the optimal bacterial concentration for compressive strength enhancement of microbial concrete. *Constr. Build. Mater.* **2018**, *183*, 202–214. [CrossRef]
90. Perito, B.; Marvasi, M.; Barabesi, C.; Mastromei, G.; Bracci, S.; Vendrell, M.; Tiano, P. A Bacillus subtilis cell fraction (BCF) inducing calcium carbonate precipitation: Biotechnological perspectives for monumental stone reinforcement. *J. Cult. Herit.* **2014**, *15*, 345–351. [CrossRef]
91. Yu, X.; Qian, C.; Xue, B.; Wang, X. The influence of standing time and content of the slurry on bio-sandstone cemented by biological phosphates. *Constr. Build. Mater.* **2015**, *82*, 167–172. [CrossRef]
92. Vijay, K.; Murmu, M. Effect of calcium lactate on compressive strength and self-healing of cracks in microbial concrete. *Front. Struct. Civ. Eng.* **2018**, *13*, 515–525. [CrossRef]
93. Kalhori, H.; Bagherpour, R. Application of carbonate precipitating bacteria for improving properties and repairing cracks of shotcrete. *Constr. Build. Mater.* **2017**, *148*, 249–260. [CrossRef]
94. Yaqoob, A.A.; Ahmad, H.; Parveen, T.; Ahmad, A.; Oves, M.; Ismail, I.M.I.; Qari, H.A.; Umar, K.; Ibrahim, M.N.M. Recent Advances in Metal Decorated Nanomaterials and Their Various Biological Applications: A Review. *Front. Chem.* **2020**, *8*, 341. [CrossRef] [PubMed]
95. Qian, C.; Ren, L.; Xue, B.; Cao, T. Bio-mineralization on cement-based materials consuming CO_2 from atmosphere. *Constr. Build. Mater.* **2016**, *106*, 126–132. [CrossRef]
96. Chen, H.; Qian, C.; Huang, H. Self-healing cementitious materials based on bacteria and nutrients immobilized respectively. *Constr. Build. Mater.* **2016**, *126*, 297–303. [CrossRef]
97. Wang, K.; Qian, C.; Wang, R. The properties and mechanism of microbial mineralized steel slag bricks. *Constr. Build. Mater.* **2016**, *113*, 815–823. [CrossRef]
98. Yaqoob, A.A.; Ibrahim, M.N.M.; Umar, K.; Ahmad, A.; Setapar, S.H.M. Applications of Supercritical Carbon Dioxide in the Rubber Industry. In *Advanced Nanotechnology and Application of Supercritical Fluids*; Springer: Cham, Switzerland, 2020; pp. 199–218.
99. Cam, N.; Benzerara, K.; Georgelin, T.; Jaber, M.; Lambert, J.-F.; Poinsot, M.; Skouri-Panet, F.; Moreira, D.; Lopez-Garcia, P.; Raimbault, E.; et al. Cyanobacterial formation of intracellular Ca-carbonates in undersaturated solutions. *Geobiology* **2017**, *16*, 49–61. [CrossRef]
100. Zhu, T.; Lin, Y.; Lu, X.; Dittrich, M. Assessment of cyanobacterial species for carbonate precipitation on mortar surface under different conditions. *Ecol. Eng.* **2018**, *120*, 154–163. [CrossRef]
101. Zhu, T.; Lu, X.; Dittrich, M. Calcification on mortar by live and UV-killed biofilm-forming cyanobacterial Gloeocapsa PCC73106. *Constr. Build. Mater.* **2017**, *146*, 43–53. [CrossRef]
102. Zhu, T.; Paulo, C.; Merroun, M.L.; Dittrich, M. Potential application of biomineralization by Synechococcus PCC8806 for concrete restoration. *Ecol. Eng.* **2015**, *82*, 459–468. [CrossRef]
103. Bundeleva, I.A.; Shirokova, L.S.; Pokrovsky, O.S.; Bénézeth, P.; Ménez, B.; Gérard, E.; Balor, S. Experimental modeling of calcium carbonate precipitation by cyanobacterium Gloeocapsa sp. *Chem. Geol.* **2014**, *374*, 44–60. [CrossRef]
104. Kumari, S.; Sarkar, P.K. Bacillus cereus hazard and control in industrial dairy processing environment. *Food Control* **2016**, *69*, 20–29. [CrossRef]
105. Li, M.; Zhu, X.; Mukherjee, A.; Huang, M.; Achal, V. Biomineralization in metakaolin modified cement mortar to improve its strength with lowered cement content. *J. Hazard. Mater.* **2017**, *329*, 178–184. [CrossRef]
106. Rozenbaum, O.; Anne, S.; Rouet, J.-L. Modification and modeling of water ingress in limestone after application of a biocalcification treatment. *Constr. Build. Mater.* **2014**, *70*, 97–103. [CrossRef]
107. Zhu, X.; Li, W.; Zhan, L.; Huang, M.; Zhang, Q.; Achal, V. The large-scale process of microbial carbonate precipitation for nickel remediation from an industrial soil. *Environ. Pollut.* **2016**, *219*, 149–155. [CrossRef] [PubMed]
108. Zhang, J.; Liu, Y.; Feng, T.; Zhou, M.; Zhao, L.; Zhou, A.; Li, Z. Immobilizing bacteria in expanded perlite for the crack self-healing in concrete. *Constr. Build. Mater.* **2017**, *148*, 610–617. [CrossRef]
109. Lors, C.; Ducasse-Lapeyrusse, J.; Gagné, R.; Damidot, D. Microbiologically induced calcium carbonate precipitation to repair microcracks remaining after autogenous healing of mortars. *Constr. Build. Mater.* **2017**, *141*, 461–469. [CrossRef]

110. Sharma, T.; Alazhari, M.; Heath, A.; Paine, K.; Cooper, R. AlkaliphilicBacillusspecies show potential application in concrete crack repair by virtue of rapid spore production and germination then extracellular calcite formation. *J. Appl. Microbiol.* **2017**, *122*, 1233–1244. [CrossRef]
111. Helmi, F.M.; Elmitwalli, H.R.; Elnagdy, S.M.; El-Hagrassy, A.F. Calcium carbonate precipitation induced by ureolytic bacteria Bacillus licheniformis. *Ecol. Eng.* **2016**, *90*, 367–371. [CrossRef]
112. Zhan, Q.; Qian, C.; Yi, H. Microbial-induced mineralization and cementation of fugitive dust and engineering application. *Constr. Build. Mater.* **2016**, *121*, 437–444. [CrossRef]
113. Erşan, Y.Ç.; Hernandez-Sanabria, E.; Boon, N.; De Belie, N. Enhanced crack closure performance of microbial mortar through nitrate reduction. *Cem. Concr. Compos.* **2016**, *70*, 159–170. [CrossRef]
114. Bai, Y.; Guo, X.; Li, Y.; Huang, T. Experimental and visual research on the microbial induced carbonate precipitation by Pseudomonas aeruginosa. *AMB Express* **2017**, *7*, 57. [CrossRef] [PubMed]
115. Lin, C.Y.; Turchyn, A.V.; Steiner, Z.; Bots, P.; Lampronti, G.I.; Tosca, N.J. The role of microbial sulfate reduction in calcium carbonate polymorph selection. *Geochim. Cosmochim. Acta* **2018**, *237*, 184–204. [CrossRef]
116. Li, M.; Fang, C.; Kawasaki, S.; Huang, M.; Achal, V. Bio-consolidation of cracks in masonry cement mortars by Acinetobacter sp. SC4 isolated from a karst cave. *Int. Biodeterior. Biodegrad.* **2019**, *141*, 94–100. [CrossRef]
117. Zhang, J.L.; Wang, C.G.; Wang, Q.L.; Feng, J.L.; Pan, W.; Zheng, X.C.; Liu, B.; Han, N.X.; Xing, F.; Deng, X. A binary concrete crack self-healing system containing oxygen-releasing tablet and bacteria and its Ca2+-precipitation performance. *Appl. Microbiol. Biotechnol.* **2016**, *100*, 10295–10306. [CrossRef]
118. Zhang, J.L.; Wu, R.S.; Li, Y.M.; Zhong, J.Y.; Deng, X.; Liu, B.; Han, N.X.; Xing, F. Screening of bacteria for self-healing of concrete cracks and optimization of the microbial calcium precipitation process. *Appl. Microbiol. Biotechnol.* **2016**, *100*, 6661–6670. [CrossRef]
119. Achal, V.; Pan, X. Influence of Calcium Sources on Microbially Induced Calcium Carbonate Precipitation by Bacillus sp. CR2. *Appl. Biochem. Biotechnol.* **2014**, *173*, 307–317. [CrossRef]
120. Lv, J.-J.; Ma, F.; Li, F.-C.; Zhang, C.-H.; Chen, J.-N. Vaterite induced by Lysinibacillus sp. GW-2 strain and its stability. *J. Struct. Biol.* **2017**, *200*, 97–105. [CrossRef]
121. Lee, Y.S.; Kim, H.J.; Park, W. Non-ureolytic calcium carbonate precipitation by Lysinibacillus sp. YS11 isolated from the rhizosphere of Miscanthus sacchariflorus. *J. Microbiol.* **2017**, *55*, 440–447. [CrossRef]
122. Xu, G.; Li, D.; Jiao, B.; Li, D.; Yin, Y.; Lun, L.; Zhao, Z.; Li, S. Biomineralization of a calcifying ureolytic bacterium Microbacterium sp. GM-1. *Electron. J. Biotechnol.* **2017**, *25*, 21–27. [CrossRef]
123. Javadi, A.S.; Badiee, H.; Sabermahani, M. Mechanical properties and durability of bio-blocks with recycled concrete aggregates. *Constr. Build. Mater.* **2018**, *165*, 859–865. [CrossRef]
124. Vashisht, R.; Attri, S.; Sharma, D.; Shukla, A.; Goel, G. Monitoring biocalcification potential of Lysinibacillus sp. isolated from alluvial soils for improved compressive strength of concrete. *Microbiol. Res.* **2018**, *207*, 226–231. [CrossRef] [PubMed]
125. Siddique, R.; Nanda, V.; Kadri, E.-H.; Khan, M.I.; Singh, M.; Rajor, A. Influence of bacteria on compressive strength and permeation properties of concrete made with cement baghouse filter dust. *Constr. Build. Mater.* **2016**, *106*, 461–469. [CrossRef]
126. Siddique, R.; Singh, K.; Singh, M.; Corinaldesi, V.; Rajor, A. Properties of bacterial rice husk ash concrete. *Constr. Build. Mater.* **2016**, *121*, 112–119. [CrossRef]
127. Krishnapriya, S.; Babu, D.L.V.; Arulaj, P.G. Isolation and identification of bacteria to improve the strength of concrete. *Microbiol. Res.* **2015**, *174*, 48–55. [CrossRef] [PubMed]
128. Hao, Y.; Cheng, L.; Hao, H.; Shahin, M.A. Enhancing fiber/matrix bonding in polypropylene fiber reinforced cementitious composites by microbially induced calcite precipitation pre-treatment. *Cem. Concr. Compos.* **2018**, *88*, 1–7. [CrossRef]
129. Montaño-Salazar, S.M.; Lizarazo-Marriaga, J.; Brandão, P.F. Isolation and Potential Biocementation of Calcite Precipitation Inducing Bacteria from Colombian Buildings. *Curr. Microbiol.* **2017**, *75*, 256–265. [CrossRef]
130. Mwandira, W.; Nakashima, K.; Kawasaki, S. Bioremediation of lead-contaminated mine waste by Pararhodobacter sp. based on the microbially induced calcium carbonate precipitation technique and its effects on strength of coarse and fine grained sand. *Ecol. Eng.* **2017**, *109*, 57–64. [CrossRef]
131. Erşan, Y.Ç.; De Belie, N.; Boon, N. Microbially induced CaCO3 precipitation through denitrification: An optimization study in minimal nutrient environment. *Biochem. Eng. J.* **2015**, *101*, 108–118. [CrossRef]

132. Daskalakis, M.I.; Rigas, F.; Bakolas, A.; Magoulas, A.; Kotoulas, G.; Katsikis, I.; Karageorgis, A.P.; Mavridou, A. Vaterite bio-precipitation induced by Bacillus pumilus isolated from a solutional cave in Paiania, Athens, Greece. *Int. Biodeterior. Biodegradation* **2015**, *99*, 73–84. [CrossRef]
133. Charpe, A.U.; Latkar, M.; Chakrabarti, T. Microbially assisted cementation—A biotechnological approach to improve mechanical properties of cement. *Constr. Build. Mater.* **2017**, *135*, 472–476. [CrossRef]
134. Liu, Z.; Zhang, Y.; Fa, K.; Zhao, H.; Qin, S.; Yan, R.; Wu, B. Desert soil bacteria deposit atmospheric carbon dioxide in carbonate precipitates. *Catena* **2018**, *170*, 64–72. [CrossRef]
135. Seifan, M.; Samani, A.K.; Berenjian, A. Induced calcium carbonate precipitation using Bacillus species. *Appl. Microbiol. Biotechnol.* **2016**, *100*, 9895–9906. [CrossRef] [PubMed]
136. Joshi, S.; Goyal, S.; Reddy, M.S. Influence of nutrient components of media on structural properties of concrete during biocementation. *Constr. Build. Mater.* **2018**, *158*, 601–613. [CrossRef]
137. Xu, J.; Yao, W. Multiscale mechanical quantification of self-healing concrete incorporating non-ureolytic bacteria-based healing agent. *Cem. Concr. Res.* **2014**, *64*, 1–10. [CrossRef]
138. Chu, J.; Ivanov, V.; Naeimi, M.; Stabnikov, V.; Liu, H.-L. Optimization of calcium-based bioclogging and biocementation of sand. *Acta Geotech.* **2014**, *9*, 277–285. [CrossRef]
139. Li, H.; Song, Y.; Li, Q.; He, J.; Song, Y. Effective microbial calcite precipitation by a new mutant and precipitating regulation of extracellular urease. *Bioresour. Technol.* **2014**, *167*, 269–275. [CrossRef]
140. Rizwan, S.A.; Khan, H.; Bier, T.A.; Adnan, F. Use of Effective Micro-organisms (EM) technology and self-compacting concrete (SCC) technology improved the response of cementitious systems. *Constr. Build. Mater.* **2017**, *152*, 642–650. [CrossRef]
141. Luo, M.; Qian, C.-X.; Li, R.-Y. Factors affecting crack repairing capacity of bacteria-based self-healing concrete. *Constr. Build. Mater.* **2015**, *87*, 1–7. [CrossRef]
142. Luo, M.; Qian, C.; Li, R.; Rong, H. Efficiency of concrete crack-healing based on biological carbonate precipitation. *J. Wuhan Univ. Technol. Sci. Ed.* **2015**, *30*, 1255–1259. [CrossRef]
143. Qian, C.; Yu, X.; Wang, X. A study on the cementation interface of bio-cement. *Mater. Charact.* **2018**, *136*, 122–127. [CrossRef]
144. Mors, R.; Jonkers, H. Feasibility of lactate derivative based agent as additive for concrete for regain of crack water tightness by bacterial metabolism. *Ind. Crop. Prod.* **2017**, *106*, 97–104. [CrossRef]
145. Gat, D.; Ronen, Z.; Tsesarsky, M. Long-term sustainability of microbial-induced CaCO3 precipitation in aqueous media. *Chemosphere* **2017**, *184*, 524–531. [CrossRef]
146. Liu, W.-T.; Lan, G.; Zeng, P. Size-dependent calcium carbonate precipitation induced microbiologically in aerobic granules. *Chem. Eng. J.* **2016**, *285*, 341–348. [CrossRef]
147. Okyay, T.O.; Nguyen, H.N.; Castro, S.L.; Rodrigues, D.F. CO_2 sequestration by ureolytic microbial consortia through microbially-induced calcite precipitation. *Sci. Total Environ.* **2016**, *572*, 671–680. [CrossRef]
148. Wiktor, V.; Jonkers, H. Field performance of bacteria-based repair system: Pilot study in a parking garage. *Case Stud. Constr. Mater.* **2015**, *2*, 11–17. [CrossRef]
149. Jroundi, F.; Gonzalez-Muñoz, M.T.; Garcia-Bueno, A.; Rodriguez-Navarro, C. Consolidation of archaeological gypsum plaster by bacterial biomineralization of calcium carbonate. *Acta Biomater.* **2014**, *10*, 3844–3854. [CrossRef]
150. Kumari, D.; Qian, X.-Y.; Pan, X.; Achal, V.; Li, Q.; Gadd, G.M. Microbially-induced Carbonate Precipitation for Immobilization of Toxic Metals. *Adv. Appl. Microbiol.* **2016**, *94*, 79–108. [CrossRef]
151. Eltarahony, M.; Zaki, S.; Abd-El-Haleem, D. Aerobic and anaerobic removal of lead and mercury via calcium carbonate precipitation mediated by statistically optimized nitrate reductases. *Sci. Rep.* **2020**, *10*, 4029. [CrossRef]
152. Jalilvand, N.; Akhgar, A.; Alikhani, H.A.; Rahmani, H.A.; Rejali, F. Removal of Heavy Metals Zinc, Lead, and Cadmium by Biomineralization of Urease-Producing Bacteria Isolated from Iranian Mine Calcareous Soils. *J. Soil Sci. Plant Nutr.* **2019**, *20*, 206–219. [CrossRef]

Publisher's Note: MDPI stays neutral with regard to jurisdictional claims in published maps and institutional affiliations.

© 2020 by the authors. Licensee MDPI, Basel, Switzerland. This article is an open access article distributed under the terms and conditions of the Creative Commons Attribution (CC BY) license (http://creativecommons.org/licenses/by/4.0/).

MDPI
St. Alban-Anlage 66
4052 Basel
Switzerland
Tel. +41 61 683 77 34
Fax +41 61 302 89 18
www.mdpi.com

Materials Editorial Office
E-mail: materials@mdpi.com
www.mdpi.com/journal/materials

www.ingramcontent.com/pod-product-compliance
Lightning Source LLC
LaVergne TN
LVHW070741100526
838202LV00013B/1280